Nanofiber Membranes for Medical, Environmental, and Energy Applications

Nanofiber Membranes for Medical, Environmental, and Energy Applications

Edited by
Ahmad Fauzi Ismail
Nidal Hilal
Juhana Jaafar
Chris J. Wright

CRC Press is an imprint of the
Taylor & Francis Group, an informa business

CRC Press
Taylor & Francis Group
6000 Broken Sound Parkway NW, Suite 300
Boca Raton, FL 33487-2742

First issued in paperback 2021

ISBN 13: 978-1-03-223985-9 (pbk)
ISBN 13: 978-0-8153-8703-9 (hbk)

DOI: 10.1201/9781351174046

Library of Congress Cataloging-in-Publication Data

Names: Ismail, Ahmad Fauzi, author. | Hilal, Nidal, author. | Jaafar, Juhana, author. | Wright, Chris J., author.
Title: Nanofiber membranes for medical, environmental, and energy applications / Ahmad Fauzi Ismail, Nidal Hilal, Juhana Jaafar, Chris Wright.
Description: Boca Raton : CRC Press, Taylor & Francis Group, 2019. | Includes bibliographical references.
Identifiers: LCCN 2019016627 | ISBN 9780815387039 (hardback : alk. paper)
Subjects: LCSH: Nanofibers. | Membranes (Technology) | Membrane separation. | Biomimetic materials. | Water--Purification--Membrane filtration.
Classification: LCC TA418.9.F5 N3525 2019 | DDC 660/.28424--dc23
LC record available at https://lccn.loc.gov/2019016627

Visit the Taylor & Francis Web site at
http://www.taylorandfrancis.com

and the CRC Press Web site at
http://www.crcpress.com

Contents

Preface

This book consists of 13 chapters with topics related to nanofiber membrane for medical, environment and energy applications. The contributors come from Asian and European countries, including Malaysia, Singapore, Turkey, Iran, and the United Kingdom. There are also contributions from Canada. The authors are experts in the development of nanofibers and membranes technology for various applications.

In particular, this book gathers numerous promising nanofiber membranes for improving and enhancing the current technologies used in drug delivery, wound healing, tissue engineering, water and wastewater treatment and purification, and fuel cells. The chapters place emphasis on the nanofiber membrane fabrication techniques and characterizations that suit the proposed applications and the way forward of the nanofiber membranes for commercial use.

This book has been separated into three major topics; medical, environmental, and energy applications. The arrangement of the content of the book is as follows.

The general information of nanofibers-based technology development for medical, environment especially water and wastewater treatment, and energy applications is presented by Chapter 1, entitled "Electrospun Polymeric Nanofiber: Production, Characterization and Application." This chapter discusses the topic to give a clear picture on the suitable technique to produce the nanofibers that meet the demand in terms of properties and performance according to applications.

It follows by Chapters 2 through 4 which cover nanofibers-based technology applications for medical. A thorough discussion to the related topic is presented by three chapters entitled "Electrospun Antimicrobial Membranes-functionality and Morphology" (Chapter 2), "Nanofiber Electrospun Membrane Based on Biodegradable Polymers for Biomedical and Tissue Engineering Application" (Chapter 3), and "Electrospun Collagen Nanofiber Membranes for Regenerative Medicine" (Chapter 4).

Due to the recent accelerated interest in highly potential nanofibers-based technology application in water and wastewater, Chapters 5 through 9 only emphasizes these particular applications. Five chapters covers this topic, which including "Nanofiber Membranes for Oily Wastewater Treatment" (Chapter 5), "Electrospun Nanofibrous Membranes for Membrane Distillation Process" (Chapter 6), "Nanofibrous Composite Membranes for Membrane Distillation" (Chapter 7), "Electrospun Nanofiber Membranes as Efficient Adsorbent for Removal of Heavy Metals" (Chapter 8), and "Surface Functionalized Electrospun Nanofibers for Removal of Toxic Pollutants in Water" (Chapter 9).

Meanwhile Chapters 10 through 13 provide the topic of the use of nanofibers in energy applications. There are four chapters provide the discussion on the said topic which are "Nanofiber Membranes for Energy Applications" (Chapter 10), "Electrospun Nanocomposite Polymer Electrolyte Membrane for Fuel Cell Applications" (Chapter 11), "Proton Transport Mechanisms in Nanofibers Ion Exchange Membrane: State of the Art and Perspectives" (Chapter 12), and

"Fabrication of Mesoporous Nanofiber Networks by Phase Separation-Based Methods" (Chapter 13).

The editors would like to highlight that apart from the growing number of research publications in membrane science and technology for the solutions of the environmental problems, the advantages of the membrane in nanofibers form is foreseen a commercially able material. With about 500 pages, this book offers the recent findings of research works from the established researchers specialized in the nanofiber membrane technologies and will definitely give full of satisfaction to the users. This book possesses its own uniqueness because established researchers from around the globe contributed to the chapters, each based on their recent research findings. Thus readers could find the most recent nanofiber membrane materials that are appropriate for medical, environment and energy applications. This could also give a clear picture on the trend of advanced materials as the way forward for the mentioned applications.

The editors would like to express their sincere thanks to all authors and co-authors for their kind support, encouragement, and their understanding of the amount of time it has taken for the book's writing. Sincere thanks also go to Barbara Knott and Danielle Zarfati from Taylor and Francis Group for their kind assistance from the beginning until the end of the publication process of this book.

Ahmad Fauzi Ismail

Nidal Hilal

Juhana Jaafar

Chris J. Wright

Editors

Professor Dr. Ahmad Fauzi Ismail is the founder and first director of Advanced Membrane Technology Research Centre (AMTEC). His research interest are in development of polymeric, inorganic and novel mixed matrix membranes for water desalination, waste water treatment, gas separation processes, membrane for palm oil refining, photocatalytic membrane for removal of emerging contaminants, development of haemodialysis membrane and polymer electrolyte membrane for fuel cell applications. His research has been published in many high impact factor journals. He also actively authored many academic books in this field, which were published by reputable international publishers. He is the author and co-author of over 600 refereed journals. He has authored 6 books, over 50 book chapters and 4 edited books, 6 patents granted, patents with the patent number of US Patent (6,521,025 B1 and WO00/27512), and has 14 patents pending. His h-index is 66 with cumulative citation of over 19,300. He won more than 150 awards national and internationally. Among the most prestigious award won is the Merdeka Award for the Outstanding Scholastic Achievement Category at 4th September 2014, Malaysia's Rising Star Award 2016 for Frontier Researcher category at 1st November 2016, and Malaysia's Research Star Award 2017 on 5 October 2017, Malaysia's Research Start Award 2018. Recently, he was appointed as UNESCO chair on groundwater arsenic within the 2030 Agenda for Sustainable Development. He is the chairman of Academy of Sciences Malaysia (Southern Region), fellow of the Academy of Sciences Malaysia, chartered engineer in the UK (CEng) and a fellow of the Institution of Chemical Engineers (FIChemE). Ahmad Fauzi also served on the Editorial Board Members of Desalination, Journal of Membrane Water Treatment, Jurnal Teknologi, Journal of Membrane Science and Research, Journal of Membrane and Separation Technology and as an advisory editorial board member of Journal of Chemical Technology and Biotechnology. He is involved extensively in R&D&C for national and multinational companies related to membrane-based processes for industrial application and currently have two spin off companies. He is the founder of Advanced Membrane Technology Research Centre (AMTEC) and now recognized as Higher Education Centre of Excellence (HICoE). Currently Ahmad Fauzi is the deputy vice chancellor of research and innovation, UTM.

His active participation in scientific research and development has also been exhibited by the recipient of research grant sponsored by government and non-government bodies. Until now, the total research grant secured is USD 11,105,360.00. Among the most recent project was Long-term Research Grant Scheme (LRGS) and European Universal Grant with the budget allocation of USD 1,466,280.00 and USD 326,535.00, respectively. Petronas Research has allocated USD 1,957,347.00 whereas SIME Darby allocated USD 247,460.00 for joint research programs with AMTEC. In 2012, he was awarded the Best Research Project award for his *Fundamental Research Grant Scheme* (FRGS) project by the Ministry of Education. In 2013, Malaysia Toray Science Foundation awarded him with a special award for his involvement and contribution in research activities and innovations. Ministry of

Education unanimously awarded him with the Innovative Action Plan for Human Capital Development Tertiary Level award.

Since 2007 until now, he has successfully generated income of about USD $1,000,000 from his invention and innovation related to membrane technology. The Innovation and Product Commercialization Award has been awarded to Fauzi in the Nation Academic Award Ceremony 2013 in recognition to his remarkable commercialization activities.

Professor Nidal Hilal is a chartered engineer in the United Kingdom (CEng), a registered European engineer (Euro Ing), an elected fellow of both the Institution of Chemical Engineers (FIChemE), and the Learned Society of Wales (FLSW). He received his bachelor's degree in chemical engineering in 1981 followed by a master's degree in advanced chemical engineering from Swansea University in 1986. He received his PhD degree from Swansea University in 1988. In 2005 he was awarded a Doctor of Science degree (DSc) from the University of Wales in recognition of an outstanding research contribution in the fields Scanning Probe Microscopy and Membrane Science and Technology. He was also awarded, by the Emir of Kuwait, the prestigious Kuwait Prize of Applied Science for the year 2005.

His research interests lie broadly in the identification of innovative and cost-effective solutions within the fields of nano-water, membrane technology, and water treatment, including desalination, colloid engineering and the nano-engineering applications of AFM. He has published 7 handbooks, 73 invited book chapters and around 450 articles in the refereed scientific literature. He has chaired and delivered lectures at numerous international conferences and prestigious organizations around the world.

He is the editor-in-chief of the international journal *Desalination*. He sits on the editorial boards of a number of international journals, is an advisory board member of several multinational organizations, and has served on/consulted for industry, government departments, research councils, and universities on an international basis.

Associate Professor Dr. Juhana Jaafar graduated with a BEng (Chemical Engineering) from Universiti Teknologi Malaysia in 2004. She was then granted with National Science Fellowship under Ministry of Higher Education (MOHE) to pursue her MSc in (Gas Engineering) from Universiti Teknologi Malaysia. In 2011, she graduated with a PhD in Gas Engineering from the same university specializing in Advanced Membrane Manufacturing for Energy Application. She has started her academic career at UTM in 2007. Currently she is a senior lecturer of, School of Chemical and Energy Engineering, Faculty of Engineering. She also holds a position as deputy direct of Advanced Membrane Technology Research Centre (AMTEC).

Her outstanding outputs in research were evident from her receiving distinguished awards at national and international levels including Asian Invention Excellent Award, 28th International Invention, Innovation and Technology Exhibition (ITEX 2017), Gold Medal, International Invention, Innovation and Design Johor 2017 (IIIDJ 2017), Most Distinguished Award – Higher Education International Conference, and Innovation Expo of Institute of Higher Education (PECIPTA 2013), Best of the Best awards at the Malaysian Technology Expo (MTE 2008) and Double Gold Medal in British Invention

Show (BIS 2008). Dr. Juhana is also active in writing for scientific publication in high-impact factor in international and national journals. To date, she has published more than 170 papers in ISI-indexed journals with H-index of 20. Until now, she had led 12 research grants and is a member of more than 60 research projects including the Long-Term Research Grant Scheme, which was awarded by Ministry of Higher Education, Malaysia, with the budget allocation of USD $1.5 million and International European Grant with the allocation of USD $700,000. Owing to her outstanding achievement in research, she has been selected as the UTMShine recipients for the year 2016.

She is actively involved with various committees at the national level, including the co-secretary general of Malaysia Membrane Society (MyMembrane), the secretary general of the National Congress on Membrane Technology 2016, steering committees for National Science Challenge 2017, International Workshop on Nanocomposite Materials for Photocatalytic Degradation of Pollutants: Advanced Opportunities for New Applications 2015, International Conference on Membrane Science and Technology 2013, Nanomaterials Technology Specialized Conference 2012, and the International Conference on Membrane Science and Technology 2009. She is also the organizing/scientific committee of 6th International Conference on Material Science and engineering technology (ICMSET 2017), International Conference on Membrane Technology and its Applications (MemSep-2017) and International Conference on Membrane Science and Technology (MST 2015).

Dr. Juhana has received several awards and recognition at the international level. Recently, she received a scholarship from AUNSEED-Net for Short-term Research Program in Japan and Short-term visit program in ASEAN country. She also has obtained a scholarship from Meiji University under the International Exchange Program 2017. She has also been listed in the Who's Who in the World, 33rd Edition 2016.

Dr. Chris J. Wright is a reader in bioprocess engineering in the College of Engineering at Swansea University. His research is focused on the biological interface and he has developed advanced characterization techniques alongside membrane fabrication in order to control bioprocess systems. His research laboratory, the Biomaterials, Biofouling and Biofilms Engineering Laboratory (B^3EL) uses electrospinning and melt electrospinning to create novel fibrous materials for application within process industries, medicine and agriculture. The underlying themes of this research are improved functionality of fibers and manufacture at scale.

Dr. Wright received his BSc in Applied Biology from the University of Wales Cardiff and completed his PhD at Swansea in Biochemical Engineering sponsored by The Brewing Research Foundation International in 1996. In 2001 he was awarded an Engineering and Physical Sciences Research Council (EPSRC) Advanced Research Fellowship. This prestigious five-year award enabled him to establish an internationally recognized research group exploiting the capabilities of atomic force microscopy (AFM) for the characterization and control of microbial interactions.

From 2006 to 2011 he was portfolio director for chemical engineering within the College of Engineering at Swansea University, and he established Medical Engineering and Environmental Engineering degree courses. He was also the director of the Multidisciplinary Nanotechnology Centre (MNC), associate director of the

Centre for Complex Fluids Processing, and founding executive member of the Centre for NanoHealth at Swansea University.

Dr. Wright's has been at the forefront in the development of AFM probe methods to measure interactions between biocolloids and surfaces. This featured the first-ever AFM cell probe in 1998 that allowed the direct measurement of the forces involved in cell-surface interactions. His research group has subsequently applied this technique for the study of bacterial, fungal and mammalian cells as well as protein-protein interactions. He has also established AFM techniques for the nanoscale characterization of the mechanical properties of biofilms and biofouling at separation membranes. To allow the translation of his research Dr. Wright has created two spinout companies, Membranology and ProColl; these companies are applying membrane technology and novel collagen fibrous materials, respectively.

Dr. Wright has over 120 international publications, 15 book chapters, and is editor of 2 books within the research area. He is on the editorial board of three international journals. His research has been sustained through major grants (in excess of £9M) from across the UK research councils, Royal Society, Royal Academy of Engineering, European funding, charity, and industry.

Contributors

Mohammed Al-Abri
Nanotechnology Research Centre
and
Department of Chemical and Petroleum Engineering
College of Engineering
Sultan Qaboos University
Al-Khoudh, Oman

Issa Sulaiman Al-Husaini
Department of Chemistry
Faculty of Science
Universiti Teknologi Malaysia
Johor Bahru, Malaysia

and

Nanotechnology Research Centre
Sultan Qaboos University
Al-Khoudh, Oman

Morteza Asghari
Separation Processes Research Group (SPRG)
Department of Chemical Engineering
and
Energy Research Institute
University of Kashan
Kashan, Iran

Nuha Awang
Advanced Membrane Technology Research Centre (AMTEC)
and
Faculty of Engineering
Universiti Teknologi Malaysia
Johor Bahru, Malaysia

Madzlan Aziz
Advanced Membrane Technology Research Centre (AMTEC)
Universiti Teknologi Malaysia
Johor Bahru, Malaysia

Brabu Balusamy
Fondazione Istituto Italiano di Tecnologia
Genova, Italy

Hasrinah Hasbullah
Advanced Membrane Technology Research Centre (AMTEC)
Universiti Teknologi Malaysia
Johor Bahru, Malaysia

Nidal Hilal
Centre for Water Advanced Technologies and Environmental Research (CWATER)
College of Engineering
Swansea University
Swansea, United Kingdom

Ahmad Fauzi Ismail
Advanced Membrane Technology Research Centre (AMTEC)
and
Faculty of Engineering
Universiti Teknologi Malaysia
Johor Bahru, Malaysia

Juhana Jaafar
Advanced Membrane Technology Research Centre (AMTEC)
and
Faculty of Engineering
Universiti Teknologi Malaysia
Johor Bahru, Malaysia

Sean James
Biomaterials, Biofouling and Biofilms Engineering Laboratory (B^{3}EL)
The Systems and Process Engineering Centre (SPEC)
College of Engineering
Swansea University
Swansea, United Kingdom

and

Hybrisan Ltd.
Port Talbot, United Kingdom

Ismail Koyuncu
Environmental Engineering Department
Civil Engineering Faculty
and
Prof. Dr. Dincer Topacik National Application and Research Center on Membrane Technologies (MEM-TEK)
Istanbul Technical University
Istanbul, Turkey

Woei-Jye Lau
Advanced Membrane Technology Research Centre (AMTEC)
Universiti Teknologi Malaysia
Johor Bahru, Malaysia

Yuan Liao
Sino-Canada Joint R&D Centre for Water and Environmental Safety
College of Environmental Science and Engineering
Nankai University
Tianjin, China

Takeshi Matsuura
Department of Chemical and Biological Engineering
University of Ottawa
Ottawa, Ontario, Canada

Lim Mim Mim
School of Biosciences and Medical Engineering
Faculty of Engineering
Universiti Teknologi Malaysia
Johor Bahru, Malaysia

Chris J. Mortimer
Biomaterials, Biofouling and Biofilms Engineering Laboratory (B^{3}EL)
The Systems and Process Engineering Centre (SPEC)
College of Engineering
Swansea University
Swansea, United Kingdom

and

Hybrisan Ltd.
Port Talbot, United Kingdom

Ali Moslehyani
Advanced Membrane Technology Research Centre (AMTEC)
Universiti Teknologi Malaysia
Johor Bahru, Malaysia

and

Department of Chemical and Biological Engineering
University of Ottawa
Ottawa, Ontario, Canada

Nor Azureen Mohamad Nor
Advanced Membrane Technology Research Centre (AMTEC)
and
Faculty of Engineering
Universiti Teknologi Malaysia
Johor Bahru, Malaysia

Faten Ermala Che Othman
Advanced Membrane Technology Research Centre (AMTEC)
Faculty of Engineering
Universiti Teknologi Malaysia
Johor Bahru, Malaysia

Mohd Hafiz Dzarfan Othman
Advanced Membrane Technology Research Centre (AMTEC)
Faculty of Engineering
Universiti Teknologi Malaysia
Johor Bahru, Malaysia

Mehmet Emin Pasaoglu
Environmental Engineering Department
Civil Engineering Faculty
and
Prof. Dr. Dincer Topacik National Application and Research Center on Membrane Technologies (MEM-TEK)
Istanbul Technical University
Istanbul, Turkey

Mukhlis A. Rahman
Advanced Membrane Technology Research Centre (AMTEC)
Faculty of Engineering
Universiti Teknologi Malaysia
Johor Bahru, Malaysia

Sadaki Samitsu
National Institute for Materials Science
Tsukuba, JAPAN

Reyhan Sengur-Tasdemir
Prof. Dr. Dincer Topacik National Application and Research Center on Membrane Technologies (MEM-TEK)
and
Nanoscience and Nanoengineering Department
Graduate School of Science, Engineering and Technology
Istanbul Technical University
Istanbul, Turkey

Anitha Senthamizhan
Fondazione Istituto Italiano di Tecnologia
Genova, Italy

Mohammad Mahdi A. Shirazi
Separation Processes Research Group (SPRG)
Department of Chemical Engineering
University of Kashan
Kashan, Iran

Naznin Sultana
Faculty of Engineering
School of Biosciences and Medical Engineering
and
Advanced Membrane Technology Research Centre (AMTEC)
Universiti Teknologi Malaysia
Johor Bahru, Malaysia

Tamer Uyar
Department of Fiber Science & Apparel Design
College of Human Ecology
Cornell University
Ithaca, New York

Rong Wang
Singapore Membrane Technology Centre
Nanyang Environment and Water Research Institute
and
School of Civil and Environmental Engineering
Nanyang Technological University
Singapore

Jonathan P. Widdowson
Biomaterials, Biofouling and Biofilms Engineering Laboratory (B^3EL)
The Systems and Process Engineering Centre (SPEC)
College of Engineering
Swansea University
Swansea, United Kingdom

Chris J. Wright
Biomaterials, Biofouling and Biofilms Engineering Laboratory (B^3EL)
The Systems and Process Engineering Centre (SPEC)
College of Engineering
Swansea University
Swansea, United Kingdom

Abdull Rahim Mohd Yusof
Department of Chemistry
Faculty of Science
and
Centre for Sustainable Nanomaterials
Ibnu Sina Institute for Scientific and Industrial Research (ISI-SIR)
Universiti Teknologi Malaysia
Johor Bahru, Malaysia

Norhaniza Yusof
Advanced Membrane Technology Research Centre (AMTEC)
Faculty of Engineering
Universiti Teknologi Malaysia
Johor Bahru, Malaysia

1 Electrospun Polymeric Nanofibers

Production, Characterization and Applications

Ali Moslehyani, Juhana Jaafar, Ahmad Fauzi Ismail, and Takeshi Matsuura

CONTENTS

1.1 INTRODUCTION

There are many different methods for nanofibrous membrane fabrication, which include melt fibrillation, island-in-sea methods, gas jet techniques, nanolithography, and self-assembly. Low cost, high production rate, and different modes of assembly are the benefits of the nanofibrous membrane. All high molecular weight polymers soluble in a solvent can be electrospun alone or by mixing with any type of miscible nanoparticles to produce polymeric nanofibrous membranes [1]. They can be fabricated from natural polymers, polymer blends, nanoparticle- or drug-impregnated polymers, and ceramic precursors in multiple structures such as beaded, ribbon-shaped, porous, and core-shell fibers [2,3].

The electrospinning process starts with the extrusion of a polymeric solution through a spinneret syringe needle (Figure 1.1). The difference between electrospinning and hollow fiberspinning is that the hollow fiber technique requires air or mechanical devices to create the extrusion force, while electrospinning needs a high voltage to charge the excluded solution. The repulsive force generated at the polymer solution surface then overcomes the surface tension of the solution and a jet erupts from the tip of the spinneret accelerating toward the region of lower electrical potential. While the solution jet is traveling through a distance, the solvent evaporates when the entanglement of the polymer chains prevents the jet from breaking up

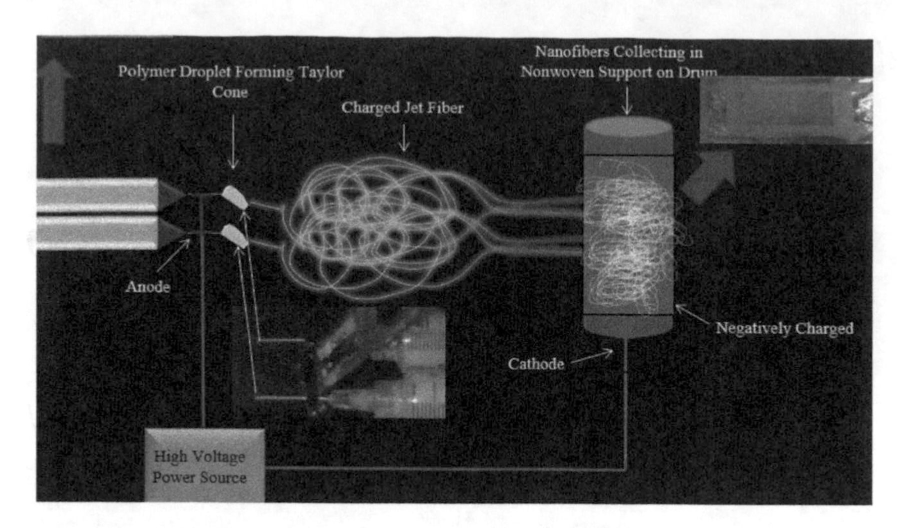

FIGURE 1.1 Electrospinning procedure.

during fiber formation. Generally, a grounded plate or a drum collector is used to collect the fibers from the spinneret [4–7].

The technology of electrospinning was invented in 1902 in the United States of America, which was however forgotten until 1990s. Then, during the last three decades researchers began new studies of electrospinning of nanofibers and their characterization among 200 universities and research institutes worldwide. The number of patents on new spinning methods and applications steeply increased and many companies such as e-Spin Technologies, Nano Technics, and KATO Tech are competitively involved to reap the unique benefits offered by electrospinning, while companies such as Donaldson and Freudenberg have been using nanofibers for air purification products for 20 years. The ability to form porous electrospun nanofibers means that the surface area of the fiber mesh can be increased tremendously [7–10].

Depending on the volatility of the solvent, porous nanofiber filaments are formed during electrospinning. Blending of dissimilar polymers is another way to fabricate porous filaments, by removing one polymer component with solvent while the other component remains insoluble [11–13]. Trajectory of the polymer solution jet is usually non uniform, which results in the formation of a nonwoven mesh. However, more ordered assembly of nanofibers is possible by managed fiber formation. Numerous approaches have been applied that produce aligned nanofibers with many degrees of order by controlling the electric field between the needle and collector plate, using a rotating mandrel, or the combination of both [5,14–16]. It has been demonstrated that a drum collector rotating at high speed can collect aligned fibers. Using a couple of conducting electrodes can also generate aligned fibers among the electrodes. To form a tubular scaffold, electrospun nanofibers were deposited on a spinning tube and the layer of nanofibers was pulled out consequently [17,18]. Another way of producing aligned nanofibers can be to use support electrodes to make an electric field profile that affects the flight of the electrospinning jet. With such versatility,

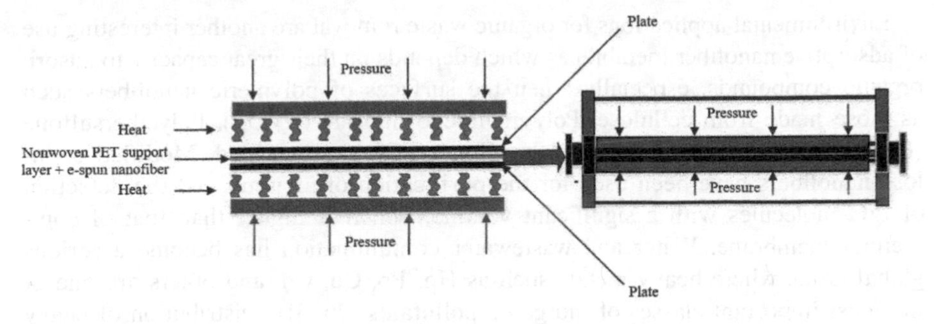

FIGURE 1.2 Illustration of the hot-pressing method.

nanofibrous membranes can be investigated for use in many applications. Presently, most tests use nonwoven fiber meshes made out of smooth fibers. Ceramic nanofibers derived from nonwoven electrospun fiber meshes have opened up many opportunities for application in different technologies. Heat-pressed nonwoven meshes have been demonstrated to be useful for high pressure applications (Figure 1.2). Therefore, the adaptability of electrospun fibers is clear and demonstrated by the proven results and current ongoing experimental studies in areas such as biotechnology, environmental engineering, healthcare, energy storage and generation, defense, and security [19–22].

1.2 ADVANTAGE OF USING ELECTROSPUN NANOFIBROUS MEMBRANE

The useful properties of electrospun nanofibrous membranes such as high absorbency, porosity, nanoscale interstitial space interconnectivity, and large surface area have meant that they have been used extensively in control of biological processes. Nanofibrous membranes with a nonwoven support have been used in various industrial biotechnology and environmental dynamic systems, because ligand molecules, bio-macromolecules, or even cells can be attached or hybridized within the nanofiber mat. Some of the examples are applications in protein filtration and wastewater purification by adsorptive barrier, or adsorptive catalyst mats, and, in the future, diagnostics and chemical detection as smart barriers [23,24].

Electrospun nanofibers have been fabricated in different sizes (50–500 nm) for dust removal in air purification by either adsorptive or physical filtration. Nanofibrous membranes have been shown to eliminate many types of contaminants from wastewater such as heavy metals, virus, and pharmaceutical byproducts. Interestingly there was no pore blockage due to the entrapment of particles inside the nanofiber mat, so the nanofibrous barrier could be efficiently reused after cleaning, which enabled the use of nanofibrous membranes during dynamic operation and allowed their use prior to many wastewater purification systems such as microfiltration, ultrafiltration, nanofiltration, and forward or reverse osmosis. These types of nanofibrous membranes are usually called adsorptive barriers that selectively capture specific target molecules by the incorporation of adsorbents in the nanofibers. Adsorptive barriers have been applied for protein and bio-products filtration in biotechnology [11,25].

Environmental applications for organic waste removal are another interesting use of adsorptive nanofiber membranes which depends on their great capacity to adsorb organic compounds, especially when the surfaces of polymeric nanofibers such as those made from cellulose, Polyvinylidene fluoride (PVDF), Polyethersulfone (PES), Polyethylenimine (PEI), Polysulfone (PSf) are modified. Modified cellulose nanofibers have been used for the purification of albumin and the detection of IgG molecules with a significant volume, which is higher than that of commercial membrane. Water and wastewater contamination has become a serious global issue, where heavy metals such as Hg, Pb, Cu, Cd, and others are one of the most important classes of inorganic pollutants [26–31]. Distribution of heavy metals in the environment is mainly attributed to the release of metal containing wastewaters from industries. For example, copper smelters release high amounts of Cd, one of the most mobile and toxic metals among the trace elements, into nearby streams It is not possible to completely remove some types of environmental contaminants such as metals by conventional water purification methods. Hence, adsorptive membranes will play a critical role in wastewater treatment to eliminate (or recycle) heavy metal ions in the future. Polymeric nanofibers, embedded with hydrated alumina/alumina hydroxide or iron oxides, were proven to be very effective for such purposes by removing toxic heavy metal ions such as As, Cr, and Pb via adsorption/chemisorption and electrostatic attraction mechanisms. Water quality level is also very sensitive to organic pollutants. Even though the concentration of the organics is usually no more than 1% of the pollutants in rivers, they tend to deplete its dissolved oxygen, making the water unable to sustain life. While the changes and pathways of metals in the environment have been studied to some extent, much less information is available on most commercial organic products due to their complex configurations [2,32–37]. Again, adsorptive membranes provide another method for eliminating organic molecules from wastewater. For example, β-cyclodextrin is a cyclic oligosaccharide composed of seven glucose units and a stereospecific toroidal structure. Due to its unique structure with a hydrophobic interior and hydrophilic exterior it can capture hydrophobic organic molecules from water by forming an inclusion complex. β-cyclodextrin has been incorporated into a poly(methyl methacrylate) nanofibrous membrane using a physical mixing method to develop an adsorptive membrane for organic waste removal [29,38–41]. Electrospun nanofibers have also received great attention for sensor applications because of their unique high surface area. This is one of the most desirable properties for improving the sensitivity of conduct metric sensors because a larger surface area will absorb more of a gas to be analyzed and change the sensor's conductivity more significantly. Nanofibers modified with a semiconductor oxide such as MoO_3, SnO_2, or TiO_2 show an electrical resistance that is sensitive to harmful chemical gases like ammonia and nitroxide. Single polypyrrole nanofibers containing SnO_2 were studied as biosensors for detecting biotin-labeled biomolecules such as DNA. Specific binding of the biomolecules to the nanofibers changes the electrical resistance of a single nanofiber. A fluorescent polymer, poly(acrylic acid)-poly(pyrene methanol), PAA-PM, was used as a sensing material for the detection of organic and inorganic waste. The fluorescence is quenched by adsorbed metal ions Fe^{3+} or Hg^{2+} or 2,4-dinitrotoluene (DNT) on the nanofiber surfaces [36,42–53].

1.3 POTENTIAL OF ELECTROSPUN NANOFIBROUS FOR WATER TREATMENT

Water purification requires several sequential steps to make collected water drinkable. These steps include removal of particulate matters, pathogens, and selected ions. Regarding the particulate removal, electrospun nanofibers are suitable for use as a size exclusion filter for particles of more than a few microns since the pore size of nanofiber ranges from submicron to few microns. A comprehensive study on the micro particles rejection using a heat-treated polysulfone (PSU) electrospun nonwoven membrane with a bubble point that corresponds to 4.6 μm showed that full flux recovery was possible for the particle size 7, 8, and 10 μm with more than 99% separation factor. Full flux recovery was also reported for 5 μm size particles with a separation factor greater than 90% using heat treated polyvinylidene fluoride (PVDF) membrane [54–58]. However, flux recovery was incomplete for particle sizes less than 3 μm. Examination of the cross-section of the membrane showed that for larger particles (more than 5 μm), the membrane acted as a size exclusion filter or screen filter with the particles trapped on the surface of the membrane. The flux could be recovered by stirring the solution inside the membrane module due to the low adhesion force between the particles and the fiber surface [58–62]. Another objective for the process of water treatment is the removal of heavy metals, which is a serious issue in many developing countries due to the rapid industrialization and impact of mining operations. Such heavy metal ions cannot be filtered out by microfiltration as the ions are too small. Instead of size exclusion, surface adsorption-based electrospun membrane has been developed. The high surface area of electrospun nanofiber meshes makes this technique suitable as more functional groups can be exposed on its surface. Open and interconnected pores formed by overlapping nanofibers allow the feed solution to pass through it. Arsenic (As) compounds are known as poisonous and cancer causing agents which are increasingly contaminating groundwater [63,64]. Chitosan electrospun nanofibers (ICS-ENF) modified by the incorporation of iron oxide were fabricated for the purpose of removing the trace amount of these arsenics from contaminated water. ICS-ENF was found to remove more than 90% of As under acidic to neutral conditions (pH 4.3–7.3). Thus its performance was better than pristine chitosan (CS) electrospun membrane which could adsorb As only little at neutral pH. When the pH was increased to 7.3, there was no adsorption of As even for ICS-ENF membrane. Adsorption by the nanofibrous membrane is dependent on the positive charge on its surface which interacts with the negatively charged As. At high pH, the surface charge of the electrospun nanofibrous membrane may turn negative and this will repel As ions, leading to their drop in adsorption capacity. In countries where there is limited available fresh water from the land, alternative source of water may come from the sea or from recycling waste water. These require technology that is able to separate salt and ions from the water source. While there is a well-established industrial purification technology such as reverse osmosis, other emerging technologies such as membrane distillation and ultra/nano-filtration are being tested with the hope of bringing down costs [65–72]. For ultrafiltration (0.1–0.01 μm) and nanofiltration (0.01–0.001 μm), the pore size of the electrospun membrane is too large to use without any modifications. Instead, the electrospun membrane can function as a supporting substrate to hold the separation layer as a thin film composite membrane. Whether the composite membrane is used for ultra-filtration or nanofiltration is dependent on the coated separation layer. High porosity and small fiber diameter of the electrospun membrane makes it an excellent supporting substrate as it gives a larger effective separation area (separation surface without underlying obstruction). Nano- and ultrafiltration membrane using electrospun membrane will theoretically give it a higher flux compared to other conventional membranes. To demonstrate effectiveness of electrospun thin film composite membrane in salt water purification, polyamide 6 nanofibers electrospun on polypropylene/polyethylene bio-component spun bond nonwoven fabric were used as the supporting substrate for interfacial *in-situ* polymerization of piperazine/trimesoyl chloride or m-phenylenediamine (MPD)/trimesoyl chloride, which function as the separation layer. An average rejection rate of 97.4% $CaCl_2$ and 96.3% NaCl was recorded for thin film nanofibrous composite (TFNC) membrane of MPD-Triethylamine (TEA)-Synferol AH (Sy-AH). Pure water flux of the TFNC was 22.5 L/m^2h while permeate flux of salt water was 12.5 L/m^2h. Using seawater, the TFNC membrane was able to retain more than 98% of the salt ions after three rounds of recirculation through the membrane [73–77].

Membrane distillation works by having a membrane to separate the salty water and pure water (permeate). There are a few parameters that affect the rate and efficiency of membrane distillation. First, there must be a temperature gradient between the feed side and permeate side. Traditionally, the separation layer is relatively thick so that heat conduction is limited from the feed to permeate, thus maintaining an optimal temperature gradient. The porosity, pore size, and tortuosity also affect the ease of water vapor passing through the membrane to form permeate. For an electrospun membrane to be used in membrane distillation, it must be able to maintain a separation between the feed and permeate water. Thus, a super hydrophobic membrane material is required [4,56,78–82]. Comparative tests on electrospun nanofibrous membrane and commercially available membrane showed that the liquid entry pressure (LEP) of electrospun PVDF membrane is less than 0.64 bar compared to 9 bar of commercial PTFE membrane. Electrospun PVDF membrane showed higher flux at a crossflow velocity of 80 mm/s than the PTFE and other commercial membranes. At a salinity lower than 50 g/kg, the thinner membrane has greater permeability and this is where the electrospun membrane has an advantage [78,82–87]. However, at higher salinity, flux for the nanofibrous membrane drops quickly while the PTFE membrane showed little difference. To increase the hydrophobicity of the membrane and maintain salt separation for longer duration, the surface of PVDF nanofibers was modified by coating silver nanoparticles through electroless silver-plating, followed by coating of membrane showed super hydrophobicity with a contact angle of 153° and water sliding angle below 10°. The flux of the membrane 32 L/m^2h was almost unchanged from the unmodified PVDF nanofibrous membrane and could be kept constant for the test duration of 8 h. Instead of super hydrophobic treatment of the surface, a commonly known super hydrophobic material polytetrafluoroethylene (PTFE) was used for fabrication of nanofibrous membrane. Since PTFE is difficult to be dissolved in most solvents, a suspension of PTFE fine particles in water was blended with water soluble polyvinyl alcohol (PVA). The PVA with PTFE particles was electrospun and the PVA component was removed through sintering up to 380°C for 30 min [4,88–93]. At this temperature the PTFE particles melted and fused together to form an interconnected nanofibrous network of PTFE. The resultant membrane showed a water contact angle of 156.7°. When tested for vacuum membrane distillation, a pure water flux of 15.8 kg/m^{2}h and a stable salt rejection of more than 98% for 10 h were recorded [85,94–96].

1.4 FUTURE PROSPECTIVE OF ELECTROSPINNING NANOFIBROUS MEMBRANE

Most of the nanofibrous composite membrane fabrication is currently based on an electrospinning approach with several weaknesses, such as difficulty of *in-situ* deposition of nanofibers, necessity for specialized equipment, low throughput, conducting targets, and high voltage. In addition, the electrospun nanofibrous mats have low mechanical strength due to the irregular and uncontrolled orientation of their component nanofibers and their low crystallinity. Therefore, more research on electrospinning is required to overcome these weaknesses. For example, nanofibrous synthesis throughput can be increased by further improvement of multi-needle and needleless electrospinning methods [39,97–102].

Recent developments in nanofiber electrospinning have meant that the application of the technology is moving at fast pace. Literature review shows that a large number of papers have been published on innovative production methods and this continues. However, to move beyond the current state of nanofibrous syntheses and applications toward technology maturation in commercial and industrial contexts, we need to address and overcome several challenges. First, it is imperative to standardize the newly developed synthetic methods, and second to bring these methods to a high engineering level such as physical properties to obtain the suitable and novel approaches of nanofibrous configurations including inter-fiber adhesion, smaller fiber dimension, and fiber surface modification. Nanofibers of more advanced configurations, such as core-shell, multilayer, and multicomponent nanofibers, may be prepared through methods similar to co-axial electrospinning [99,103–107].

Combining nanofibrous synthesis with other techniques, such as heat treatment, plasma treatment, chemical grafting, and control of fiber arrangement, is adaptable to further improve the surface properties of nanofibers. Eventually, the development of integrated nanofibrous synthesis strategies which are able to offer the best production procedures, such as the advanced nanofibrous structures of co-axial electrospinning, the high throughput of centrifugal jet spinning, and the *in situ* nanofibrous deposition of solution blow spinning, is highly desirable and should be the ultimate goal in the synthesis of nanofibers. Further, moving beyond synthesis, the next challenge is the identification of applications, those for which nanofibers are capable of revolutionizing a current practice. Nanofiber applications can be summarized in three important fields: energy, water, and healthcare. Most of these areas of application are still at the stage of emergence. In the case of energy application, despite the fact that nanofiber-based energy storage devices have demonstrated enhanced electrochemical properties and cycling performance, the potential of nanofiber-based technology in the area has not been fully exploited [108–113].

In biotechnology and environmental engineering, it is important to note that there are various efficient methods applicable for separation and purification, which include chemical precipitation, electrochemical processes, ion exchange, adsorption, and membrane filtration. Membrane adsorption (MA) combines the advantages of both adsorption and membrane filtration, offering an alternative option for effluent treatment. In particular, MA by electrospun adsorptive nanofiber membranes (EANMs) looks promising due to EANM's high permeation rate and adsorption capacity, and the possibility of their reuse over multiple times via appropriate desorption. For this reason, EANMs have attracted much interest recently as one of the most efficient techniques for biotechnological and environmental engineering via low operation cost. However, the use of suitable and compatible adsorbent particles is the most challenging parameter in the preparation of highly efficient EANMs, which needs more investigation [114–117].

Healthcare has a massive interest in the nanofibrous electrospinning process for both wound dressing and tissue engineering, since nanofiber mats are one of the most attractive options for helping the renewal of different kinds of tissues and cells. While the relationship between a wide range of nanofibrous supports and cells has been studied, the application of this research needs further proof-of-concept in terms of proliferation, cellular adhesion, and differentiation within the 3D structure of the electrospun mat. Additional analysis needs to be considered for the changes in cellular behaviors that are influenced by the topographical signals delivered by the nanofibrous supports. Obviously, more general *in vivo* investigations in human clinical trials are needed to demonstrate the influences and importance of nanofibrous process in biomedical engineering and healthcare [39,118–120].

1.5 CONCLUSIONS

Electrospinning is an old technology, which has been existing in the literature for more than 60 years. Although it is still immature, electrospinning is thought to be the best possible method for the fabrication of continuous nanofibers. A comprehensive as well as state-of-the-art review on electrospinning together with the applications of polymer nanofibers produced by this technique was made by this chapter. Relatively few polymers have been electrospun into nanofibers until now. Moreover, the understanding of the electrospinning process, the characterization of nanofibers, and the exploration of nanofiber applications specifically as nanocomposites seem to be very limited. Extensive researches and developments in all these three areas are required in the future.

REFERENCES

1. S.L. Shenoy, W.D. Bates, H.L. Frisch, G.E. Wnek, Role of chain entanglements on fiber formation during electrospinning of polymer solutions: Good solvent, non-specific polymer–polymer interaction limit, *Polymer*, 46 (2005) 3372–3384.
2. S. Ramakrishna, K. Fujihara, W.-E.Teo, T. Yong, Z. Ma, R. Ramaseshan, Electrospun nanofibers: Solving global issues, *Materials Today*, 9 (2006) 40–50.
3. R. Erdem, M. İlhan, E. Sancak, Analysis of EMSE and mechanical properties of sputter coated electrospun nanofibers, *Applied Surface Science*, 380 (2016) 326–330.
4. F.E. Ahmed, B.S. Lalia, R. Hashaikeh, A review on electrospinning for membrane fabrication: Challenges and applications, *Desalination*, 356 (2015) 15–30.

5. I. Sas, R.E. Gorga, J.A. Joines, K.A. Thoney, Literature review on superhydrophobic self-cleaning surfaces produced by electrospinning, *Journal of Polymer Science Part B: Polymer Physics*, 50 (2012) 824–845.
6. S. Nataraj, K. Yang, T. Aminabhavi, Polyacrylonitrile-based nanofibers—A state-of-the-art review, *Progress in Polymer Science*, 37 (2012) 487–513.
7. A.L. Andrady, *Science and Technology of Polymer Nanofibers*, Hoboken, NJ: John Wiley & Sons, 2008.
8. S. Agarwal, S. Jiang, Nanofibers and electrospinning, in: *Encyclopedia of Polymeric Nanomaterials* (S. Kobayashi and K. Müllen, Eds.), Springer, Berlin, Germany, 2015, pp. 1323–1337.
9. K. Chawla, *Fibrous Materials*, Cambridge, UK: Cambridge University Press, 2016.
10. A. Gittsegrad, Heterogeneous Structural Organization of Polystyrene Fibers Prepared by Electrospinning, (2018).
11. W.E. Teo, S. Ramakrishna, A review on electrospinning design and nanofibre assemblies, *Nanotechnology*, 17 (2006) R89.
12. A. Patanaik, R.D. Anandjiwala, R. Rengasamy, A. Ghosh, H. Pal, Nanotechnology in fibrous materials–a new perspective, *Textile Progress*, 39 (2007) 67–120.
13. Z. Li, C. Wang, *One-Dimensional Nanostructures: Electrospinning Technique and Unique Nanofibers*, Heidelberg, Germany: Springer, 2013.
14. Z. Kurban, A. Lovell, S.M. Bennington, D.W. Jenkins, K.R. Ryan, M.O. Jones, N.T. Skipper, W.I. David, A solution selection model for coaxial electrospinning and its application to nanostructured hydrogen storage materials, *The Journal of Physical Chemistry C*, 114 (2010) 21201–21213.
15. F.K. Ko, Y. Wan, *Introduction to Nanofiber Materials*, Cambridge, UK: Cambridge University Press, 2014.
16. B. Bera, Literature review on electrospinning process (A fascinating fiber fabrication technique), *Imperial Journal of Interdisciplinary Research*, 2 (2016) 8.
17. E. Llorens Domenjó, Advanced electrospun scaffolds based on biodegradable polylactide and poly (butylene succinate) for controlled drug delivery and tissue engineering, doctoral thesis, Universitat Politècnica de Catalunya (2014).
18. Z.-K. Xu, X.-J. Huang, L.-S. Wan, *Surface Engineering of Polymer Membranes*, Berlin, Germany: Springer Science & Business Media, 2009.
19. S. Honarbakhsh, *Polymer-Based Nanofibers Impregnated with Drug Infused Plant Virus Particles as a Responsive Fabric for Therapeutic Delivery*, Raleigh, NC: North Carolina State University, 2010.
20. C. Wang, T.-T. Hsiue, Core–shell fibers electrospun from phase-separated blend solutions: Fiber formation mechanism and unique energy dissipation for synergistic fiber toughness, *Biomacromolecules*, 18 (2017) 2906–2917.
21. P. Heikkilä, *Nanostructured Fibre Composites, and Materials for Air Filtration*, Tampere, Finland: Tampere University of Technology, 2008.
22. A.K. Alves, C.P. Bergmann, F.A. Berutti, *Novel Synthesis and Characterization of Nanostructured Materials*, Berlin, Germany: Springer, 2013.
23. G. Cicala, S. Mannino, Applications of epoxy/thermoplastic blends, in: *Handbook of Epoxy Blends*, 2016, pp. 1–25.
24. M.Z.B. Mukhlish, Development of flexible ceramic nanofiber membranes for energy and environmental applications. Dissertation, Kagoshima University, 2018.
25. D. Liang, B.S. Hsiao, B. Chu, Functional electrospun nanofibrous scaffolds for biomedical applications, *Advanced Drug Delivery Reviews*, 59 (2007) 1392–1412.
26. S. Wu, F. Li, H. Wang, L. Fu, B. Zhang, G. Li, Effects of poly (vinyl alcohol)(PVA) content on preparation of novel thiol-functionalized mesoporous PVA/SiO_2 composite nanofiber membranes and their application for adsorption of heavy metal ions from aqueous solution, *Polymer*, 51 (2010) 6203–6211.

27. P. Kampalanonwat, P. Supaphol, Preparation and adsorption behavior of aminated electrospun polyacrylonitrile nanofiber mats for heavy metal ion removal, *ACS Applied Materials & Interfaces*, 2 (2010) 3619–3627.
28. L.-L. Min, Z.-H. Yuan, L.-B. Zhong, Q. Liu, R.-X. Wu, Y.-M. Zheng, Preparation of chitosan based electrospun nanofiber membrane and its adsorptive removal of arsenate from aqueous solution, *Chemical Engineering Journal*, 267 (2015) 132–141.
29. E. Salehi, P. Daraei, A.A. Shamsabadi, A review on chitosan-based adsorptive membranes, *Carbohydrate Polymers*, 152 (2016) 419–432.
30. A. Moslehyani, T. Matsuura, A.F. Ismail, M.H.D. Othman, A review on application of electrospun polymeric nanofibers, in: *National Congress on Membrane Technology 2016* (*NATCOM 2016*), 2016.
31. A. Celebioglu, T. Uyar, Electrospinning of polymer-free nanofibers from cyclodextrin inclusion complexes, *Langmuir*, 27 (2011) 6218–6226.
32. E. Pehlivan, T. Altun, S. Parlayıcı, Utilization of barley straws as biosorbents for Cu^{2+} and Pb^{2+} ions, *Journal of Hazardous Materials*, 164 (2009) 982–986.
33. P. Brown, S. Gill, S. Allen, Metal removal from wastewater using peat, *Water Research*, 34 (2000) 3907–3916.
34. T. Hua, R. Haynes, Y.-F. Zhou, A. Boullemant, I. Chandrawana, Potential for use of industrial waste materials as filter media for removal of Al, Mo, As, V and Ga from alkaline drainage in constructed wetlands–adsorption studies, *Water Research*, 71 (2015) 32–41.
35. K. Rao, M. Mohapatra, S. Anand, P. Venkateswarlu, Review on cadmium removal from aqueous solutions, *International Journal of Engineering, Science and Technology*, 2 (2010).
36. A. Kulkarni, V. Bambole, P. Mahanwar, Electrospinning of polymers, their modeling and applications, *Polymer-Plastics Technology and Engineering*, 49 (2010) 427–441.
37. A.I. Zouboulis, E.N. Peleka, P. Samaras, Removal of toxic materials from aqueous streams, in: *Mineral Scales and Deposits* (Z. Amjad and K. Demadis, Eds.), Elsevier, Amsterdam, the Netherlands, 2015, pp. 443–473.
38. C. Manta, G. Peralta-Altier, L. Gioia, M.F. Méndez, G. Seoane, K. Ovsejevi, Synthesis of a thiol-β-cyclodextrin, a potential agent for controlling enzymatic browning in fruits and vegetables, *Journal of Agricultural and Food Chemistry*, 61 (2013) 11603–11609.
39. N. Bhardwaj, S.C. Kundu, Electrospinning: A fascinating fiber fabrication technique, *Biotechnology Advances*, 28 (2010) 325–347.
40. R. Zhao, Y. Wang, X. Li, B. Sun, C. Wang, Synthesis of β-cyclodextrin-based electrospunnanofiber membranes for highly efficient adsorption and separation of methylene blue, *ACS Applied Materials & Interfaces*, 7 (2015) 26649–26657.
41. T. Uyar, R. Havelund, J. Hacaloglu, F. Besenbacher, P. Kingshott, Functional electrospun polystyrene nanofibers incorporating α-, β-, and γ-cyclodextrins: Comparison of molecular filter performance, *ACS Nano*, 4 (2010) 5121–5130.
42. P. Suja, C. Reshmi, P. Sagitha, A. Sujith, Electrospun nanofibrous membranes for water purification, *Polymer Reviews*, 57 (2017) 467–504.
43. D. Noreña-Caro, M. Álvarez-Láinez, Functionalization of polyacrylonitrile nanofibers with β-cyclodextrin for the capture of formaldehyde, *Materials & Design*, 95 (2016) 632–640.
44. J. Kang, J. Han, Y. Gao, T. Gao, S. Lan, L. Xiao, Y. Zhang, G. Gao, H. Chokto, A. Dong, Unexpected enhancement in antibacterial activity of N-halamine polymers from spheres to fibers, *ACS Applied Materials & Interfaces*, 7 (2015) 17516–17526.
45. J. Weiss, K. Kanjanapongkul, S. Wongsasulak, T. Yoovidhya, Electrospunfibers: Fabrication, functionalities and potential food industry applications, in: *Nanotechnology in the Food, Beverage and Nutraceutical Industries*, Elsevier, Cambridge, UK, 2012, pp. 362–397.

46. S. Torres-Giner, R. Pérez-Masiá, J.M. Lagaron, A review on electrospun polymer nanostructures as advanced bioactive platforms, *Polymer Engineering & Science*, 56 (2016) 500–527.
47. R. Das, S.B.A. Hamid, M.E. Ali, A.F. Ismail, M. Annuar, S. Ramakrishna, Multifunctional carbon nanotubes in water treatment: The present, past and future, *Desalination*, 354 (2014) 160–179.
48. B. Satilmis, T. Uyar, Superhydrophobic hexamethylene diisocyanate modified hydrolyzed polymers of intrinsic microporosity electrospun ultrafine fibrous membrane for the adsorption of organic compounds and oil/water separation, *ACS Applied Nano Materials*, 1 (2018) 1631–1640.
49. X. Wang, C. Drew, S.-H. Lee, K.J. Senecal, J. Kumar, L.A. Samuelson, Electrospunnanofibrous membranes for highly sensitive optical sensors, *Nano Letters*, 2 (2002) 1273–1275.
50. W. Wang, N.-K. Wong, M. Sun, C. Yan, S. Ma, Q. Yang, Y. Li, Regenerable fluorescent nanosensors for monitoring and recovering metal ions based on photoactivatable monolayer self-assembly and host–guest interactions, *ACS Applied Materials & Interfaces*, 7 (2015) 8868–8875.
51. N. Brandon, V. Di Noto, K. More, R. Unocic, D. Cullen, Program—Symposium M: Fuel cells, electrolyzers and other electrochemical energy systems, Materials Research Society, Warrendale, PA, 2014, https://mafiadoc.com/programasymposium-m-fuel-cells-electrolyzers-_598e00dd1723ddcd6988ad69.html (October 30, 2017).
52. M.A. Khan, F. Qazi, Z. Hussain, M.U. Idrees, S. Soomro, S. Soomro, Recent Trends in Electrochemical Detection of NH_3, H_2S and NO_x Gases, *International Journal of Electrochemical Science*, 12 (2017) 1711–1733.
53. G. Wypych, *Handbook of Fillers*, ChemTec Publishing, Scarborough, Canada, 2016.
54. R. Gopal, S. Kaur, C.Y. Feng, C. Chan, S. Ramakrishna, S. Tabe, T. Matsuura, Electrospun nanofibrous polysulfone membranes as pre-filters: Particulate removal, *Journal of Membrane Science*, 289 (2007) 210–219.
55. S. Kaur, Z. Ma, R. Gopal, G. Singh, S. Ramakrishna, T. Matsuura, Plasma-induced graft copolymerization of poly (methacrylic acid) on electrospun poly (vinylidene fluoride) nanofiber membrane, *Langmuir*, 23 (2007) 13085–13092.
56. C. Feng, K. Khulbe, T. Matsuura, R. Gopal, S. Kaur, S. Ramakrishna, M. Khayet, Production of drinking water from saline water by air-gap membrane distillation using polyvinylidene fluoride nanofiber membrane, *Journal of Membrane Science*, 311 (2008) 1–6.
57. S.A.A.N. Nasreen, S. Sundarrajan, S.A.S. Nizar, R. Balamurugan, S. Ramakrishna, Advancement in electrospun nanofibrous membranes modification and their application in water treatment, *Membranes*, 3 (2013) 266–284.
58. S. Homaeigohar, M. Elbahri, Nanocomposite electrospun nanofiber membranes for environmental remediation, *Materials*, 7 (2014) 1017–1045.
59. S. Kaur, S. Sundarrajan, D. Rana, R. Sridhar, R. Gopal, T. Matsuura, S. Ramakrishna, The characterization of electrospun nanofibrous liquid filtration membranes, *Journal of Materials Science*, 49 (2014) 6143–6159.
60. C. Feng, K. Khulbe, T. Matsuura, S. Tabe, A. Ismail, Preparation and characterization of electro-spun nanofiber membranes and their possible applications in water treatment, *Separation and Purification Technology*, 102 (2013) 118–135.
61. C. Feng, K. Khulbe, T. Matsuura, Recent progress in the preparation, characterization, and applications of nanofibers and nanofiber membranes via electrospinning/interfacial polymerization, *Journal of Applied Polymer Science*, 115 (2010) 756–776.
62. L. Zhang, L.-G. Liu, F.-L. Pan, D.-F. Wang, Z.-J. Pan, Effects of heat treatment on the morphology and performance of PSU electrospun nanofibrous membrane, *Journal of Engineered Fabrics & Fibers (JEFF)*, 7 (2012).
63. X. Wang, B.S. Hsiao, Electrospunnanofiber membranes, *Current Opinion in Chemical Engineering*, 12 (2016) 62–81.

64. J.C. Halbur, *Inorganic Surface Modification of Nonwoven Polymeric Substrates*, Raleigh, NC: North Carolina State University, 2014.
65. L.-L. Min, L.-B. Zhong, Y.-M. Zheng, Q. Liu, Z.-H. Yuan, L.-M. Yang, Functionalized chitosan electrospunnanofiber for effective removal of trace arsenate from water, *Scientific Reports*, 6 (2016) 32480.
66. G. Ognibene, C.M. Gangemi, A. D'Urso, R. Purrello, G. Cicala, M.E. Fragalà, Combined approach to remove and fast detect heavy metals in water based on PES–TiO_2 electrospun mats and porphyrin chemosensors, *ACS Omega*, 3 (2018) 7182–7190.
67. S. Muthu Prabhu, C. Chuaicham, K. Sasaki, A mechanistic approach for the synthesis of carboxylate-rich carbonaceous biomass-doped lanthanum-oxalate nanocomplex for arsenate adsorption, *ACS Sustainable Chemistry & Engineering*, 6 (2018) 6052–6063.
68. W.A. Sarhan, H.M. Azzazy, High concentration honey chitosan electrospun nanofibers: Biocompatibility and antibacterial effects, *Carbohydrate Polymers*, 122 (2015) 135–143.
69. Y.-T. Jia, J. Gong, X.-H. Gu, H.-Y. Kim, J. Dong, X.-Y. Shen, Fabrication and characterization of poly (vinyl alcohol)/chitosan blend nanofibers produced by electrospinning method, *Carbohydrate Polymers*, 67 (2007) 403–409.
70. J. Fang, H. Niu, T. Lin, X. Wang, Applications of electrospun nanofibers, *Chinese Science Bulletin*, 53 (2008) 2265.
71. K. Ohkawa, D. Cha, H. Kim, A. Nishida, H. Yamamoto, Electrospinning of chitosan, *Macromolecular Rapid Communications*, 25 (2004) 1600–1605.
72. S. Haider, S.-Y. Park, Preparation of the electrospun chitosan nanofibers and their applications to the adsorption of Cu (II) and Pb (II) ions from an aqueous solution, *Journal of Membrane Science*, 328 (2009) 90–96.
73. J.-S. WU, X.-Y. JIN, Influence factors analysis for bio-component thermo-bonded nonwovens strength ration in machine and cross direction [J], *Nonwovens*, 6 (2008) 013.
74. E.B. Bond, Bicomponent fibers comprising a thermoplastic polymer surrounding a starch rich core, United States patent US 6,783,854, 2004.
75. A. Carbonell, T. Boronat, E. Fages, S. Gironés, E. Sanchez-Zapata, J. Perez-Alvarez, L. Sanchez-Nacher, D. Garcia-Sanoguera, Wet-laid technique with *Cyperus esculentus*: Development, manufacturing and characterization of a new composite, *Materials & Design*, 86 (2015) 887–893.
76. K. Shalumon, N. Binulal, N. Selvamurugan, S. Nair, D. Menon, T. Furuike, H. Tamura, R. Jayakumar, Electrospinning of carboxymethyl chitin/poly (vinyl alcohol) nanofibrous scaffolds for tissue engineering applications, *Carbohydrate Polymers*, 77 (2009) 863–869.
77. D. Wang, Rapid, controllable and environmentally benign fabrication of thermoplastic nanofibers and applications, University of California, Davis, 2008.
78. B.S. Lalia, E. Guillen-Burrieza, H.A. Arafat, R. Hashaikeh, Fabrication and characterization of polyvinylidenefluoride-co-hexafluoropropylene (PVDF-HFP) electrospun membranes for direct contact membrane distillation, *Journal of Membrane Science*, 428 (2013) 104–115.
79. J. Prince, G. Singh, D. Rana, T. Matsuura, V. Anbharasi, T. Shanmugasundaram, Preparation and characterization of highly hydrophobic poly (vinylidene fluoride)–Clay nanocomposite nanofiber membranes (PVDF–clay NNMs) for desalination using direct contact membrane distillation, *Journal of Membrane Science*, 397 (2012) 80–86.
80. R. Gopal, S. Kaur, Z. Ma, C. Chan, S. Ramakrishna, T. Matsuura, Electrospunnanofibrous filtration membrane, *Journal of Membrane Science*, 281 (2006) 581–586.
81. M. Essalhi, M. Khayet, Self-sustained webs of polyvinylidene fluoride electrospun nanofibers at different electrospinning times: 1. Desalination by direct contact membrane distillation, *Journal of Membrane Science*, 433 (2013) 167–179.

82. E. Drioli, A. Ali, F. Macedonio, Membrane distillation: Recent developments and perspectives, *Desalination*, 356 (2015) 56–84.
83. E. Drioli, A. Ali, S. Simone, F. Macedonio, S. Al-Jlil, F. Al Shabonah, H. Al-Romaih, O. Al-Harbi, A. Figoli, A. Criscuoli, Novel PVDF hollow fiber membranes for vacuum and direct contact membrane distillation applications, *Separation and Purification Technology*, 115 (2013) 27–38.
84. N. Hamzah, C. Leo, Membrane distillation of saline with phenolic compound using superhydrophobic PVDF membrane incorporated with TiO_2 nanoparticles: Separation, fouling and self-cleaning evaluation, *Desalination*, 418 (2017) 79–88.
85. M.M.A. Shirazi, A. Kargari, M. Tabatabaei, Evaluation of commercial PTFE membranes in desalination by direct contact membrane distillation, *Chemical Engineering and Processing: Process Intensification*, 76 (2014) 16–25.
86. Y. Huang, Q.-L. Huang, H. Liu, C.-X. Zhang, Y.-W. You, N.-N. Li, C.-F. Xiao, Preparation, characterization, and applications of electrospun ultrafine fibrous PTFE porous membranes, *Journal of Membrane Science*, 523 (2017) 317–326.
87. M. Essalhi, M. Khayet, Self-sustained webs of polyvinylidene fluoride electrospun nanofibers at different electrospinning times: 2. Theoretical analysis, polarization effects and thermal efficiency, *Journal of Membrane Science*, 433 (2013) 180–191.
88. M. Wang, G. Liu, H. Yu, S.-H. Lee, L. Wang, J. Zheng, T. Wang, Y. Yun, J.K. Lee, ZnO nanorod array modified PVDF membrane with superhydrophobic surface for vacuum membrane distillation application, *ACS Applied Materials & Interfaces*, 10 (2018) 13452–13461.
89. Z.-M. Huang, Y.-Z. Zhang, M. Kotaki, S. Ramakrishna, A review on polymer nanofibers by electrospinning and their applications in nanocomposites, *Composites Science and Technology*, 63 (2003) 2223–2253.
90. W. Qing, X. Shi, Y. Deng, W. Zhang, J. Wang, C.Y. Tang, Robust superhydrophobic-superoleophilic polytetrafluoroethylene nanofibrous membrane for oil/water separation, *Journal of Membrane Science*, 540 (2017) 354–361.
91. S. Morimune, M. Kotera, T. Nishino, K. Goto, K. Hata, Poly (vinyl alcohol) nanocomposites with nanodiamond, *Macromolecules*, 44 (2011) 4415–4421.
92. B.L. Anneaux, R. Ballard, D.P. Garner, Electrospinning of PTFE with high viscosity materials, U.S. Patent No. 8,178,030, 2012.
93. Y. Liao, C.-H. Loh, M. Tian, R. Wang, A.G. Fane, Progress in electrospun polymeric nanofibrous membranes for water treatment: Fabrication, modification and applications, *Progress in Polymer Science*, 77 (2017) 69–94.
94. B.L. Anneaux, R. Ballard, D.P. Garner, Electrospinning of PTFE with high viscosity materials, U.S. Patent Application No. 13/446,300, 2013.
95. W. Kang, H. Zhao, J. Ju, Z. Shi, C. Qiao, B. Cheng, Electrospun poly (tetrafluoroethylene) nanofiber membranes from PTFE-PVA-BA-H 2 O gel-spinning solutions, *Fibers and Polymers*, 17 (2016) 1403–1413.
96. J. Li, Q. Zhong, Y. Yao, S. Bi, T. Zhou, X. Guo, M. Wu, T. Feng, R. Xiang, Electrochemical performance and thermal stability of the electrospun PTFE nanofiber separator for lithium-ion batteries, *Journal of Applied Polymer Science*, 135 (2018) 46508.
97. S. Mohammadzadehmoghadam, Y. Dong, S. Barbhuiya, L. Guo, D. Liu, R. Umer, X. Qi, Y. Tang, Electrospinning: Current Status and Future Trends, in: *Nano-size Polymers* (S. Fakirov, Ed.), Springer, Cham, Switzerland, 2016, pp. 89–154.
98. S.W. Choi, S.M. Jo, W.S. Lee, Y.R. Kim, An electrospun poly (vinylidene fluoride) nanofibrous membrane and its battery applications, *Advanced Materials*, 15 (2003) 2027–2032.
99. Q.P. Pham, U. Sharma, A.G. Mikos, Electrospinning of polymeric nanofibers for tissue engineering applications: A review, *Tissue Engineering*, 12 (2006) 1197–1211.

100. Z. Ma, M. Kotaki, R. Inai, S. Ramakrishna, Potential of nanofiber matrix as tissue-engineering scaffolds, *Tissue Engineering*, 11 (2005) 101–109.
101. K. Yoon, B.S. Hsiao, B. Chu, High flux ultrafiltration nanofibrous membranes based on polyacrylonitrile electrospun scaffolds and crosslinked polyvinyl alcohol coating, *Journal of Membrane Science*, 338 (2009) 145–152.
102. H. Ma, C. Burger, B.S. Hsiao, B. Chu, Fabrication and characterization of cellulose nanofiber based thin-film nanofibrous composite membranes, *Journal of Membrane Science*, 454 (2014) 272–282.
103. A. Frenot, I.S. Chronakis, Polymer nanofibers assembled by electrospinning, *Current Opinion in Colloid & Interface Science*, 8 (2003) 64–75.
104. K. Jayaraman, M. Kotaki, Y. Zhang, X. Mo, S. Ramakrishna, Recent advances in polymer nanofibers, *Journal of Nanoscience and Nanotechnology*, 4 (2004) 52–65.
105. Y. Zhang, C.T. Lim, S. Ramakrishna, Z.-M. Huang, Recent development of polymer nanofibers for biomedical and biotechnological applications, *Journal of Materials Science: Materials in Medicine*, 16 (2005) 933–946.
106. T. Subbiah, G. Bhat, R. Tock, S. Parameswaran, S. Ramkumar, Electrospinning of nanofibers, *Journal of Applied Polymer Science*, 96 (2005) 557–569.
107. Y. Dzenis, Spinning continuous fibers for nanotechnology, *Science*, 304 (2004) 1917–1919.
108. V. Beachley, X. Wen, Polymer nanofibrous structures: Fabrication, biofunctionalization, and cell interactions, *Progress in Polymer Science*, 35 (2010) 868–892.
109. S. Zhong, Y. Zhang, C.T. Lim, Fabrication of large pores in electrospun nanofibrous scaffolds for cellular infiltration: A review, *Tissue Engineering Part B: Reviews*, 18 (2011) 77–87.
110. C. Burger, B.S. Hsiao, B. Chu, Nanofibrous materials and their applications, Annual Review of Materials Research, 36 (2006) 333–368.
111. S. Cheng, D. Shen, X. Zhu, X. Tian, D. Zhou, L.-J. Fan, Preparation of nonwoven polyimide/silica hybrid nanofiberous fabrics by combining electrospinning and controlled in situ sol–gel techniques, *European Polymer Journal*, 45 (2009) 2767–2778.
112. R. Nayak, R. Padhye, I.L. Kyratzis, Y.B. Truong, L. Arnold, Recent advances in nanofibre fabrication techniques, *Textile Research Journal*, 82 (2012) 129–147.
113. L. Melone, L. Altomare, I. Alfieri, A. Lorenzi, L. De Nardo, C. Punta, Ceramic aerogels from TEMPO-oxidized cellulose nanofibre templates: Synthesis, characterization, and photocatalytic properties, *Journal of Photochemistry and Photobiology A: Chemistry*, 261 (2013) 53–60.
114. R. Ramaseshan, S. Sundarrajan, Y. Liu, R. Barhate, N.L. Lala, S. Ramakrishna, Functionalized polymer nanofibre membranes for protection from chemical warfare stimulants, *Nanotechnology*, 17 (2006) 2947.
115. R. Zhang, Y. Liu, M. He, Y. Su, X. Zhao, M. Elimelech, Z. Jiang, Antifouling membranes for sustainable water purification: Strategies and mechanisms, *Chemical Society Reviews*, 45 (2016) 5888–5924.
116. S. Thomas, Y. Grohens, N. Ninan, *Nanotechnology Applications for Tissue Engineering*, Oxford, UK: William Andrew, 2015.
117. D. Grafahrend, Nanofibers and nanoparticles by electrostatic processing for medical and pharmaceutical applications, in Lehrstuhl für Textilchemie und Makromolekulare Chemie, 2010.
118. T. Xu, J.M. Miszuk, Y. Zhao, H. Sun, H. Fong, Bone tissue engineering: Electrospun polycaprolactone 3D nanofibrous scaffold with interconnected and hierarchically structured pores for bone tissue engineering (Adv. Healthcare Mater. 15/2015), *Advanced Healthcare Materials*, 4 (2015) 2237–2237.

119. H. Chen, P. Jia, H. Kang, H. Zhang, Y. Liu, P. Yang, Y. Yan, G. Zuo, L. Guo, M. Jiang, Upregulating Hif-1α by hydrogel nanofibrous scaffolds for rapidly recruiting angiogenesis relative cells in diabetic wound, *Advanced Healthcare Materials*, 5 (2016) 907–918.
120. P. Zahedi, I. Rezaeian, S.O. Ranaei-Siadat, S.H. Jafari, P. Supaphol, A review on wound dressings with an emphasis on electrospun nanofibrous polymeric bandages, *Polymers for Advanced Technologies*, 21 (2010) 77–95.

2 Electrospun Antimicrobial Membranes-Functionality and Morphology

Chris J. Mortimer, Sean James, Nidal Hilal, and Chris J. Wright

CONTENTS

2.1 INTRODUCTION

Electrospinning is a commonly used technique for the fabrication of micron and sub-micron non-woven materials. It is an area that has been subject to a vast amount of research due to the technique's facile nature and ability to rapidly develop polymer fibers with a diameter on the nanoscale. Scientific interest in nanotechnology and the relatively recent ability to image and measure fibers using Scanning Electron Microscopy (SEM) has led to an increase in the use of electrospinning.

Membrane-based technologies are increasing in popularity, particularly over the last few decades. This is due to their high separation efficiencies, relatively low costs and ease of operation (Ahmed et al. 2015). A membrane acts as a barrier, typically between two phases that allow selected substances to be transported from one side to another (Ulbricht 2006). Membranes can be classified as either porous or dense depending on its structure (Takht Ravanchi et al. 2009). The selectivity and transport properties of a membrane are strongly dependent on its pore structure. Membranes can be broken down into classification based on their pore size, including microfiltration (MF), ultrafiltration (UF) and nanofiltration (NF). Electrospinning is becoming a more commonly used technique for the fabrication of membranes, particularly focused on micro- and ultrafiltration. The ability to electrospin fibers of varying diameters with a tuneable pore size make it a very suitable method for their fabrication.

2.2 ELECTROSPINNING

Electrospinning as a technique was first documented by Rayleigh in 1878 (Strutt and Rayleigh 1878), with the first patents filed by Cooley (1900) and a series of patents were filed by Anton Formhals during the period from 1934 to 1944 (Formhals 1934). Electrospinning uses a high voltage power supply to create a large potential difference between a grounded "collector" structure and a polymer solution or melt being delivered at a constant rate through an aperture, such as a blunt end needle. As the voltage is increased the like charges within the polymer fluid directly oppose surface tension, resulting in the normally spherical droplet at the aperture distending into a conical shape. This cone is referred to as the "Taylor" cone, after Sir Geoffrey Taylor who first mathematically modeled the phenomenon (Taylor 1964, 1969; Yarin et al. 2001). At a critical voltage the electrostatic attractive force between the solution and the collector causes a jet of the polymer solution to be expelled from the cone tip towards the grounded collector surface. This jet then undergoes a whipping instability and dries in flight, depositing the nanofibers on the collector (Bhardwaj and Kundu 2010) (Figure 2.1).

Electrospun fiber morphology can be finely tuned by varying a number of processing and solution parameters. An increase in polymer concentration in a given solution will result in an increase in the solution viscosity. It has been shown by a number of research groups that fiber diameter decreases with polymer concentration (Katti et al. 2004; Mit-uppatham et al. 2004; Mo et al. 2004; Tan et al. 2005). Although this offers a certain level of control as the polymer concentration decreases, it also leads to beaded fibers and a wider size distribution. Mit-uppatham et al. showed this with Polyamide-6 showing at low concentrations a large number of droplets.

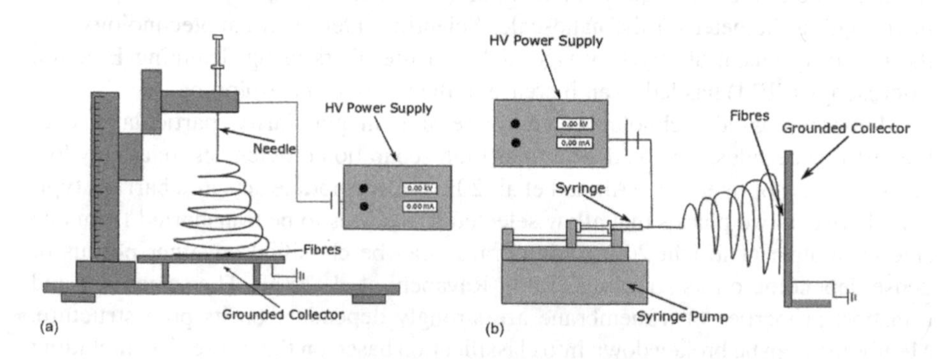

FIGURE 2.1 A schematic diagram of electrospinning apparatus in (a) a vertical setup and (b) a horizontal setup.

As solution concentration increased the number of beads decreased and fiber morphology improved (Mit-uppatham et al. 2004). The conductivity of the solution is a crucial parameter. When the conductivity is low this results in the production of fibers with a greater diameter. This is most likely due to poor jet elongation (Tan et al. 2005). As the conductivity increases, the fiber diameter tends to decrease. This shows a clear relationship between the conductivity of a solution and the level of jet elongation. As the conductivity of the solution is increased, the jet undergoes a higher degree of elongation along its axis due to the repulsion of charges under the electric field (Son et al. 2004). The viscosity of a polymer solution will increase with increased molecular weight, meaning a lower concentration of polymer is required to form non-beaded fibers with tight size distributions. At higher molecular weights there are a greater number of chain entanglements and therefore a higher viscosity at equivalent concentration compared to a lower molecular weight. As a result even at a low polymer concentration a high molecular weight polymer can provide sufficient chain entanglements to overcome the effects of surface tension and result in a uniform jet (Tan et al. 2005). A solvent with a high vapor pressure at normal temperatures can be referred to as being volatile. During the electrospinning process this results in the jet spending less time in the elongation stage undergoing stretching as the solvent evaporates off. Therefore, a more volatile solvent (higher vapor pressure) will result in fibers of a larger average diameter. This allows the researcher to control the fiber diameter to a certain degree through the choice of solvent based on its vapor pressure, although a solvent with a particularly low vapor pressure may be unsuitable and result in the deposition of wet fibers, with coalescence at fiber junctions. It has been shown by a number of research groups that the applied voltage does not have a great effect on the fiber diameter (Buchko et al. 1999; Katti et al. 2004; Tan et al. 2005). Electrospinning will only occur at a certain range of voltage although if the voltage is too high this can result in multiple jets forming which can decrease the diameter of fibers but widen the size distribution (Tan et al. 2005). Generally, fiber diameter decreases as applied voltage increases due to the greater columbic forces causing greater stretching of the solution and resulting in a smaller fiber diameter with fibers drying more quickly (Mo et al. 2004; Bhardwaj and Kundu 2010). Baumgarten showed that as voltage increases jet diameter initially decreases and then increases as the voltage continues to increase (Baumgarten 1971). Because of this flow rate must increase as the voltages increases. It has also been reported that deposition rates increase with increased voltage (Buchko et al. 1999). This could explain the inconsistent reports of the effect that voltage has on fiber diameter and other parameters such as the feed rate and tip to collector distance must also be considered when observing the effects of applied voltage (Andrady 2008). The flow rate is the rate at which the solution is pumped through the needle to maintain a droplet of solution available for Taylor cone formation. The flow rate of the solution has not been shown to have a significant effect on the fiber diameter or morphology (Tan et al. 2005), although some studies have shown an increase in flow rate can result in an increase in fiber diameter (Kidoaki et al. 2006). The ideal flow rate should match the rate at which the solution is being ejected from the tip (Andrady 2008). At lower flow rates electrospinning tends to be intermittent, with the Taylor cone often receding into the needle. If the distance between the spinneret and the collector is either too small or too large, the result is beaded fibers. The ideal distance is the minimum distance required to allow the fibers enough time to dry between the spinneret and the collector (Baumgarten 1971; Buchko et al. 1999). At distances too small, fibers are deposited wet, with insufficient drying time resulting in coalescence at fiber junctions. Tip to collector distance is usually selected according to the vapor pressure of the solvent. When increasing the distance, the voltage applied to the spinneret needs to be increased to maintain the electric field.

Although needle electrospinning allows excellent control over both fiber diameter and their composition, it has an extremely low throughput and is often limited to flow rates of less than 0.5 mL per hour. In contrast, a free-surface electrospinning set up, such as the Elmarco NanoSpider™ system, is capable of forming fibers with a throughput significantly greater than the conventional needle based systems (Wang et al. 2009). In the free-surface electrospinning approach the spinning solution is held in a bath, rather than being delivered through an aperture; i.e. the needle, with the whole solution then being connected to a high voltage power supply. In the specific case of the NanoSpider™ the solution is held in a reservoir and fed onto a wire spinning electrode. Thus, concentrating the electric field on the thin layer of polymer coating the wire. In this process many Taylor cones are formed along the length of the wire and electrospinning occurs upwards towards a collector located above the electrode. This increases the throughput of the process many thousands of times above the conventional needle-based system. However, much higher voltages, up to 82 kV in the case of the NanoSpider™, are required and solution properties such as viscosity, conductivity and surface tension must be more tightly controlled (Thoppey et al. 2011; Niu and Lin 2012). The Elmarco NanoSpider™ is available in many forms, from a small lab system through to pilot and production line systems offering extremely high throughputs and an easy transition from lab to commercial scale (Figure 2.2).

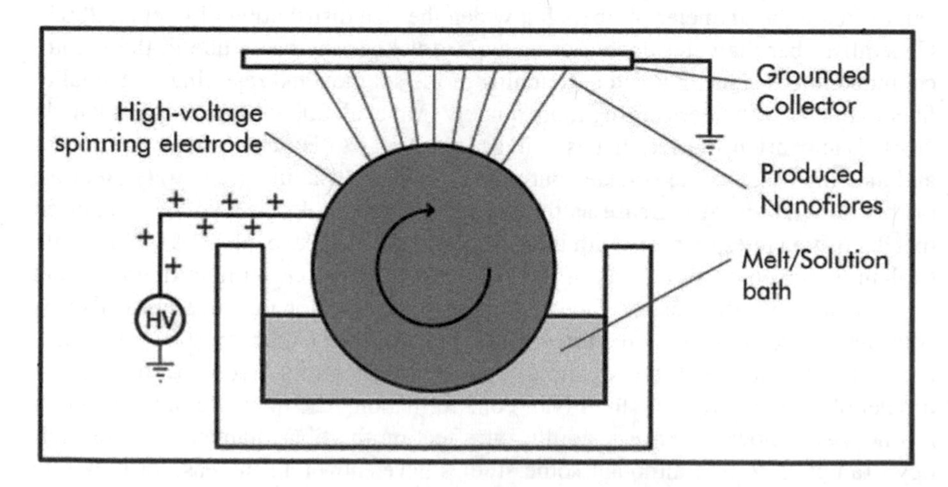

FIGURE 2.2 A schematic diagram showing a free-surface electrospinning setup. A polymer solution/melt is held in a bath and a spinning electrode connected to a high voltage power supply is utilized to form multiple jets. Nanofibers are electrospun upwards and collected on a grounded collector plate.

A nozzle-less (free-surface) system can overcome the challenges faced with a multi-needle in terms of reliability, cleaning and consistency (Petrík 2009). Lukas et al. studied the process of multiple jet formation from a free liquid surface in an electric field (Lukas et al. 2008). Their study showed how the hypothesis (based on a profound analysis of a dispersion law) explains that the system begins to self-organize in mesoscopic scale due to the mechanism of the "fastest forming instability" above a critical value of applied electric field intensity/field strength. Using Euler's equations for liquid surface waves they were able to derive an expression for the critical spatial period (wavelength) which is the average distance between individual jets emerging from a liquid. This is a major advantage of free-surface electrospinning with the number and location of jets along the electrode set up naturally in their optimal position. In the case of multiple needle systems, the jet distribution is made artificially with a mismatch between "natural" jet distribution and the actual structure leading to instabilities and a heterogeneous nanofiber layer.

2.3 BACTERIAL FOULING OF MEMBRANES

Electrospun nanofiber membranes have shown considerable promise for water purification applications (Burger et al. 2006; Yoon et al. 2008; Barhate and Ramakrishna 2007). However, as with all materials utilized in the aqueous environment, the predisposition of these materials to adsorption of organic matter, and the subsequent increase in biofouling of these materials is well documented (Neihof and Loeb 1974; Gómez-Suárez et al. 2002; Schneider and Leis 2003; Hwang et al. 2012a, 2012b, 2013; El-Kirat-Chatel et al. 2013; James et al. 2017). Hence, remediation of organic adsorption characteristics, and therefore biofouling, is of paramount importance to their design. A major contributing factor to the formation of an organic conditioning layer is the deposition of extracellular polymeric substance (EPS), a temporally dynamic mix of polysaccharides, DNA, lipids, proteins, and environmentally derived humic substances (Wingender et al. 2001; Conrad et al. 2003; Flemming and Wingender 2003; Flemming and Wingender 2010).

Numerous studies have been conducted on the deposition of model organic foulants; BSA, alginates, and humic acids, and industrially relevant foulants; purified EPS, membrane permeates, and cellular fragments (Hwang et al. 2013; Barnes et al. 2014; Liu and Mi 2014; Suwarno et al. 2016). The aforementioned studies concluded that the physiochemical changes, in the form of changes in wettability, zeta potential, and surface roughness induced a significant increase in the overall rate of bacterial adhesion (reviewed here [James et al. 2017]). Hence, innate physiochemical properties that prevent adsorption of foulants or the incorporation of self-cleaning elements offers a significant advantage over the current technologies. In this regard nanofibers excel, studies conducted by Zhifeng et al. demonstrated that blending of electrospun polyaniline (PANI) membranes with existing polysulfone membranes exhibited a significant rejection of common model proteins, BSA and albumin egg albumin, while retaining the high porosity associated with nanofibers and a decreased rate pf flux reduction (Fan et al. 2008). Furthermore Abrigo et al. (2015b) demonstrated the through control of the fiber diameter significant control over the microbial adhesion rate can be achieved. During this study it was found that as the fiber diameter tended towards the microbial cell diameter of the target organism an increase in the total adhesion rate was observed. As these studies were conducted over an extended incubation period ~6 h, it can be concluded that the effect due to fiber diameter may be able to supersede that caused by organic fouling.

2.4 FIBER MORPHOLOGY

The growing use of electrospinning to fabricate nanofibrous structures for use in wound dressings, tissue engineering and filtration processes has increased the need for an understanding of the interactions between bacteria and nanostructures. The adhesion characteristics and colonization of bacteria on these materials are still not completely understood but are essential to aid their future development (Abrigo et al. 2014).

Fibrous materials have been used extensively in the fabrication of synthetic membranes used in separation processes (Khulbe et al. 2015). By controlling the fiber diameter different pore sizes can be obtained making them useful for many applications. Focusing on air filtration where removal of microorganisms is an important efficiency indicator, electrospun nanofiber mats can be used due to their small diameter fibers and small pore size allowing them to trap particles while allowing sufficient air flow. Filters have been electrospun from many synthetic and biopolymers such as nylon 6 for use as high efficiency air particulate (HEPA) filters and ultra-low particulate air (ULPA) filters (Zhang et al. 2009), polyethylene oxide (PEO) for use in composite filters for long range applications (Patanaik et al. 2010), polyacrylonitrile (PAN) for nanoparticle filtration (Yun et al. 2007), and a composite of polylactide/polyhydroxybutyrate (PLA/PHB) for aerosol particle filtration (Nicosia et al. 2015). One major problem with the very small pore size of nanofibrous filters is their contamination leading to a short life span for the membrane filter. Bacterial contamination can therefore become a problem with bacterial adhesion resulting in colonization and biofilm formation. An understanding of bacterial interaction with nanofibers is therefore essential to progress the development of nanofibrous air filtration systems and to aid the design and addition of antimicrobials.

Adhesion of bacteria is one of the early stages of biofilm formation occurring after or alongside the laying down of a conditioning film depending on the process environment. Biofilms are now widely regarded as the predominant mode of microbial life in both nature and disease due to their inherent resistance to antimicrobials and ability to adapt and "protect" themselves from attack (Costerton et al. 1999; Donlan 2002; Hall-Stoodley et al. 2004; Xu and Siedlecki 2012). How the cells initially interact in surface adhesion is therefore a very important factor when developing antimicrobial and antibiofilm surfaces for application in medicine and the process industries. In recent years there has been extensive research into the adhesion mechanisms of bacteria with more recent studies focusing on the interaction between bacteria and textured surfaces; surfaces that have a controllable nanoscale morphology. The use of surfaces with micro and nanosurfaces, which have the potential to control bacterial adhesion and subsequent biofilm formation, is another technique in the portfolio of methods available to control microbial attachment. The mode of action of such an approach is to reduce the area of surface interaction and the number of participating bonds, thus rendering the initial bacterial attachment more labile and vulnerable to physical methods of disruption such as hydrodynamic shear (Guélon et al., 2011). Prevention of attachment through physical means compliments or removes the need for chemical methods.

The advent of nanotechnology and improved fabrication techniques triggered extensive research on the control of microbial adhesion by nanostructured surfaces. However, conspicuous by its absence, is a research focus on microbial adhesion and subsequent biofilm formation at nanofiber structures, given their industrial and clinical importance. Indeed, there are very few quantitative studies examining the adhesion of micro-organisms to nanofiber textured surfaces. As at any surface there are a number of mechanisms that govern the adhesion of bacteria to nanofibers. These can be physio-chemical or specific and are influenced by the surface chemistry and the morphology of the fibers and the fiber mat. Physio-chemical interactions will have a contribution from electrostatic and van der Waals forces and hydrophobicity as described by Derjaguin, Landau, Verwey, and Overbeek (DLVO) and extended DLVO theories. Specific interactions involve macromolecules at the bacterial surfaces that bind to specific sites at the host and material surfaces. Surface appendages such as slime layers, fimbriae and pili can also contribute to bacterial surface interactions. For a more detailed discussion on bacterial adhesion to surfaces and the influence of material see Ploux et al. (2010) and Ubbink and Schar-Zammaretti (2007). The area of contact between the interacting cell and the surface is an important determinant of the size of the interaction forces; the greater the area of contact the larger number of interacting surface groups or molecular structures contributing to the bacterial cell attachment. When considering adhesion at planar surfaces the surface roughness and the deformation of the cell boundaries in contact with the surface will influence the area of contact between the cell and the surface. At surfaces covered in nanofibers the actual surface area of contact may be significantly reduced due to the void areas between the fibers. However, as previously stated the interaction of bacteria with fibrous structures requires more research. Kargar et al. (2012) investigated the interaction of *Pseudomonas aeruginosa* with nanofiber-textured surfaces. They used highly aligned polystyrene nanofibers of different diameters and varying gap size fabricated using a fiber extrusion technique. They selected three fiber diameters, one smaller than the diameter of the organism, one equal to the diameter of the organism and one greater in diameter to that of the organism. Similarly, the gap size between fibers was selected with a gap sizer smaller than, equal to and greater than that of the organism. Samples were submerged in a suspension of *P. aeruginosa* and SEM was used to image the samples. The images were analyzed, using the total number of bacteria per fiber length to quantify the total adhesion density. To further analyze the state of adhesion, bacteria were characterized according to their alignment with the fibers and spacing. Kargar et al. found that the total adhesion density increases both with fiber diameter and spacing distance [58]. This study used aligned fibers to easier understand the interaction with the fibers themselves. The absence of a control sample with a smooth surface limits the findings to how the bacteria interact with the nanofiber textured surface with no comparison to smooth surfaces (Figure 2.3).

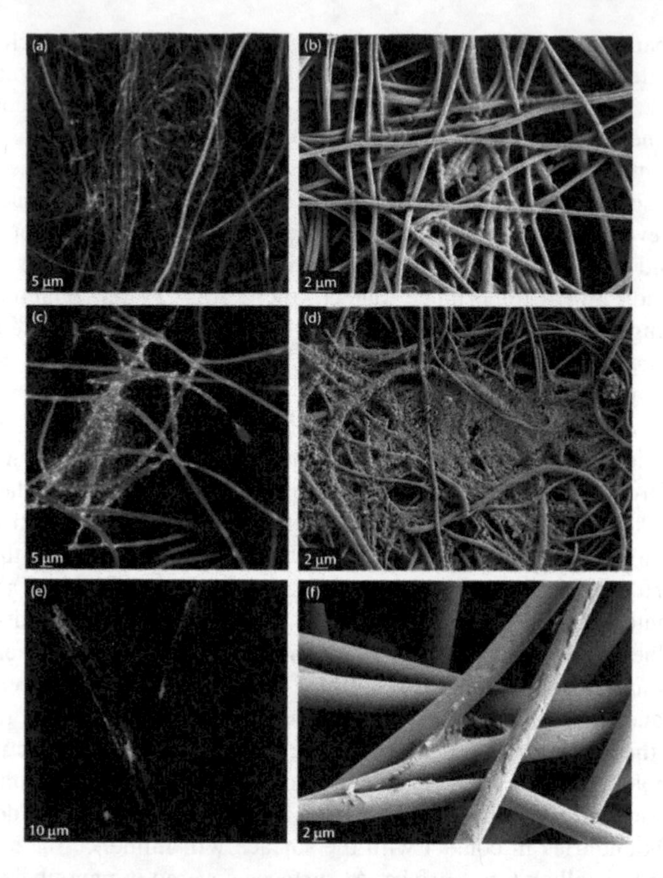

FIGURE 2.3 Confocal (a, c and e) and SEM (b, d and f) images of *E. coli* cells interacting with electrospun polystyrene meshes with varying fiber diameters. (a, b) Average diameter = 500 ± 200 nm; (c, d) average diameter = 1000 ± 100 nm; (e, f) average diameter = 3000 ± 1000 nm. (Reprinted with permission from Abrigo, M. et al., *ACS Appl. Mater. Interfaces*, 7, 7644–7652, 2015b. Copyright 2015b American Chemical Society.)

Abriago et al. investigated the effect of fiber diameter on bacterial attachment, proliferation and growth at electrospun fiber constructs (Figure 2.4) (Abrigo et al. 2015b). In their study varying concentrations of polystyrene (PS) in DMF were used to electrospin fibers of different diameter from 300 to 3000 nm. Electrospun meshes were then tested against *Escherichia coli*, *P. aeruginosa* and *Staphylococcus aureus* both in solution and on agar plates. Their work demonstrated that the fiber diameter influences bacterial proliferation. An average fiber diameter close to that of the bacteria offered the best support for bacterial adhesion and proliferation. Rod shaped cells tended to wrap themselves around fibers with a smaller diameter than their length limiting the ability of the cells to bridge gaps between fibers and form colonies (Figure 2.4). Round cells tended to proliferate through nanofibrous substrates yet when the diameter was larger they were found to have adhered to the surface. Again, these findings were limited by the absence of a smooth control

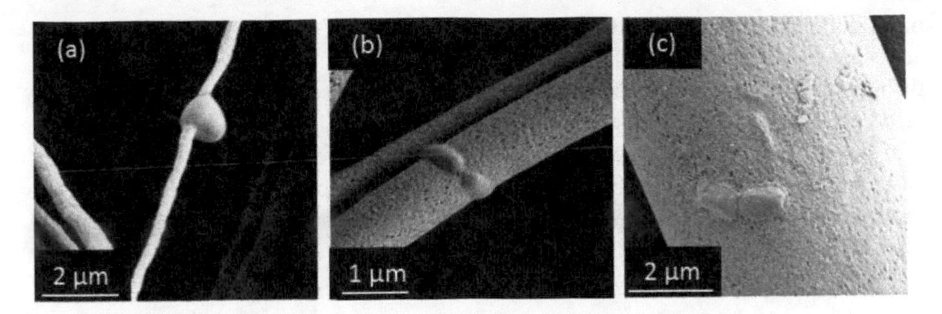

FIGURE 2.4 SEM images of *E. coli* cells adhered to polystyrene fibers of average diameter: (a) 0.3 µm, (b) 1 µm, (c) 5 µm. (Reprinted with permission from Abrigo, M. et al., *ACS Appl. Mater. Interfaces*, 7, 7644–7652, 2015b. Copyright 2015b American Chemical Society.)

sample. In a further study Abriago et al. (2015a) investigated the bacterial response to different surface chemistries of electrospun nanofibers. Polystyrene (PS) nanofibers were electrospun and plasma coated with a number of different monomers including allylamine (ppAAm), acrylic acid (ppAAc), 1,7-octadiene (ppOct) and 1,8-cineole (ppCo). The same techniques as the previous paper were used to characterize bacterial interactions with the fibers (Abrigo et al. 2015b). The plasma coating did not induce a significant change in fiber morphology. The surface chemistry was found to have a significant effect on bacterial adhesion and proliferation. A ppAAm coating (hydrophilic and rich in amine positively charged groups) resulted in the highest attraction of viable *E. coli* cells forming colonies and clusters across the interstices of the mesh. There was a significantly lower number of *E. coli* cells found on fibers with a hydrophilic, negatively charged ppAAc coating. The cells spread throughout the fibrous network. Fibers with a hydrophobic ppOct coating were found to have a higher proportion of live cells when compared to untreated PS fibers forming clusters at fiber crossovers. The ppCo coating had no inhibitory effect although a high proportion of dead isolated bacterial cells were found to have adhered to the fibers. Cells were wrapped around fibers with no clusters at fiber crossovers or across the interstices of the mesh. The results demonstrate the effect of surface chemistry on the interaction between bacteria and nanofibers offering a further parameter to be altered during electrospinning fiber fabrication to meet a specific anti-microbial attachment application.

To further increase the understanding of bacterial interaction with nanofibers the studies of Abriago et al. need to be extended for different materials with different stiffness' and mechanical properties. Different inoculation techniques and fluidic environments can be used such as a static or parallel plate flow chambers and rotating disk systems. By applying shear stress while microbial colonization is occurring the understanding of the influence of fiber morphology on fouling can be extended to biofilm formation. These findings can be compared with the interactions of bacteria with planar surfaces and other nanostructured surfaces. This research will be able to guide the design of electrospun non-wovens and improve their microbial control in a broad range of applications. Tissue engineering scaffolds can be optimized to reduce microbial colonization with the tantalizing possibility of promoting mammalian cell adhesion while minimizing the risk of infection by inhibiting bacterial cell adhesion, through fine tuning of the nanofibrous material structure. Similar characteristics are also desirable in wound dressings. A more rigorous understanding of the influence of nanofiber structure on microbial adhesion will also help with the development of systems susceptible to biofilm formation such as filtration systems, whereby reducing the adherent cell number, or controlling biofilm morphology and mechanical properties, formation of biofilm can be inhibited or slowed.

2.5 FIBER FUNCTIONALITY

Further to control of fiber morphology, additives can be used to improve antimicrobial and antifouling properties of nanofibers. Several studies have outlined potential antimicrobials for applications requiring microbial control such as silver and zinc oxide nanoparticles with most studies preferring the former (Shalumon et al. 2011; Gopiraman et al. 2016; Zhang et al. 2016).

2.5.1 Metallic Nanoparticles

Metal nanoparticles have been found to exhibit antimicrobial activity. This has proven to be attractive due to their large specific surface area, low toxicity, excellent stability and long-lasting action (Song et al. 2016). A number of different types of metallic nanoparticles such as copper, zinc, titanium, magnesium, gold and silver have been shown to have antimicrobial efficacy against bacteria, viruses and simpler eukaryotic organisms (Moritz and Geszke-Moritz 2013). Incorporation of metallic nanoparticles, particularly silver, has been shown to be effective in a number of fields including tissue engineering and water treatment. Electrospinning nanoparticles that have already been synthesized is the most basic and therefore most commonly used technique for the fabrication of nanoparticle-nanofiber composites. However, this process can often be multi-stage and time consuming requiring particles to be pre-synthesized and subsequently functionalized to reduce the effects of particle agglomeration and allow homogenous distribution throughout the nanofibers. There are a number of examples in the literature where nanofibers have been electrospun with a number of different nanoparticles including iron oxide, silver, silica, titanium oxide and zinc oxide. Mehrasa et al. (2015). synthesized silica nanoparticles by a template removal method and added them to a solution of PLGA/Gelatin before electrospinning using a single needle electrospinning set up with a rotating collector to obtain aligned fibers. It was reported that the incorporation of silica nanoparticles into PLGA nanofibers enhanced hydrophobicity, improved tensile mechanical properties and led to improved cell attachment and proliferation of PC12 cells. Razzaz et al. (2016) purchased titanium oxide (TiO_2) nanoparticles, which were added to a chitosan/acetic acid solution and electrospun using a single needle electrospinning setup. The nanocomposite fibers were suitable for the removal of heavy metal ions, with maximum adsorption capacities of 710.3 mg/g (Cu(II)) and 526.5 mg/g (Pb(II)). They also fabricated chitosin nanofibers and immersed them in a solution of nanoparticles for 1 h to allow the nanoparticles to coat the fibers. They found that not only were the adsorption capacities lower for the coating method but they also showed a loss of adsorption performance after five adsorption/desorption cycles.

Alternative options include post treatment where the precursor is electrospun into the nanofibers and subsequently treated to form nanoparticles on the surface. Post treatment techniques involve the inclusion of a precursor in the electrospun nanofibers, which undergoes post-electrospinning processing technique to form nanoparticles within the nanofibers. This can also be achieved by immersing nanofibers in a nanoparticle solution, as presented by Razzaz et al. (2016) with TiO_2 nanoparticles but this requires pre-synthesis of nanoparticles. Formo et al. (2008) presented a method for the preparation of TiO_2 nanofibers with platinum nanoparticles and nanowires. Titanium oxide nanoparticles were prepared from a solution containing PVP and titanium tetraisopropoxide followed by calcination at 510°C. The fibers were then immersed in a polyol reduction bath containing an ethylene glycol solution with a H_2PtCl_6 and PVP for 7 h, resulting in the deposition of nanofibers on the surface of the nanofibers. Xiao et al. (2009) prepared nanofibers from PVA/PAA before immersion in an aqueous solution of ferrous tichloride to allow ferric cations to complex with free carboxyl groups on PAA through ion exchange. Sodium borohydride was then added to the mats to reduce the ferric ions to IONPs. Another example of post treatment nanocomposite fabrication is the work of Barakat who prepared nanofibers from a solution of PVA and Iron (II) acetate (FeAc) (Barakat 2012). The nanofiber mats were dried for 24 h before undergoing calcination at 700°C for 5 h in an argon atmosphere to form IONPs. XRD confirmed there was no hematite present and FTIR confirmed there was no magnetite present, showing that the IONPs present were maghemite. Wang et al. (2005) presented a method to prepare silver nanoparticles dispersed in polyacrylonitrile (PAN) nanofibers combining a reduction reaction with electrospinning. They electrospun a solution of PAN in DMF containing 10 wt.% silver nitrate ($AgNO_3$). The nanofibers were then immersed in a hydrazinium hydroxide (N_2H_5OH) aqueous solution where the silver nitrate in the nanofibers was reduced to silver nanoparticles. Nataraj et al. (2010) presented a three-stage *in-situ* synthesis technique where the chemical reagents ($FeSO_4$ and $FeCl_3$) were added to a PAN solution and electrospun. The electrospun mat was then immersed in KOH for 4 h, stabilized at 280°C in air for 2 h and carbonized at 800°C in nitrogen atmosphere. XPS was used to identify the phase of the IONPs, which were identified as Fe_2O_3.

In recent year's research has focused on the development of *in-situ* synthesis techniques combined with electrospinning with less steps, more simplicity and lower production costs. This is an area that has been investigated in more depth for metal nanoparticles. For example, Jin et al. (2005) presented a one-step technique to prepare silver nanoparticles in Poly(vinylpyrrolidone) PVP nanoparticles. In their study silver nitrate ($AgNO_3$) was reduced in a PVP/DMF solution with DMF as the reducing agent. Solutions were then electrospun resulting in PVP nanofibers containing silver nanoparticles. Saquing et al. (2009) presented a facile one-step technique to synthesize and incorporate silver nanoparticles into electrospun nanofibers. They chose PEO as the electrospinning polymer, which is also used as a reducing agent for the metal salt precursor and protects the formed nanoparticles from agglomeration. The fiber quality was improved with the addition of the silver nanoparticles and fiber diameter was reduced due to an increase in the electrical conductivity of the solution. Rujitanaroj et al. (2008) prepared a solution of $AgNO_3$ and gelatin in acetic acid/distilled water. This solution was allowed to age before electrospinning using a needle based setup. The fibers were cross-linked using glutaraldehyde vapor and antimicrobial activity was confirmed against common bacteria found on burn wounds using a disc diffusion method. Faridi-Majidi et al. present a one-stage *in-situ* synthesis technique for the electrospinning of PEO nanofibers containing IONPs (Faridi-Majidi and Sharifi-Sanjani 2007). In their technique, $FeCl_3$ and $FeSO_4$ were added to the PEO/distilled water electrospinning solution. Electrospinning was carried out in an ammonia atmosphere to reduce the iron compounds to IONPs. SEM and TEM confirmed the nanofibrous morphology and presence of nanoparticles. XRD was used to identify the phase of iron oxide nanoparticles as maghemite. In a recent study, we have developed a novel one-stage *in-situ* synthesis technique to fabricate PEO and PVP nanofibers containing magnetite magnetic nanoparticles (MNPs) (Burke et al. 2017). We also demonstrated an ability to scale up the process from laboratory to industrial scale using a commercially available free-surface electrospinning setup. In our technique a 2:1 molar ratio of ferric and ferrous chloride was added to a PEO solution in deionized water containing sodium borohydride, and used to reduce the ions to nanoparticles. The reaction was allowed to progress before being electrospun (Figure 2.4). Nanofiber mats were cross-linked using UV irradiation, EDX was used to confirm the presence of iron, DLS showed the average nanoparticle diameter to range from 8 nm (PVP) to 26 nm (PEO), XRD confirmed the phase of the nanoparticles to be magnetite and NMR showed a shortening in both T1 and T2 relaxation times confirming the nanoparticles could provide a suitable relaxation channel, verifying the magnetic behavior of the MNPs (Figure 2.5).

Silver nanoparticles are the most commonly used for antimicrobial applications with various silver loaded wound dressings commercially available. Silver nanoparticles have been widely used as an antimicrobial agent for many years due to their easy production, high antimicrobial activity and their ability to be incorporated into a diverse range of products, including nanofibers. There are many examples of silver nanoparticle-nanofiber composite, with nanofibers both encapsulating and coated with silver nanoparticles. A range of different biocompatible natural and synthetic polymers have been used to produce antimicrobial scaffolds for a number of different applications. Kharaghani et al. (2018) presented electrospun polyacrylonitrile (PAN) nanofibers with a surface functionalized with silver nanoparticles by a facile wetting process. The functionalized nanofibers were shown to exhibit antimicrobial activity against *E. coli* and *S. aureus* with considerable sustainability as well as showing good biocompatibility with mammalian cells making them suitable for application as a disinfecting membrane filter for biomedical and hygienic applications. Zhao et al. (2017) presented a fabrication method for a membrane bioreactor for waste-water treatment consisting of an antimicrobial three-dimensional woven fabric filters of weft yarn wrapped with electrospun PAN nanofibers containing 2 wt.% silver nanoparticles. Disc diffusion assays showed that the fabrics could suppress the growth of bacterial colonies and long-term filtration performance tests showed the filters containing

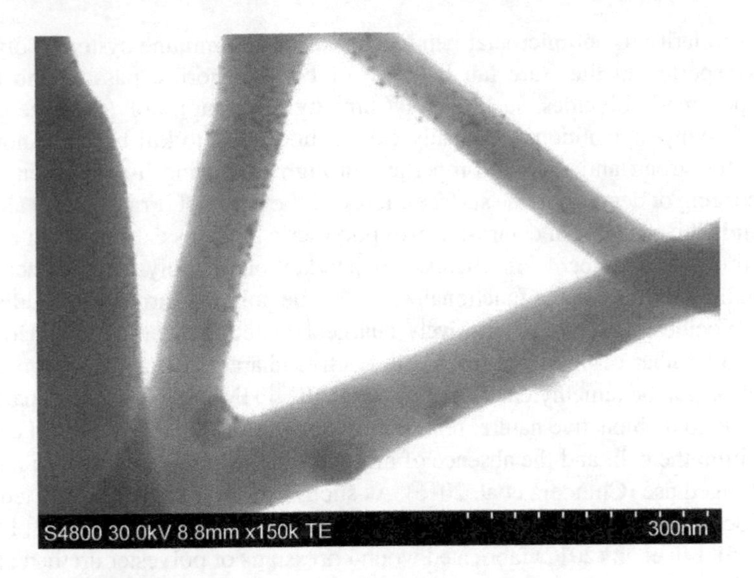

FIGURE 2.5 High magnification scanning transmission electron microscopy image of PEO nanofiber containing MNPs.

nanoparticles had a 40%–50% higher flux and substantially larger flux recovery proportions than the control filter without nanoparticles.

Another metal nanoparticle with known antimicrobial properties is copper (Cu). Zhang et al. (2015) presented a method to fabricate PAN nanofibers with zero-valent copper nanoparticles. They were shown to present antimicrobial efficacy against *S. aureus* when tested using a disc diffusion technique. Titanium dioxide (titania) and zinc oxide are also reported to be antimicrobial although these are less commonly used than silver. However, there are still reports of these in nanoparticle-nanofiber composites. Lubasova and Barbora (2014) compared the antimicrobial properties of electrospun PVA nanofiber membranes with zinc oxide, zirconium dioxide, tin oxide and titanium dioxide nanoparticles. An increase in nanoparticle concentration of all particles resulted in an increase in antimicrobial efficacy when tested using the modified AATCC method (100-2004) tested against *E. coli* and *S. aureus*. Zinc oxide and zirconium oxide proved to be the most effective with percentage reductions up to 100% for zinc oxide and 90% for zirconium oxide against *E. coli*. Tin oxide and titanium dioxide proved to be less effective with reductions of up 81% (*S. aureus*) and 50% (*E. coli*), respectively.

2.5.2 Polymeric Biocides

As a result of overuse silver resistance is rising with some researchers calling for tighter control (Percival et al. 2005) and fabrication of such fibers can become complex with the heterogeneity of the silver antimicrobial distribution within such fibers (Maillard and Hartemann 2013). This has made polymeric biocides of interest due to their similarity to antimicrobial peptides used by the immune system. Polymeric biocides reported in literature fall into one of two categories, passive and active. Passive polymeric biocides, such as poly(dimethyl siloxane), poly(ethylene glycol) and poly(*n*-vinyl-pyrrolidone), typically exhibit no ability to kill bacteria; however, they exhibit strong antiadhesive properties through increasing hydrophilicity, positively charging or decreasing the surface energy of the material (Francolini et al. 2014; Zhang and Chiao 2015). In contrast, active polymeric biocides exhibit direct biocidal activity through a number of mechanisms dependant on the polymer. This activity is traditionally the result of the functionalization of the polymer through the addition of cationic biocides, peptides or positively charged quaternary compounds. However, there are a number of polymers that exhibit active characteristics without modification, such as polyhexamethylene biguanide (PHMB). PHMB has become of particular interest due to its bioactive nature, reported to be the result of chromosomal condensation within the cell, and the absence of any reported resistance despite widespread and prolonged use (Chindera et al. 2016). As such, PHMB is the most reported polymeric biocide used in electrospinning (Lin et al. 2012; Dilamian et al. 2013; Llorens et al. 2015). Liu et al. (2012) fabricated wound dressings of polyester urethane (PEU) and cellulose acetate (CA) containing PHMB. Their results show bacterial reduction rates of up to 99.88% (~3 log) when tested according to the AATCC Quantitative Test Method (100-1999) against *E. coli*. Other groups have successfully reported the electrospinning of antimicrobial nanofibers of various other polymers containing PHMB. Llorens et al. (2015) electrospun PHMB loaded Polylactic Acid (PLA) scaffolds which were shown to inhibit the growth of both *E. coli* and *Microccus luteus* along with a decrease in cell adhesion percentages. Dilamian et al. (2013) electrospun membrane of chitosan/poly (ethylene oxide) incorporating PHMB showing inhibitory effects against *E. coli* and *S. aureus*.

Other antimicrobials to be recently incorporated in nanofibers include quarternary ammonium compounds (QACs). Nicosia et al. (2015) incorporated QACs into Polylactide/polyhydroxybutyrate (PLA/PHB) nanofibers which exhibited antimicrobial properties for aerosol particle filtration. They found that the addition of QACs to the electrospinning solution had no adverse effect on the electrospinning process or subsequent filtration efficiency. Antimicrobial nanofibers of poly(vinyl alcohol) (PVA) were synthesized by Park et al. (2017) incorporating benzyl triethylammonium chloride (BTEAC). They used a heat-methanol treatment to improve their water stability with bacterial removal being enhanced by incorporation of BTEAC into the PVA nanofibers. A very low level of BTEAC was shown to leach from the fibers, which was proven to be below toxicity levels.

2.6 CONCLUSIONS

Nanotechnology has come a long way in the past two decades with electrospinning emerging as a strong candidate for the commercial production of nanofibers for medical, filtration and textile applications. It has been shown that by controlling solution properties, processing parameters, and electrospinning modifications nanofiber diameter can be controlled. This is crucial to the design of a membrane to select appropriate pore size and permeability. Membrane fouling still remains a significant barrier to the application of nanofibers to membrane processes. However, recent research suggests that control of fiber morphology alongside the inclusion of antimicrobial materials can reduce biofouling of membrane materials. The effect of fiber diameter on bacterial adhesion has been discussed along with discussion of antimicrobials commonly used and incorporated in nanofibers, with a focus on filtration applications. Future research is expected to focus on the understanding of the adhesion to fibers of different diameters and further investigation of polymeric biocides as an alternative to metallic nanoparticles.

REFERENCES

Abrigo, Martina, Peter Kingshott, and Sally L. McArthur. 2015a. "Bacterial Response to Different Surface Chemistries Fabricated by Plasma Polymerization on Electrospun Nanofibers." *Biointerphases* 10 (4): 04A301. doi:10.1116/1.4927218.

Abrigo, Martina, Peter Kingshott, and Sally L. McArthur. 2015b. "Electrospun Polystyrene Fiber Diameter Influencing Bacterial Attachment, Proliferation, and Growth." *ACS Applied Materials and Interfaces* 7 (14): 7644–52. doi:10.1021/acsami.5b00453.

Abrigo, Martina, Sally L. McArthur, and Peter Kingshott. 2014. "Electrospun Nanofibers as Dressings for Chronic Wound Care: Advances, Challenges, and Future Prospects." *Macromolecular Bioscience* 14 (6): 772–792. doi:10.1002/mabi.201300561.

Ahmed, Farah Ejaz, Boor Singh Lalia, and Raed Hashaikeh. 2015. "A Review on Electrospinning for Membrane Fabrication: Challenges and Applications." *Desalination* 356: 15–30. doi:10.1016/j.desal.2014.09.033.

Andrady, Anthony L. 2008. *Science and Technology of Polymer Nanofibers*. Hoboken, NJ: John Wiley & Sons. doi:10.1002/9780470229842.

Barakat, Nasser A. M. 2012. "Synthesis and Characterization of Maghemite Iron Oxide (γ-Fe_2O_3) Nanofibers: Novel Semiconductor with Magnetic Feature." *Journal of Materials Science* 47 (17): 6237–6245. doi:10.1007/s10853-012-6543-7.

Barhate, Rajendrakumar Suresh, and Seeram Ramakrishna. 2007. "Nanofibrous Filtering Media: Filtration Problems and Solutions from Tiny Materials." *Journal of Membrane Science* 296 (1–2): 1–8. doi:10.1016/j.memsci.2007.03.038.

Barnes, Robert J., Ratnaharika R. Bandi, Felicia Chua, Jiun Hui Low, Theingi Aung, Nicolas Barraud, Anthony G. Fane, Staffan Kjelleberg, and Scott A. Rice. 2014. "The Roles of Pseudomonas Aeruginosa Extracellular Polysaccharides in Biofouling of Reverse Osmosis Membranes and Nitric Oxide Induced Dispersal." *Journal of Membrane Science* 466: 161–72. doi:10.1016/j.memsci.2014.04.046.

Baumgarten, Peter K. 1971. "Electrostatic Spinning of Acrylic Microfibers." *Journal of Colloid and Interface Science* 36 (1): 71–79. doi:10.1016/0021-9797(71)90241-4.

Bhardwaj, Nandana, and Subhas C. Kundu. 2010. "Electrospinning: A Fascinating Fiber Fabrication Technique." *Biotechnology Advances* 28 (3): 325–47. doi:10.1016/j.biotechadv.2010.01.004.

Buchko, Christopher J., Loui C. Chen, Yu Shen, and David C. Martin. 1999. "Processing and Microstructural Characterization of Porous Biocompatible Protein Polymer Thin Films." *Polymer* 40 (26): 7397–7407.

Burger, Christian, Benjamin S. Hsiao, and Benjamin Chu. 2006. "Nanofibrous Materials and Their Applications." *Annual Review of Materials Research*. doi:10.1146/annurev.matsci.36.011205.123537.

Burke, Luke, Chris J. Mortimer, Daniel J. Curtis, Aled R. Lewis, Rhodri Williams, Karl Hawkins, Thierry G.G. Maffeis, and Chris J. Wright. 2017. "In-Situ Synthesis of Magnetic Iron-Oxide Nanoparticle-Nanofibre Composites Using Electrospinning." *Materials Science and Engineering: C* 70: 512–519. doi:10.1016/j.msec.2016.09.014.

Chindera, Kantaraja, Manohar Mahato, Ashwani Kumar Sharma, Harry Horsley, Klaudia Kloc-Muniak, Nor Fadhilah Kamaruzzaman, Satish Kumar et al. 2016. "The Antimicrobial Polymer PHMB Enters Cells and Selectively Condenses Bacterial Chromosomes." *Scientific Reports* 6: 23121. doi:10.1038/srep23121.

Conrad, Arnaud, Merja Kontro, Minna M. Keinänen, Aurore Cadoret, Pierre Faure, Laurence Mansuy-Huault, and Jean-Claude Block. 2003. "Fatty Acids of Lipid Fractions in Extracellular Polymeric Substances of Activated Sludge Flocs." *Lipids* 38 (10): 1093–1105. doi:10.1007/s11745-006-1165-y.

Cooley, John Francis. 1900. Improved Methods of and Apparatus for Electrically Separating the Relatively Volatile Liquid Component from the Component of Relatively Fixed Substances of Composite Fluids. Patent GB 6385, p. 19.

Costerton, J. William, Philip. S. Stewart, and E. Peter Greenberg. 1999. "Bacterial Biofilms: A Common Cause of Persistent Infections." *Science* 284 (5418): 1318–1322. doi:10.1126/science.284.5418.1318.

Dilamian, Mandana, Majid Montazer, and J. Masoumi. 2013. "Antimicrobial Electrospun Membranes of Chitosan/Poly (Ethylene Oxide) Incorporating Poly (Hexamethylene Biguanide) Hydrochloride." *Carbohydrate Polymers* 94 (1): 364–371. doi:10.1016/j.carbpol.2013.01.059.

Donlan, Rodney M. 2002. "Biofilms: Microbial Life on Surfaces." *Emerging Infectious Diseases* 8 (9): 881–890. doi:10.3201/eid0809.020063.

El-Kirat-Chatel, Sofiane, Dalila Mil-Homens, Audrey Beaussart, Arsenio M. Fialho, and Yves F. Dufrêne. 2013. "Single-Molecule Atomic Force Microscopy Unravels the Binding Mechanism of a Burkholderia Cenocepacia Trimeric Autotransporter Adhesin." *Molecular Microbiology* 89 (4): 649–659. doi:10.1111/mmi.12301.

Fan, Zhifeng, Zhi Wang, Ning Sun, Jixiao Wang, and Shichang Wang. 2008. "Performance Improvement of Polysulfone Ultrafiltration Membrane by Blending with Polyaniline Nanofibers." *Journal of Membrane Science* 320 (1–2): 363–371. doi:10.1016/j.memsci.2008.04.019.

Faridi-Majidi, Reza, and Naser Sharifi-Sanjani. 2007. "In Situ Synthesis of Iron Oxide Nanoparticles on Poly(Ethylene Oxide) Nanofibers through an Electrospinning Process." *Journal of Applied Polymer Science* 105 (3): 1351–1355. doi:10.1002/app.26230.

Flemming, Hans-Curt, and Jost Wingender. 2003. "Extracellular Polymeric Substances (EPS): Structural, Ecological and Technical Aspects." In *Encyclopedia of Environmental Microbiology*. Hoboken, NJ: John Wiley & Sons. doi:10.1002/0471263397.env292.

Flemming, Hans-Curt, and Jost Wingender. 2010. "The Biofilm Matrix." *Nature Reviews Microbiology* 8 (9): 623–633. doi:10.1038/nrmicro2415.

Formhals, Anton. 1934. Process and Apparatus for Preparing Artificial Threads: US, 1975504.

Formo, Eric, Eric Lee, Dean Campbell, and Younan Xia. 2008. "Functionalization of Electrospun TiO_2 Nanofibers with Pt Nanoparticles and Nanowires for Catalytic Applications." *Nano Letters* 8 (2): 668–672. doi:10.1021/nl073163v.

Francolini, Iolanda, Gianfranco Donelli, Fernanda Crisante, Vincenzo Taresco, and Antonella Piozzi. 2014. "Antimicrobial Polymers for Anti-Biofilm Medical Devices: State-of-Art and Perspectives." In *Biofilm-Based Healthcare Associated Infections II* (Gianfranco Donelli Ed.), pp. 93–117. Springer, Cham, Switzerland.

Gómez-Suárez, Cristina, Jos Pasma, Arnout J. van der Borden, Jost Wingender, Hans Curt Flemming, Henk J. Busscher, and Henny C. van der Mei. 2002. "Influence of Extracellular Polymeric Substances on Deposition and Redeposition of Pseudomonas Aeruginosa to Surfaces." *Microbiology* 148 (4): 1161–1169. doi:10.1099/00221287-148-4-1161.

Gopiraman, Mayakrishnan, Abdul Wahab Jatoi, Seki Hiromichi, Kyohei Yamaguchi, Han Yong Jeon, Ill Min Chung, and Kim Ick Soo. 2016. "Silver Coated Anionic Cellulose Nanofiber Composites for an Efficient Antimicrobial Activity." *Carbohydrate Polymers* 149 (September): 51–59. doi:10.1016/j.carbpol.2016.04.084.

Guélon, Thomas, Jean-Denis Mathias, and Paul Stoodley. 2011. "Advances in Biofilm Mechanics." In *Biofilm Highlights* (Hans-C Flemming, Jost Wingender, Ulrich Szewzyk Eds.), pp. 111–139. Springer, Berlin, Germany.

Hall-Stoodley, Luanne, J. William Costerton, Paul Stoodley, Montana State, and Biofilm Engineering. 2004. "Bacterial Biofilms: From the Natural Environment to Infectious Diseases." *Nature Reviews Microbiology* 2 (2): 95–108. doi:10.1038/nrmicro821.

Hwang, Geelsu, Seoktae Kang, Mohamed Gamal El-Din, and Yang Liu. 2012a. "Impact of an Extracellular Polymeric Substance (EPS) Precoating on the Initial Adhesion of Burkholderia Cepacia and Pseudomonas Aeruginosa." *Biofouling* 28 (6): 525–538. doi:10.1080/08927014.2012.694138.

Hwang, Geelsu, Seoktae Kang, Mohamed Gamal El-Din, and Yang Liu. 2012b. "Impact of Conditioning Films on the Initial Adhesion of Burkholderia Cepacia." *Colloids and Surfaces B: Biointerfaces* 91 (1): 181–188. doi:10.1016/j.colsurfb.2011.10.059.

Hwang, Geelsu, Jiaming Liang, Seoktae Kang, Meiping Tong, and Yang Liu. 2013. "The Role of Conditioning Film Formation in Pseudomonas Aeruginosa PAO1 Adhesion to Inert Surfaces in Aquatic Environments." *Biochemical Engineering Journal* 76: 90–98. doi:10.1016/j.bej.2013.03.024.

James, Sean A., Nidal Hilal, and Chris J. Wright. 2017. "Atomic Force Microscopy Studies of Bioprocess Engineering Surfaces – Imaging, Interactions and Mechanical Properties Mediating Bacterial Adhesion." *Biotechnology Journal* 12 (7): 1–11. doi:10.1002/biot.201600698.

Jin, Wen-Ji, Hwang Kyu Lee, Eun Hwan Jeong, Won Ho Park, and Ji Ho Youk. 2005. "Preparation of Polymer Nanofibers Containing Silver Nanoparticles by Using Poly(N-Vinylpyrrolidone)." *Macromolecular Rapid Communications* 26 (24): 1903–1907. doi:10.1002/marc.200500569.

Kargar, Mehdi, Ji Wang, Amrinder S. Nain, and Bahareh Behkam. 2012. "Controlling Bacterial Adhesion to Surfaces Using Topographical Cues: A Study of the Interaction of Pseudomonas Aeruginosa with Nanofiber-Textured Surfaces." *Soft Matter* 8 (40): 10254. doi:10.1039/c2sm26368h.

Katti, Dhirendra S, Kyle W Robinson, Frank K. Ko, and Cato T. Laurencin. 2004. "Bioresorbable Nanofiber-based Systems for Wound Healing and Drug Delivery: Optimization of Fabrication Parameters." *Journal of Biomedical Materials Research Part B: Applied Biomaterials* 70 (2): 286–96.

Kharaghani, Davood, Yun Kee Jo, Muhammad Qamar Khan, Yeonsu Jeong, Hyung Joon Cha, and Ick Soo Kim. 2018. "Electrospun Antibacterial Polyacrylonitrile Nanofiber Membranes Functionalized with Silver Nanoparticles by a Facile Wetting Method." *European Polymer Journal* 108 (July): 69–75. doi:10.1016/j.eurpolymj.2018.08.021.

Khulbe, Kailash Chandra, Chaoyang Feng, Ahmad Fauzi Ismail, and Takeshi Matsuura. 2015. "Development of Large-Scale Industrial Applications of Novel Membrane Materials: Carbon Nanotubes, Aquaporins, Nanofibers, Graphene, and Metal-Organic Frameworks." In *Membrane Fabrication* (Nidal Hilal, Ahmad Fauzi Ismail, Chris Wright Eds.), pp. 383–436. CRC Press, London, UK. doi:10.1201/b18149-15.

Kidoaki, Satoru, Il Keun Kwon, and Takehisa Matsuda. 2006. "Structural Features and Mechanical Properties of in Situ–Bonded Meshes of Segmented Polyurethane Electrospun from Mixed Solvents." *Journal of Biomedical Materials Research Part B: Applied Biomaterials* 76B (1): 219–229. doi:10.1002/jbm.b.30336.

Lin, Tong, Xin Liu, Tong Lin, Yuan Gao, Zhiguang Xu, Chen Huang, Gang Yao, Linlin Jiang, Yanwei Tang, and Xungai Wang. 2012. "Antimicrobial Electrospun Nanofibers of Cellulose Acetate and Polyester Urethane Composite for Wound Dressing Urethane Composite for Wound Dressing." *Journal of Biomedical Materials Research Part B: Applied Biomaterials*. doi:10.1002/jbm.b.32724.

Liu, Yaolin, and Baoxia Mi. 2014. "Effects of Organic Macromolecular Conditioning on Gypsum Scaling of Forward Osmosis Membranes." *Journal of Membrane Science* 450: 153–161. doi:10.1016/j.memsci.2013.09.001.

Llorens, Elena, Silvia Calderón, Luis J. Del Valle, and Jordi Puiggalí. 2015. "Polybiguanide (PHMB) Loaded in PLA Scaffolds Displaying High Hydrophobic, Biocompatibility and Antibacterial Properties." *Materials Science and Engineering C* 50: 74–84. doi:10.1016/j.msec.2015.01.100.

Lubasova, Daniela, and Smykalova Barbora. 2014. "Antibacterial Efficiency of Nanofiber Membranes with Biologically Active Nanoparticles." *International Conference on Agriculture, Biology and Environmental Sciences (ICABES'14)*, December 8–9, Bali, Indonesia, pp. 55–58.

Lukas, David, Arindam Sarkar, and Pavel Pokorny. 2008. "Self-Organization of Jets in Electrospinning from Free Liquid Surface: A Generalized Approach." *Journal of Applied Physics* 103 (8). doi:10.1063/1.2907967.

Maillard, Jean-Yves, and Philippe Hartemann. 2013. "Silver as an Antimicrobial: Facts and Gaps in Knowledge." *Critical Reviews in Microbiology* 39 (4): 373–383. doi:10.3109/1040841X.2012.713323.

Mehrasa, Mohammad, Mohammad Ali Asadollahi, Kamran Ghaedi, Hossein Salehi, and Ayyoob Arpanaei. 2015. "Electrospun Aligned PLGA and PLGA/Gelatin Nanofibers Embedded with Silica Nanoparticles for Tissue Engineering." *International Journal of Biological Macromolecules* 79: 687–695. doi:10.1016/j.ijbiomac.2015.05.050.

Mit-uppatham, Chidchanok, Manit Nithitanakul, and Pitt Supaphol. 2004. "Ultrafine Electrospun Polyamide-6 Fibers: Effect of Solution Conditions on Morphology and Average Fiber Diameter." *Macromolecular Chemistry and Physics* 205 (17): 2327–2338.

Mo, Xiumei, Chengyu Xu, Masaya Kotaki, and Seeram Ramakrishna. 2004. "Electrospun P (LLA-CL) Nanofiber: A Biomimetic Extracellular Matrix for Smooth Muscle Cell and Endothelial Cell Proliferation." *Biomaterials* 25 (10): 1883–1890.

Moritz, Michał, and Małgorzata Geszke-Moritz. 2013. "The Newest Achievements in Synthesis, Immobilization and Practical Applications of Antibacterial Nanoparticles." *Chemical Engineering Journal* 228: 596–613. doi:10.1016/j.cej.2013.05.046.

Nataraj, Sanna Kotrappanavar, Bo-hye Kim, Kap Seung Yang, and Hee-gweon Woo. 2010. "*In-Situ* Deposition of Iron Oxide Nanoparticles on Polyacrylonitrile-Based Nanofibers by Chemico-Thermal Reduction Method." *Journal of Nanoscience and Nanotechnology* 10 (5): 3530–3533. doi:10.1166/jnn.2010.2290.

Neihof, Rex, and George Loeb. 1974. "Dissolved Organic Matter in Seawater and the Electric Charge of Immersed Surfaces." *Journal of Marine Research* 32 (1): 5–12.

Nicosia, Alessia, Weronika Gieparda, Joanna Foksowicz-Flaczyk, Judyta Walentowska, Dorota Wesołek, B. Vazquez, Francesco Prodi, and Franco Belosi. 2015. "Air Filtration and Antimicrobial Capabilities of Electrospun PLA/PHB Containing Ionic Liquid." *Separation and Purification Technology* 154: 154–160. doi:10.1016/j.seppur.2015.09.037.

Niu, Haitao, and Tong Lin. 2012. "Fiber Generators in Needleless Electrospinning." *Journal of Nanomaterials* 2012: 12.

Park, Jeong Ann, and Song Bae Kim. 2017. "Antimicrobial Filtration with Electrospun Poly(Vinyl Alcohol) Nanofibers Containing Benzyl Triethylammonium Chloride: Immersion, Leaching, Toxicity, and Filtration Tests." *Chemosphere* 167: 469–477. doi:10.1016/j.chemosphere.2016.10.030.

Patanaik, Asis, Valencia Jacobs, and Rajesh D. Anandjiwala. 2010. "Performance Evaluation of Electrospun Nanofibrous Membrane." *Journal of Membrane Science* 352 (1–2): 136–142. doi:10.1016/j.memsci.2010.02.009.

Percival, Steven L., P. G. Bowler, and D. Russell. 2005. "Bacterial Resistance to Silver in Wound Care." *Journal of Hospital Infection* 60 (1): 1–7. doi:10.1016/j.jhin.2004.11.014.

Petrík, Stanislav. 2009. *Industrial Production Technology for Nanofibers.* IntechOpen.

Ploux, Lydie, Arnaud Ponche, and Karine Anselme. 2010. "Bacteria/Material Interfaces: Role of the Material and Cell Wall Properties." *Journal of Adhesion Science and Technology* 24 (13–14): 2165–2201. doi:10.1163/016942410X511079.

Razzaz, Adib, Shima Ghorban, Layla Hosayni, Mohammad Irani, and Majid Aliabadi. 2016. "Chitosan Nanofibers Functionalized by TiO_2 Nanoparticles for the Removal of Heavy Metal Ions." *Journal of the Taiwan Institute of Chemical Engineers* 58: 333–343. doi:10.1016/j.jtice.2015.06.003.

Rujitanaroj, Pim-on, Nuttaporn Pimpha, and Pitt Supaphol. 2008. "Wound-Dressing Materials with Antibacterial Activity from Electrospun Gelatin Fiber Mats Containing Silver Nanoparticles." *Polymer* 49 (21): 4723–4732.

Saquing, Carl D., Joshua L Manasco, and Saad A Khan. 2009. "Electrospun Nanoparticle–Nanofiber Composites via a One-Step Synthesis." *Small* 5 (8): 944–951. doi:10.1002/smll.200801273.

Schneider, Ren P., and Andrew Leis. 2003. "Conditioning Films in Aquatic Environments." In *Encyclopedia of Environmental Microbiology (Gabriel Bitton Ed.)*, Hoboken, NJ: John Wiley & Sons. doi:10.1002/0471263397.env036.

Shalumon, K. T., K. H. Anulekha, Sreeja V. Nair, S. V. Nair, Krishnaprasad Chennazhi, and R. Jayakumar. 2011. "Sodium Alginate/Poly(Vinyl Alcohol)/Nano ZnO Composite Nanofibers for Antibacterial Wound Dressings." *International Journal of Biological Macromolecules* 49 (3): 247–254. doi:10.1016/j.ijbiomac.2011.04.005.

Son, Won Keun, Ji Ho Youk, Taek Seung Lee, and Won Ho Park. 2004. "The Effects of Solution Properties and Polyelectrolyte on Electrospinning of Ultrafine Poly(Ethylene Oxide) Fibers." *Polymer* 45 (9): 2959–2966. doi:10.1016/j.polymer.2004.03.006.

Song, K., Q. Wu, Y. Qi, and T. Kärki. 2016. "Electrospun Nanofibers with Antimicrobial Properties." *Electrospun Nanofibers* 1: 551–569. doi:10.1016/B978-0-08-100907-9.00020-9.

Strutt, John William, and Lord Rayleigh. 1878. "On the Instability of Jets." *Proceedings of the London Mathematical Society* 10: 4–13.

Suwarno, Stanislaus Raditya, S. Hanada, Tzyy Haur Chong, S. Goto, Masahiro Henmi, and A. G. Fane. 2016. "The Effect of Different Surface Conditioning Layers on Bacterial Adhesion on Reverse Osmosis Membranes." *Desalination* 387: 1–13. doi:10.1016/j.desal.2016.02.029.

Takht Ravanchi, Maryam, Tahereh Kaghazchi, and Ali Kargari. 2009. "Application of Membrane Separation Processes in Petrochemical Industry: A Review." *Desalination* 235 (1–3): 199–244. doi:10.1016/j.desal.2007.10.042.

Tan, S. H., Ryuji Inai, Masaya Kotaki, and Seeram Ramakrishna. 2005. "Systematic Parameter Study for Ultra-Fine Fiber Fabrication via Electrospinning Process." *Polymer* 46 (16): 6128–6134. doi:10.1016/j.polymer.2005.05.068.

Taylor, Geoffrey. 1964. "Disintegration of Water Drops in an Electric Field." *Proceedings of the Royal Society of London. Series A. Mathematical and Physical Sciences* 280 (1382): 383–397.

Taylor, Geoffrey. 1969. "Electrically Driven Jets." *Proceedings of the Royal Society of London. A. Mathematical and Physical Sciences* 313 (1515): 453–475.

Thoppey, Nagarajan Muthuraman, Jason Bochinski, Laura Clarke, and Russell E. Gorga. 2011. "Edge Electrospinning for High Throughput Production of Quality Nanofibers." *Nanotechnology* 22 (34): 345301. doi:10.1088/0957-4484/22/34/345301.

Ubbink, Job, and Prisca Schär-Zammaretti. 2007. "Colloidal Properties and Specific Interactions of Bacterial Surfaces." *Current Opinion in Colloid and Interface Science* 12 (4–5): 263–270. doi:10.1016/j.cocis.2007.08.004.

Ulbricht, Mathias. 2006. "Advanced Functional Polymer Membranes." *Polymer* 47 (7): 2217–2262. doi:10.1016/j.polymer.2006.01.084.

Wang, Xin, Haitao Niu, Tong Lin, and Xungai Wang. 2009. "Needleless Electrospinning of Nanofibers with a Conical Wire Coil." *Polymer Engineering & Science* 49 (8): 1582–1586.

Wang, Yongzhi, Qingbiao Yang, Guiye Shan, Ce Wang, Jianshi Du, Shugang Wang, Yaoxian Li, Xuesi Chen, Xiabin Jing, and Yen Wei. 2005. "Preparation of Silver Nanoparticles Dispersed in Polyacrylonitrile Nanofiber Film Spun by Electrospinning." *Materials Letters* 59 (24–25): 3046–3049. doi:10.1016/j.matlet.2005.05.016.

Wingender, Jost, Martin Strathmann, Alexander Rode, Andrew Leis, and Hans-Curt Flemming. 2001. "Isolation and Biochemical Characterization of Extracellular Polymeric Substances from Pseudomonas Aeruginosa." *Methods in Enzymology* 336: 302–314. doi:10.1016/S0076-6879(01)36597-7.

Xiao, Shili, Mingwu Shen, Rui Guo, Shanyuan Wang, and Xiangyang Shi. 2009. "Immobilization of Zerovalent Iron Nanoparticles into Electrospun Polymer Nanofibers: Synthesis, Characterization, and Potential Environmental Applications." *The Journal of Physical Chemistry C* 113 (42): 18062–18068. doi:10.1021/jp905542g.

Xu, Li Chong, and Christopher A. Siedlecki. 2012. "Submicron-Textured Biomaterial Surface Reduces Staphylococcal Bacterial Adhesion and Biofilm Formation." *Acta Biomaterialia* 8 (1): 72–81. doi:10.1016/j.actbio.2011.08.009.

Yarin, Alexander L, Sureeporn Koombhongse, and Darrell Hyson Reneker. 2001. "Taylor Cone and Jetting from Liquid Droplets in Electrospinning of Nanofibers." *Journal of Applied Physics* 90 (9): 4836–46.

Yoon, Kyunghwan, Benjamin S. Hsiao, and Benjamin Chu. 2008. "Functional Nanofibers for Environmental Applications." *Journal of Materials Chemistry* 18 (44): 5326. doi:10.1039/b804128h.

Yun, Ki Myoung, Christopher J. Hogan, Yasuko Matsubayashi, Masaaki Kawabe, Ferry Iskandar, and Kikuo Okuyama. 2007. "Nanoparticle Filtration by Electrospun Polymer Fibers." *Chemical Engineering Science* 62 (17): 4751–4759. doi:10.1016/j.ces.2007.06.007.

Zhang, Chenglin, Chunping Li, Jie Bai, Junzhong Wang, and Hongqiang Li. 2015. "Synthesis, Characterization, and Antibacterial Activity of Cu NPs Embedded Electrospun Composite Nanofibers." *Colloid and Polymer Science* 293 (9): 2525–2530. doi:10.1007/s00396-015-3640-6.

Zhang, Hongbin, and Mu Chiao. 2015. "Anti-Fouling Coatings of Poly(Dimethylsiloxane) Devices for Biological and Biomedical Applications." *Journal of Medical and Biological Engineering* 35 (2): 143–155.

Zhang, Shu, Woo Sub Shim, and Jooyoun Kim. 2009. "Design of Ultra-Fine Nonwovens via Electrospinning of Nylon 6: Spinning Parameters and Filtration Efficiency." *Materials and Design* 30 (9): 3659–3666. doi:10.1016/j.matdes.2009.02.017.

Zhang, Shu, Yongan Tang, and Branislav Vlahovic. 2016. "A Review on Preparation and Applications of Silver-Containing Nanofibers." *Nanoscale Research Letters* 11 (1): 80. doi:10.1186/s11671-016-1286-z.

Zhao, Fang, Si Chen, Qiaole Hu, Gang Xue, Qingqing Ni, Qiuran Jiang, and Yiping Qiu. 2017. "Antimicrobial Three Dimensional Woven Filters Containing Silver Nanoparticle Doped Nanofibers in a Membrane Bioreactor for Wastewater Treatment." *Separation and Purification Technology* 175: 130–139. doi:10.1016/j.seppur.2016.11.024.

3 Nanofiber Electrospun Membrane Based on Biodegradable Polymers for Biomedical and Tissue Engineering Application

Lim Mim Mim, Naznin Sultana, Hasrinah Hasbullah, and Madzlan Aziz

CONTENTS

3.1 INTRODUCTION

Conventionally, autograft, allograft, and xenograft are the ways for replacing diseased or damaged tissues or organs. However, these treatments have limitations including lack of healthy cells for biopsy, lack of organ donors, the patients are not strong enough to withstand surgery, and the high possibilities of graft rejection and disease transmittance. Due to these limitations, tissue engineering has emerged to provide a new medical approach to regenerate almost every tissue or organ in human body. To be successful, tissue engineering needs a wide range of knowledge and a team of experts such as engineers, material scientists, mathematicians, cell biologists, clinicians, and geneticists.

The term "tissue engineering" is defined by Langer and Vacanti as "*an interdisciplinary field that applies the principles of engineering and life sciences towards the development of biological substitutes that restore, maintain or improve tissue function or a whole organ*" (Langer and Vacanti 1993). The function of tissue engineering is to restore, replace, improve, or maintain diseased or damaged tissue or organs by developing biological substitutes such as tissue analogue scaffolds or recreating the tissue. The objective of tissue engineering is to create a suitable environment for cells that mimic the natural environment in our body and help to control and regulate the function of cells.

Generally, there are three strategies of tissue engineering (Langer and Vacanti 1993): (i) implantation into the body of cell substitutes or cells that have been previously isolated and cultured *in vitro* (cell therapy), (ii) insertion of growth factors to induce the growth of tissue, and (iii) cell seeding on scaffolds (nanofiber membrane). The strategy of cell seeding on nanofiber membrane has been the most frequently related to the concept of tissue engineering. In this strategy, there are three important elements: (i) cells, (ii) scaffold (nanofiber membrane), and (iii) bioreactor. It is crucial to choose the right cell source which is compatible with the patient and without contamination to avoid rejection and infection. The nanofiber membrane is a biocompatible, biodegradable and temporary supporting material for cellular growth. The bioreactor provides an environment for cells that imitates the natural environment in the body. Cells seeded on the nanofiber membrane and placed in the bioreactor are expected to behave similarly as in the body. First, cells are isolated from the patient and cultured *in vitro* to expand the cells. Then, cells are seeded on an appropriate scaffold with desired properties together with growth factors and signaling molecules to promote cell adhesion and proliferation. These cells-seeded nanofiber membranes are located in a cell-specific bioreactor with growth media and a supportive environment for cells to proliferate and generate extracellular matrix (ECM) because every type of cells has different pH, oxygen, mechanical, and temperature requirements (Pörtner and Giese 2007). Lastly, after sufficient *in vitro* growth of the cells on and within the cell-seeded nanofiber membrane the construct is transplanted into the body of the patient to restore or regenerate the damaged or diseased tissue or organ.

3.2 NANOFIBER MEMBRANE MATERIAL SELECTION

The selection of material plays an important role in fabricating tissue engineering scaffold. More than 200 polymers have been used in the process of nanofiber membrane fabrication. Biocompatible and biodegradable polymers are more advantageous and preferable. Polymers are divided into two groups: (1) synthetic polymers and (2) natural polymers. Synthetic polymers have better mechanical strength, but slower degradation rate than natural polymers. They are biocompatible, but hydrophobic and lack of cell-recognition signals. However, synthetic polymers are highly useful as their properties can be tailored for targeted applications by controlling porosity, degradation rate, and mechanical strength. They are low in cost, have longer storage time and can be produced in a large uniform amount (Brahatheeswaran et al. 2011). A variety of synthetic polymers have been explored and evaluated for use in nanofiber membrane fabrication in tissue engineering. Examples of synthetic polymers are polycaprolactone (PCL), poly(lactic acid-co-glycolic acid) (PLAGA), poly(lactide-co-glycolide) (PLGA), poly(3-hydroxybutyrate-co-3-hydroxyvalerate) (PHBV), polyurethane (PU), poly(ethylene oxide) (PEO), and poly(lactic acid) (PLA).

PCL is a synthetic aliphatic polyester, which is an implantable device material that has been approved by the Food and Drug Administration (FDA) (Woodruff and Hutmacher 2010). It has advantages of low cost, ease of shaping and manufacture, biodegradability, biocompatibility, low toxicity, and good mechanical strength. PCL is considered as a soft and hard tissue compatible bioresorbable material. However, it has slow degradation rates of up to 3–4 years (Woodruff and Hutmacher 2010). Due to the slow degradation rate of PCL, it is more suitable to be used as long-term drug delivery carrier.

Apart from synthetic polymers, natural polymers are also widely used. Natural polymers are hydrophilic, have low toxicity and excellent biocompatibility, and contain cell-recognition signals which lead to a better interaction with cells and enhances cell behavior on the scaffold. However, natural polymers are more challenging to be used in fiber fabrication than synthetic polymers. Natural polymers have some disadvantages such as fast degradation rates and low reproducibility. Natural polymers are obtained from natural sources and purified to avoid foreign body response when implanted into the body. They are categorized into three groups: (1) proteins (collagen, gelatin [Ge], silk, elastin, fibrinogen, myosin, actin, and keratin), (2) polynucleotides (deoxyribonucleic acid [DNA] and ribonucleic acid [RNA]), and (3) polysaccharides (cellulose, chitin, dextran, glycosaminoglycans, and amylose).

Ge is obtained by partial hydrolysis of collagen. Ge that is derived from an acid treated process is known as Type A Ge and Ge that is derived from an alkaline process is known as Type B Ge (GMIA 2012). Collagen is extracted from the skin, connective tissues (tendon and cartilage), and bones of cattle, fish, poultry or fish. Ge is an FDA-approved polymer which has been used widely in various fields including pharmaceutical and food industry (GMIA 2012). Ge is hydrophilic, favorable to cells, low in cost, biocompatible, biodegradable, non-immunogenic, and anti-thrombogenic. It has the potential to be used as wound dressing material, drug delivery carrier, and development of biomimicking artificial ECM for tissue engineering applications. However, due to its water soluble and less stable properties as well as week mechanical strength, it limits the function of Ge for potential applications. Ge is a water-soluble polymer due to the presence of amino group that form hydrogen bonds with water molecules. Its fiber structure would be lost in an aqueous environment such as that found in a wound. To overcome this, Ge requires chemical cross-linking.

Polymer blending is one of the most effective methods to fabricate nanofiber membranes with desired properties for specific application. Nanofiber membrane fabricated from a single polymer might have "fishnet effect" (high fiber density) which will reduce cellular penetration. Synthesizing scaffolds by polymer blending can avoid fishnet effect and the need for cross-linking. Polymer blending of synthetic-natural polymers has been widely investigated compared to polymer blending of natural-natural and synthetic-synthetic polymers. It gives advantages of providing the scaffold with mechanical strength and durability from the synthetic polymer as well as the biological functionality of natural polymer. By blending of natural and synthetic polymers, it improves the degradation rate of the synthetic polymer, too (Lim and Sultana 2017).

The combination of hydrophobic and hydrophilic polymers is widely used. The hydrophobic polymer can serve as a backbone for the scaffold while the hydrophilic polymer enhances cell attachment and proliferation. Examples of polymer blending are collagen/poly(L-lactic acid)–co-poly(ε-caprolactone) (collagen/PLLCL), PCL/collagen, PCL/Ge, Chitosan/PCL, and poly(L-lactic acid)-co-poly(ε-caprolactone) (PLACL)/Ge.

3.3 NANOFIBER MEMBRANE FABRICATION

The fabrication technique is a key factor in the production of scaffolds with desired properties. Many techniques have been used, including freeze-drying, gas foaming, rapid prototyping, solvent casting/particulate leaching, and melt molding. To fabricate scaffolds of nanofiber membrane, various methods have been explored, including phase separation, drawing, self-assembly, electrospinning, and template synthesis.

The phase separation technique consists of five steps: (1) polymer dissolution, (2) gelation, (3) solvent extraction from gel with water, (4) freezing, and (5) freeze-drying under vacuum. First, a polymer that can gel is dissolved in a solvent and stirred at a fixed temperature until the solution is mixed homogeneously. The solution is transferred to the refrigerator and set at the gelation temperature of the polymer followed by immersing the resultant gel in distilled water for solvent exchange and then transferred to a freezer at −18°C. Lastly, the frozen gel is freeze-dried under vacuum for 1 week. The phase separation technique is able to form fibers with diameters in the range of 50–500 nm and length of few micrometers (Zhang et al. 2005a). However, this technique is time consuming. It needs more than 1 week to form a fibrous scaffold. Moreover, it is unable to produce homogeneous fibers and difficult to control the scaffold morphology.

Self-assembly is a technique of reorganizing existing components into desired designs and has been used for fiber fabrication. Atoms and molecules reorganize and rearrange themselves via weak and non-covalent bonding including hydrogen bonding, electrostatic, and hydrophobic forces into a well-defined and stable structure which is in nano-sized. Self-assembly technique produces fibrous scaffolds with diameters less than 100 nm and length of few micrometers (Ndreu 2007). Self-assembly is a time-consuming technique as it takes a few days to fabricate a fibrous scaffold. Moreover, it is an expensive technique with complex design parameters and low productivity (Rajput 2012).

Template synthesis technique utilizes nano porous membrane as a template to form nanofibers or hollow shaped tubules. A polymer solution is extruded through the nano porous membrane by water pressure. The polymer solution is solidified when in contact with a solidification solution and forms fibers. The diameter of the fibers is dependent on the pore size of the template. This technique is advantageous where various raw materials can be used in the fabrication of fibrous scaffold such as metals, polymers, carbons, and semiconductors (Ndreu 2007). However, this technique has low productivity and is unable to produce single and continuous fibers (Rajput 2012). The fibers produced are only in the length of a few micrometers (Ndreu 2007).

The drawing technique uses a micropipette for fabricating nanofibers. The micropipette is dipped into a droplet near the solution solid surface contact line through a micromanipulator and withdrawn from the droplet to produce nanofibers. In this technique, less equipment is needed. However, instabilities occur during the process will lead to fiber breaking. The fiber size is dependent on the opening of the extrusion mold and it is difficult to fabricate fibers with diameter less than 100 nm. The solution used in this technique needs to be cohesive to withstand the stress during pulling and can undergo strong deformations (Ndreu 2007).

Electrospinning, a technique that utilizes electrostatic forces to produce fibers, was developed 100 years ago. The electrospinning technique is a simple and cost-effective technique for fiber membrane scaffold fabrication (Zhang et al. 2005b). This technique is versatile and can spin a wide range of materials such as polymers, composites, and ceramics into extremely thin fibers ranging from micrometers to a few nanometers (Goh et al. 2013). These thin fibers mimic the structural dimension of ECM of tissues and organs (Zhang et al. 2005b). Moreover, the small dimension of fibers provides high surface area to volume ratio for cellular growth (Zhang et al. 2005b). This technique is also able to produce fibers with high porosity which is ideal for gene, drug, and cell delivery. It can be integrated with biomolecules and living cells for different applications (Zhang et al. 2005b). The characteristics of fiber membrane scaffolds produced by the electrospinning technique are favorable for cell attachment, expression, proliferation as well as transportation of nutrient, oxygen, and waste products. Among the techniques mentioned above, electrospinning is the most versatile and promising technique to fabricate fibrous scaffolds with controlled porosity, pore size, and fiber diameter (Brahatheeswaran et al. 2011).

Electrospinning passes a fluid solution, through a millimeter diameter needle, which then forms into a solid fiber that is four or five times smaller in diameter by using an electrostatic force (Huang et al. 2003). The three basic components of electrospinning setup are: (1) high-voltage power supply, (2) syringe pump, and (3) grounded collector. There are three stages of electrospinning process: (1) jet initiation, (2) whipping, and (3) elongation of a jet and solidification of the jet into fibers. The electrospinning process is controlled by the high voltage electric field between two opposite polarity electrodes. The positive (red) electrode is connected to the needle to electrify the polymer solution and the negative (black) electrode is connected to the grounded collector which is usually a metal screen. A schematic diagram of electrospinning is shown in Figure 3.1.

In electrospinning a polymer solution is placed in a syringe. A syringe pump is used to force the polymer solution with a controlled rate through a needle, which is connected to high voltage power supply. When the voltage increases, it induces a charge on the polymer solution, forming an electrostatic force which causes the repulsion of similar charges in the solution. When increasing voltage, the electrostatic force is also increasing. A conical shape of the pendent droplet (Taylor cone) (Taylor 1969) at the tip of the needle is formed when electrostatic force is balanced with surface tension. The formation of the Taylor cone is at a semi-vertical angle, $\varphi = 49.3°$ (Taylor 1969). By increasing the voltage jet initiation occurs. A jet of polymer is ejected when the electrostatic force is larger than the surface tension, followed by whipping and elongation. At this stage, there might be splitting or branching of

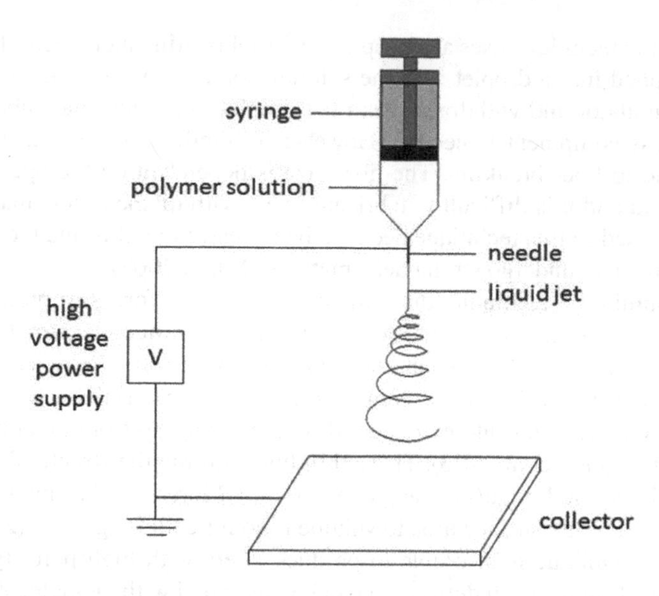

FIGURE 3.1 Schematic diagram of electrospinning.

the primary jet into multiple sub-jets called fiber splaying which has been experienced by a few research groups (Shkadov and Shutov 2001). This phenomenon is due to the instability of the jet. During the elongation process the volatile solvent is evaporated. Evaporation of solvent reduces the diameter of the fiber and leads to the formation of pores. Lastly is the solidification of fiber which is attracted to the opposite charged grounded collector.

3.4 NANOFIBER MEMBRANE CHARACTERIZATION

An important step to evaluate the suitability of nanofiber membrane for tissue engineering application is characterization. To illustrate this we now describe a study within our laboratory, where PCL and PCL/Ge were electrospun and characterized for their physical and chemical properties including average fiber diameter, pore size, wettability, porosity and density, surface roughness, and *in vitro* degradation (Lim et al. 2015). *In vitro* human skin fibroblast (HSF) cells culture was used to investigate the suitability of HSF cells growth on these nanofiber membranes. HSF cells morphology was characterized using cell fixation followed by observing the cells under the scanning electron microscope (SEM). HSF cells also underwent cytotoxicity testing on both nanofiber membranes by using an MTT (3-(4,5-dimethylthiazol-2-yl)-2,5-diphenyltetrazolium bromide/ethylthiazoletetrazolium) assay.

PCL solution was prepared by dissolving PCL in formic acid while PCL/Ge solution was prepared by dissolving PCL and Ge in the ratio of 70:30 in formic acid. Na Bond Nanofiber Electrospinning Unit, China, was used for electrospinning process. A 23 gauge stainless steel needle was attached to the syringe as a nozzle and the syringe was placed on a syringe pump (NE-300, New Era Pump Systems, Inc.).

High voltage was applied for 2 h at a distance of 10 cm between capillary tip to the collector. Solution and processing parameters such as polymer concentration, types of solvent, and flow rate were optimized during the electrospinning process.

Wettability (surface hydrophobicity) of nanofiber membranes was determined by measuring the contact angle using video contact angle system (VCA Optima, AST Products, Inc.). Sessile drop method was used. Surface hydrophobicity is a crucial factor that will influence cell responses to a scaffold. An angle of more than 90° is considered hydrophobic and lower than 90° is hydrophilic. Studies have shown that cells are more favorable on hydrophilic surfaces and show good spreading and proliferation on these surfaces (Yildirim et al. 2010).

Porosity and density of nanofiber membranes were measured by liquid displacement method. In this study, ethanol was used as displacement liquid (Sun et al. 2014). Initial weight (W) of scaffolds was noted and immersed in a known volume (V_1) of ethanol for 5 min. The volume of ethanol and ethanol-impregnated fiber was recorded as V_2. Residual volume after ethanol-impregnated fiber was removed and recorded as V_3. Total volume, density, and porosity of the fibers were calculated as the equations below (Sun et al. 2014).

Total volume of fiber was

$$V = V_2 - V_3 \tag{3.1}$$

Density (d) of the fiber was

$$d = \frac{W}{V_2 - V_3} \tag{3.2}$$

Porosity (ε) of the fiber was

$$= \frac{(V_1 - V_3)}{(V_2 - V_3)} \tag{3.3}$$

Pores are important for cell migration, infiltration, nutrient supply, oxygen, and waste product exchange. Decreasing fiber diameter of fibrous scaffold will decrease the size of pores. The average size of fibroblasts, osteoblasts, and chondrocytes is 10 μm. According to a study done by Lowery et al. (2010), fibroblasts proliferated at a faster rate on scaffold with pore diameters bigger than 6 μm.

Porosity is the percentage of void volume in a scaffold which is also an important factor for cell infiltration. Inconsistencies of porosity will lead to poor cellular distribution and penetration. In a study of bone cells, it was found that increasing porosity increased the growth of bone cells (Toth et al. 1995). The preferred porosity for cell penetration is within the range of 60%–90% (Chong et al. 2007). Hence, scaffolds have to be fabricated with high porosity and optimal pore size, which varies with different tissue engineering applications.

A study has been conducted to investigate HSF cell behavior on PCL/chitosan and nanohydroxyapatite (nHA)/PCL/chitosan porous scaffolds fabricated using a freeze-drying technique (Jin et al. 2015). The addition of nHA decreased the hydrophilicity of PCL/chitosan porous scaffolds from 70.93 ± 2.99° to 59.25 ± 2.22°. The results showed that a more hydrophilic scaffold of nHA/PCL/chitosan enhances the growth and proliferation of HSF cells (Jin et al. 2015). Another study was conducted to evaluate porous poly (L-lactic acid)/PLGA (PLLA/PLGA) (70:30) scaffolds with different pore structures using HSF cells. The results showed that scaffolds with pore size smaller than 160 μm were more suitable for the growth of HSF cells (Yang et al. 2002). Another study fabricated PLAGA electrospunfibers with different fiber diameters and characterized their influence on cell behavior by seeding with HSF cells (Kumbar et al. 2008). HSF cells showed significant progressive growth on fiber diameters in the range of 350–1100 nm.

Surface topography and surface roughness (Ra) of nanofiber membranes were performed by atomic force microscope (AFM) (TT-AFM, AFM Workshop) using pre-set scanning area of 15 micron by 15 micron. Surface roughness is divided into macroroughness (100 μm–millimeters), microroughness (100 nm–100 μm), and nanoroughness (less than 100 nm). Cell responses toward surface roughness are variable for different cells. Primary rat osteoblasts cells showed high proliferation rates and high alkaline phosphatase activity on microscale rough surfaces of 0.81 μm (Hatano et al. 1999). Nanoscale rough surfaces (6.26–49.38 nm) have been shown to support neurite growth and increase axonal length (De Bartolo et al. 2008). For human vein endothelial cells, nanoscale rough surfaces (10–102 nm) have enhanced cell adhesion and growth (Chung et al. 2003).

Degradation was evaluated by morphological changes and water uptake. Nanofiber membranes were sectioned into 1 × 1 cm², weighted, and incubated at 37°C in 10 mL phosphate buffer saline (PBS). After 4, 8, and 12 weeks of incubation, nanofiber membranes were washed with distilled water to remove residual buffer salts. Nanofiber membranes were blotted dry with tissue paper to remove excess water and weighted. Water uptake of nanofiber membranes was calculated using the following equation (Sultana and Khan 2012):

$$\text{Water Uptake}(\%) = \frac{W_w - W_d}{W_d} \times 100, \qquad (3.4)$$

where W_w is weight of fiber after incubation, W_d is weight of fiber before incubation.

Nanofiber membranes were dried at room temperature until a constant weight had achieved and coated with gold for the preparation to be viewed under FESEM. Morphological changes of the fibers after degradation were determined by FESEM images. Degradation rate has to be coincided with the rate of new tissue formation. Nanofiber fabrication by blending of natural and synthetic polymers can have a better control of desired degradation rate.

PCL and PCL/Ge nanofiber membranes were cut into circular discs (1.6 cm in diameter) and washed three times using PBS with 1 wt.% Pen Strep followed by exposure to ultraviolet (UV) radiation for 2 h. HSF cells were used to investigate cell behavior on both PCL and PCL/Ge nanofiber membranes. HSF cells were cultured in Dulbecco's Modified Eagle Medium (DMEM) containing 10% fetal bovine serum (FBS) and 1% Pen Strep in a CO_2 incubator (37°C, 5% CO_2). DMEM was replaced every 2 days. When HSF cells reached 80% confluence, they were trypsinized and counted using a hemocytometer (Counting chambers, Neubauer-improved, Hirschmann). A density of 10 × 10^3 cells/well were seeded on each scaffold in 24-well plate for cell counting, morphology, and cell cytotoxicity tests.

HSF cells morphology on PCL and PCL/Ge nanofiber membranes was observed by using SEM. Scaffolds were washed with PBS and fixed with 4% glutaraldehyde for 1 h at 4°C. After 1 h, scaffolds underwent gradual dehydration in 30%, 50%, 70%, 95%, and 100% of ethanol and air-dried in biosafety cabinet (BSC) The dried samples were sputter coated with platinum for 20 s and observed under SEM.

MTT assay was used to test for cell cytotoxicity on PCL, PCL/Ge nanofiber membrane. HSF cell cultured in wells without membrane was used as control. Nanofiber membranes were sectioned into circular discs (1.6 cm in diameter) and sterilized by washing three times using PBS with 1% Pen Strep followed by exposing to UV radiation for 2 hours. 10 × 10^3 HSF cells were seeded on the scaffolds for 3 days in each well in 24-well plates with DMEM as medium. After 3 days, DMEM was removed. New DMEM and MTT solution (5 g/L) were added to each well followed by incubation for 4 hours. Twenty-four well plates were wrapped with aluminum foil as MTT is sensitive to light. After 4 hours, purple formazan was formed. Medium was removed and added with 1 mL of dimethyl sulfoxide (DMSO) to dissolve the formazan. After all formazan was dissolved, 100 μL aliquot was transferred to 96 well-plate, 5 replicates per each sample. DMSO was set as blank. Absorbance was measured at 570 nm using a microplate spectrophotometer (Epoch, BioTek).

Before applying a nanofiber membrane on an injured area, it needs to go through several tests to ensure the scaffold is safe to be used and meets the requirements for the application. Initially, *in vitro* study needs to be conducted, followed by *in vivo* and clinical study. Cell viability/cytotoxicity, cell attachment, cell morphology, and cell proliferation studies are the most common *in vitro* cell behavior studies done by most research groups. There are various methods of measuring cell viability/cytotoxicity such as direct cell counting by trypan blue dye exclusion method, MTT, live/dead cell viability, and 4′,6-diamidino-2-phenylindole (DAPI) staining.

Cell morphology on nanofiber membranes can be viewed under SEM or FESEM. Cells that are attached and growing on the nanofiber membrane surface are fixed using glutaraldehyde at 4°C following by dehydration with a series of graded ethanol concentrations. Cells are dead but remain with relatively intact morphology on the nanofiber membrane. A higher end SEM, FESEM can also be used. In addition, there is a more convenient method to study cell morphology without cell fixing by using fluorescein diacetate and chloromethylfluoresceindiacetate (CMFDA) staining and using fluorescent microscopy or confocal laser scanning microscopy. Fluorescein diacetate and CMFDA are non-fluorescent lipophilic dyes that can pass through cell membranes. Hydrolyzing enzymes in cells will hydrolyze it into hydrophilic fluorescent products which are not able to diffuse out of the cells. CMFDA is non-toxic to cells and does not affect cellular growth. Hence, stained cells can be monitored in a long-term culture.

Cell attachment and proliferation can be observed by determining the number of cells on the scaffold. Counting cell number after cells are trypsinized is the most straightforward method for cell attachment and proliferation. However, some cells are difficult to be detached especially from a rough surface. There are some colorimetric methods to determine the number of cells including the MTT assay and 3-(4,5-dimethylthiazol-2-yl)-5-(3-carboxymethoxyphenyl)-2-(4-sulfophenyl)-2H-tetrazolium (MTS) assay. Both these assays measure the cellular metabolic activity of cells. Viable cells contain dehydrogenase enzymes which will reduce the tetrazolium salt of MTT and MTS to purple formazan. The formazan can be dissolved in dimethyl sulfoxide and quantified by optical absorbance. The absorbance reflects the number of viable cells. The advantage of using MTT and MTS methods is cells do not need to be detached from the scaffold. However, it also has a disadvantage that different cells contain different amount of dehydrogenase enzyme (Ma et al. 2007). Several studies have used MTT and MTS for cell attachment and proliferation studies (Ndreu 2007; Kumbar et al. 2008; Sultana and Wang 2012). In addition, DAPI staining can also be used to investigate cell attachment on scaffolds.

An ideal nanofiber membrane provides a framework and supports cell attachment, proliferation, differentiation, and promotes the formation of ECM by the cells. When a cell comes into contact with a scaffold, it interacts with the scaffold by formation of focal adhesion points. This provides signals to cells through integrin (cell adhesion receptors) and mediates cell adhesion. After cell adhesion, active compounds will be released to give signals for cell proliferation and differentiation. Cell behavior on the scaffold is greatly influenced by the surface topography, chemistry, and architecture of the scaffold.

3.5 APPLICATIONS

Various applications in tissue engineering have been explored, including vascular tissue engineering, nerve tissue engineering, bone tissue engineering, tendon/ligament tissue engineering, heart tissue engineering, and skin tissue engineering. Skin tissue engineering has been the focus of much research. The skin is the largest organ of the human body, covering 16% of total body weight with an area approximately 1.8 m^2 in an adult. Thickness of the skin is in the range of 1.5–4.0 mm depending on parts of the body and skin maturity. The main function of the skin is to serve as a physical barrier to protect the body from microbial invasion. In addition, it also protects against UV light, mechanical, chemical, thermal, and dehydration. Skin is also capable of thermoregulation and sensation. Skin is divided into three layers: epidermis, dermis, and hypodermis. Due to the exposure of skin to the external environment, it can be injured easily. Skin defects are caused by burning, tumors, chronic wounds, and other dermatological conditions. The most common skin defect is a burn, which has been categorized as a global health problem by World Health Organization (WHO). According to WHO, there are 300,000 deaths every year from burn injuries. Burns are categorized into three degrees: first degree (damage of epidermis layer), second degree (damage of epidermis and some dermis layer or damage of all dermis layer), and third degree (damage of all three layers of skin). Second- and third-degree burns are considered as deep dermal injuries. Skin has a natural healing ability in the sequence of, inflammation, migration, proliferation, maturation, and hemostasis of skin tissue. However, deep dermal injuries are difficult to heal naturally, and any loss of skin that is more than 4 cm depth needs a graft for healing. Conventionally, autografting and allografting are the methods for treating deep dermal injuries.

Skin tissue engineering has been developed recently due to the limitations of autografts and allografts such as availability of healthy autologous skin, infections, and scarring. Skin tissue engineering facilitates the treatment of patients with deep dermal injuries, deep skin burns, and skin related disorders. Skin tissue engineering aims to promote the healing of wounds, repair, and the regeneration of skin tissue with reduced scarring. A successful tissue engineered skin is a combination of a wound dressing and a skin graft. Tissue engineered skin substitutes can be divided into epidermal, dermal, and dermo-epidermal substitutes.

The three essential components of skin tissue engineering are types of cells, scaffold, and environmental condition. Cells are grown *in vitro*, seeded on the tissue engineered scaffold, and implanted *in vivo* at the injury site. There are three main types of cells found in skin: (1) keratinocytes, (2) melanocytes, and (3) fibroblasts. Keratinocytes and melanocytes are found in the epidermis layer. Keratinocytes form a stratified squamous epithelium to generate the epidermal layer and melanocytes are located at a lower layer of the epidermis and provide pigments for skin color. Fibroblasts are located in the dermis layer and there are no cells in hypodermis layer. In skin tissue engineering the types of cells need to be selected carefully for investigating different types of skin substitute. For study on epidermal substitutes, keratinocytes have to be selected and for dermal substitute, fibroblasts have to be selected. An ideal scaffold should mimic the physical and functional characteristics of natural skin. Before *in vivo* implantation, cells seeded on scaffolds have to be cultured in an appropriate bioreactor to provide an environment that mimics the skin tissue environment.

A study was done to investigate *in vitro* cell cytotoxicity and behavior of HSF cells on the fabricated PCL and PCL/Ge nanofiber membranes to determine the suitability of these scaffolds for skin tissue engineering application (Lim et al. 2015). PCL and PCL/Ge nanofiber membranes were tested for cell cytotoxicity after 3 days of culture using a quantitative MTT assay.

In this study, HSF cells cultured on a well plate without any scaffold was used as control. Cytotoxicity results are shown in Figure 3.2.

Both PCL and PCL/Ge nanofiber membranes gave high absorbance inferring that both membranes were non-toxic to cells. This result was expected as PCL and Ge have already been approved by FDA and there is no evidence to demonstrate hazard to human beings. Moreover, the absorbance was higher than the control. Cells were growing and proliferating well on both scaffolds. This showed that nanofiber membranes enhanced the growth of cells as compared to the surface of the well plate. Both membranes formed favorable substrates for cells as the fiber membranes mimic ECM and provided a high surface area to volume ratio for cell attachment and proliferation. The absorbance of PCL/Ge nanofiber membrane was slightly higher than that of the PCL nanofiber membrane. This was due to the hydrophilicity of PCL/Ge nanofiber membrane. Cells grow more favorably on the hydrophilic surface compared to the hydrophobic surface. Furthermore, there is a biochemical interaction between the cells and Ge (denatured collagen). This interaction is the binding of collagen I components to the receptors on the cell membrane which is mediated

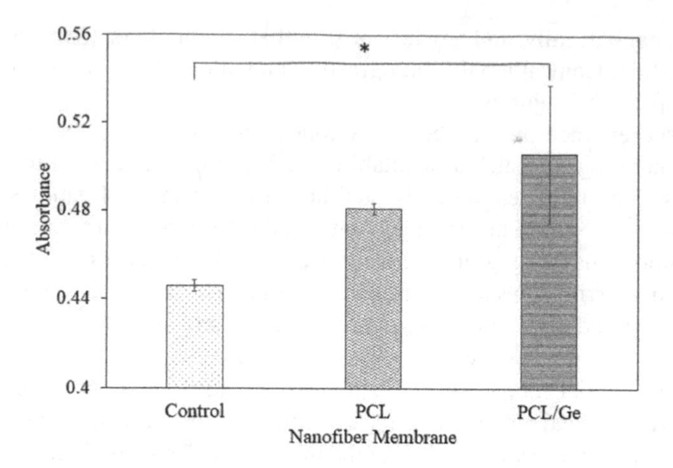

FIGURE 3.2 Cell cytotoxicity on control, PCL, and PCL/Ge nanofiber membrane. Data are representative of three independent experiments and plotted as mean ± SD ($n = 3$); $^*p < 0.05$. (From Lim, M.M. et al., *J. Nanomater*, 2015, 10, 2015.)

by ECM glycoprotein fibronectin (Schnell et al. 2007). Ge is derived from collagen by hydrolysis and its structure is similar to collagen. Therefore, it can be concluded that blending of Ge with PCL enhances the interaction between cells and PCL/Ge nanofiber membrane.

In this study the number of viable cells attached to the nanofiber membrane was also counted by using a dye exclusion method and trypan blue (Lim et al. 2015). Trypan blue is a dye that can differentiate between living and dead cells. It selectively colors dead cells in blue and living cells will refract brightly. Figure 3.3 shows the result of the viable cell count. Tissue culture plate wells with only cells were used as control. No dead cell were observed, but the growth of cells on each sample was different. According to the growth curve, cells started to attach after 4 h and were well attached after 24 h. Cells were in the proliferation state at 72 h. With increasing time, the number of cells on each sample increased, demonstrating cell attachment occurs successfully and cells have proliferated. The high porosity of both nanofiber membranes provides an optimum growth environment for cells by giving more spaces for nutrient and metabolic waste exchange. All of the data were statistically significant.

For the PCL nanofiber membrane, the number of cells increased with time, but the cell number was less than the control, indicating that cells preferred to adhere to the well plate compared to the PCL nanofiber membrane. Hydrophilic surface enhances effective adhesion of cells and enhances cell migration and the PCL nanofiber membrane was hydrophobic. However, others have observed that hepatocytes attach and proliferate on PCL nanofibers membrane (Lubasová et al. 2010). Cells were still growing on the hydrophobic fibrous PCL scaffold due to the fiber structure that mimics ECM, but they showed slower growth rate. For the hydrophilic PCL/Ge nanofiber membrane, cell number was less at 24 h, but it increased rapidly and significantly in number within 72 h. This showed that PCL/Ge nanofiber membrane has accelerated the proliferation and differentiation of HSF.

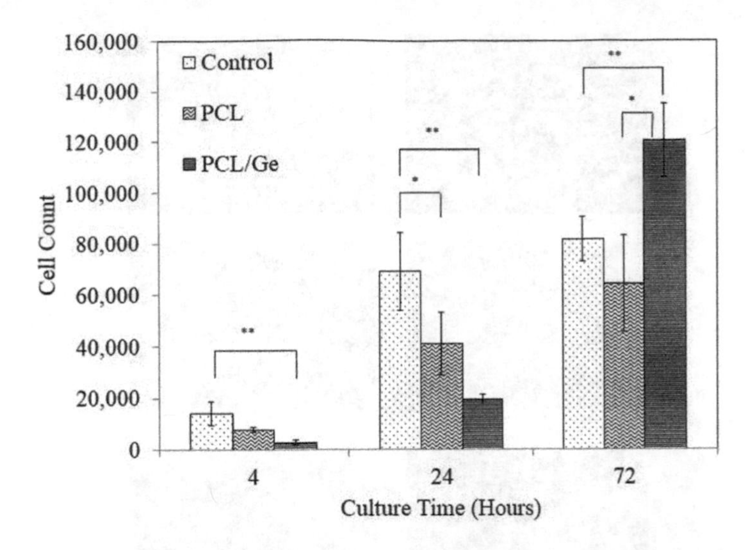

FIGURE 3.3 Cell counting result of control, PCL, and PCL/Ge nanofiber membranes at 4, 24, and 72 h using dye exclusion method (trypan blue). Data are representative of three independent experiments and plotted as mean ± SD ($n = 3$); $^{*}p < 0.05$, $^{**}p < 0.01$. (From Lim, M.M. et al., *J. Nanomater*, 2015, 10, 2015.)

The morphology of HSF cells cultured on PCL and PCL/Ge nanofiber membranes were studied by SEM after culture for 4 h, 1, 3, and 7 days. These results are shown in Figure 3.4. PCL and PCL/Ge nanofiber membranes were tested for cell cytotoxicity and the result showed that both membranes are non-toxic to cells. Since both membranes were not harmful to cells, it was also observed from the SEM images that cells were growing on the scaffolds. However, these cells were randomly orientated. With increasing culturing time, cells spread on the nanofiber membranes and increased in size. At day 7, the surface of membranes was almost completely covered by cells and cells had grown into a confluent state, forming a monolayer.

A shown in Figure 3.4 (PCL), cells aggregated and formed a globular morphology with extended filopodia on the PCL nanofiber membrane. The SEM images showed poor attachment of cells on the PCL membrane. The aggregation of cells is due to the hydrophobicity of the PCL membrane. This result is similar to a study that was reported (Khil et al. 2005). It was also found that a hydrophobic surface will lead to poor attachment of cells (Ma et al. 2007). Although there was poor attachment of cells on the PCL nanofiber membrane, the number of cells was increased; cells were still growing but with poor attachment.

For PCL/Ge nanofiber membrane (Figure 3.4h and j), cells were more flattened, but some globular morphology can be seen in the SEM images of 4 hours and day 1. This demonstrated slow attachment of cells on the membrane. In the SEM image of day 3 (Figure 3.4h), cells no longer maintained the globular morphology but exhibited flattened and stretched morphology with an increasing surface area of cells. It showed good cell to cell and cell to scaffold interaction. On day 7 (Figure 3.4j), a population of cells was spreading well with flattened morphology on the PCL/Ge nanofiber membrane.

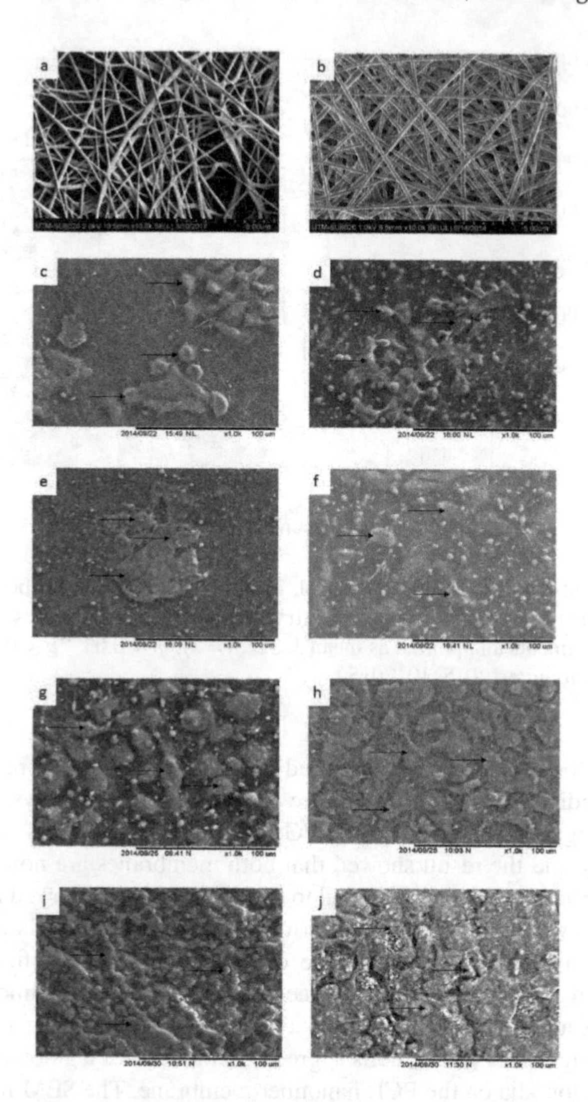

FIGURE 3.4 Morphology of HSF cells cultured on PCL (left) and PCL/Ge (right) nanofiber membranes at different time points (a, b) before culturing; magnification 10000x, (c, d) 4 h, (e, f) day 1, (g, h) day 3, (i, j) day 7; magnification 1000x. (Arrows indicate cells.) (From Lim, M.M. et al., *J. Nanomater*, 2015, 10, 2015.)

The hydrophilic surface of PCL/Ge nanofiber membranes were more favorable for the attachment of cells, as compared to PCL nanofiber membranes as at the PCL/Ge surface cells had flattened morphology instead of a globular morphology. For both membranes, cells did not penetrate into the scaffold due to the relatively small pore size (PCL: 811.38 ± 343.42 nm, PCL/Ge: 802.52 ± 351.37 nm) compared to the size of fibroblast cells (10–15 μm). To enhance cell infiltration, membranes have to be equipped with micropores. There are two proposed methods to overcome this problem including fiber leaching to create micropores and layered scaffolds by sequential electrospinning and combining nanofiber membranes with microfiber membranes.

PCL and PCL/Ge membranes have nano-sized fiber diameters with a high porosity. This gives advantages in providing an optimum growth environment to cells by having a larger surface area to volume ratio and giving more spaces for nutrient and metabolic waste exchange. The main difference between PCL and PCL/Ge nanofiber membranes were the surface wettability and degradation rate. PCL nanofiber membranes had slow degradation rate with hydrophobic properties while PCL/Ge nanofiber membrane has faster degradation rate with hydrophilic properties. The observations in this section demonstrated that surface wettability is a crucial factor for cell attachment and proliferation. There was good cell to cell and cell to scaffold interaction on PCL/Ge nanofiber membrane. Hence, PCL/Ge nanofiber membrane has a greater potential to be used for skin tissue engineering application.

Silver-coated nanofiber membranes has the potential to be used as a scaffold with antibacterial properties for skin tissue engineering application to prevent bacterial infections and also promote skin tissue regeneration (Lim and Sultana 2016). We found that silver-coated PCL/Ge nanofiber membranes were non-toxic to HSF cells and demonstrated great antibacterial activity toward Gram positive (*Bacillus cereus*) and Gram negative (*Escherichia coli*) bacteria (Lim and Sultana 2016).

Apart from skin tissue engineering application, drug-loaded nanofiber electrospun membrane can be act as drug release system to give a controlled drug release. An *in vitro* drug release study was done on PCL electrospun membrane with incorporation of *fluorescein isothiocyanate*-dextran (FD70S) drug (Lim and Sultana 2014). FD70S was successfully released from the PCL electrospun membrane and degraded after the release.

Bone tissue engineering using nanofibers is widely investigated by researchers. Bone has a hierarchical and complex structure. Nanofiber scaffold has the features to provide suitable extracellular matrix environment for osteoblasts adhesion, proliferation and differentiation and suitable for bone tissue regeneration. A study by Linh and Lee (2012) holds promises in the field of artificial bone implant using a three-dimensional polymer composite system. The three-dimensional polymer composite system was fabricated by cross-linking polyvinyl alcohol/gelatin (PVA/Ge) using methanol treatment. Osteoblasts were proliferated on the PVA/Ge nanofibers and expressing philopodial extensions (Linh and Lee 2012).

Three-dimensional PCL nanofibers have also been fabricated for bone regeneration application. This fabrication used thermally induced nanofiber self-agglomeration technique and followed by freeze drying technique. By using this fabrication technique, high porosity (≈96.4%) of nanofibers which was similar to natural extracellular matrix was achieved and showed high cell viability. *In vivo* results demonstrated three-dimensional PCL nanofibers as a favorable synthetic extracellular matrix for bone regeneration (Xu et al. 2015).

In heart tissue engineering application, the growth of cardiomyocytes (CMs) needs a nature-mimicking environment which provides a suitable biochemical and structural properties for the maturation of CMs. Nanofibrous scaffolds fabricated using electrospinning technique have been widely used for tissue engineering application as nanofibers mimic the architecture of native ECM of the myocardium. Nanofibrous scaffolds could also enhance migration along with cell-cell junction and induce cellular alignment which results in the formation of vascularized tissue constructs (Kharaziha et al. 2013). Several studies have been done in investigating nanofibrous scaffolds for heart tissue engineering.

A study showed human adipose-derived stem cells were significantly more preferable to proliferate and differentiate on aligned nanofibrouspolycaprolactone (PCL) scaffold as compared to random nanofibrous PCL scaffold. There was also an increase in cardiovascular-related gene markers (Myo-D, α-MHC, and GATA-4) on aligned nanofibrous PCL scaffold (Safaeijavan et al. 2014). Moreover, aligned polyester urethane urea (PEUU) nanofibrous scaffold blended with gelatin (PEUU/G) at ratio 70:30 meets the specifications of cardio-bioengineering such as biodegradable, cytocompatible, similar anisotropic mechanical properties as myocardium, hydrophilicity, and enhances CMs attachment as well as proliferation. Biodegradation rate of PEUU/G 70:30 nanofibrous scaffold was also investigated. It is believed that the degradation of nanofibrous scaffold will lead to a decrease in its mechanical properties and varnished after new CM formation (Jamadi et al. 2016). Blending of 7:3 ratio poly (L-lactide-co-caprolactone) (PLCL) and poly (2-ethyl-2-oxazoline) (PEOz) nanofibrous scaffold with an incorporation of dual growth factors, VEGF and bFGF, has improved biological response of ischemic tissue without edema and reduced scar tissue formation. This growth factor embedded nanofibrous patch was studied on a rabbit with acute myocardial infarction. It resulted in restoration of cardiac function after ischemia (Lakshmanan et al. 2016).

Another study investigated restoration of infarcted myocardium by culturing human pluripotent stem cell-derived cardiac tissue-like constructs (hPSC-derived CMs) on low-thickness aligned biodegradable poly(D, L-lactic-co-glycolic acid) (PLGA) nanofibrous scaffolds. Low thickness of nanofibrous scaffold made the cells infiltrate and envelop easily in the scaffold. Multi-layered and elongated CMs showed upregulated cardiac biomarkers, robust drug response, and improved cardiac functions. *In vivo* study of rat model showed enhanced cell survival after 4 weeks post-surgery (Li et al. 2017). Although nanofiber is a promising scaffold for cardiac regeneration strategy, several characteristics need to take note such as fiber degradability, fiber layer thickness, and porosity of the fibers to enhance cell infiltration, attachment, and proliferation.

3.6 CONCLUSIONS

Nanofiber electrospun membranes fabricated using electrospinning have a great potential to be used for medical applications. Electrospinning is able to fabricate nanoscale fibrous scaffolds that mimic the natural ECM of our body and enhance cell attachment, drug loading, and mass transfer. Several characteristics need to be taken into account when designing electrospun membranes, which includes fiber diameter, pore size, wettability, biodegradable, degradation rate, and biocompatibility. Present and future work must focus on characterizing these properties to fabricate an optimized nanofiber membrane for specific medical applications. This offers hope for the improved care of millions of patients who suffer from inadequate current medical treatments.

REFERENCES

Brahatheeswaran D, Y Yasuhiko, M Toru, and D Sakthi Kumar. 2011. "Polymeric scaffolds in tissue engineering application: A review." *International Journal of Polymer Science* 2011: Article ID 290602, 1–19.

Chong EJ, TT Phan, IJ Lim, YZ Zhang, BH Bay, S Ramakrishna, and CT Lim. 2007. "Evaluation of electrospun PCL/gelatin nanofibrous scaffold for wound healing and layered dermal reconstitution." *Acta Biomaterialia* 3 (3):321–330.

Chung T-W, D-Z Liu, S-Y Wang, and S-S Wang. 2003. "Enhancement of the growth of human endothelial cells by surface roughness at nanometer scale." *Biomaterials* 24 (25):4655–4661.

De Bartolo, L, M Rende, S Morelli, G Giusi, S Salerno, A Piscioneri, A Gordano, A Di Vito, M Canonaco, and E Drioli. 2008. "Influence of membrane surface properties on the growth of neuronal cells isolated from hippocampus." *Journal of Membrane Science* 325 (1):139–149.

GMIA. 2012. *Gelatin Handbook*. Gelatin Manufacturers Institute of America. http://www.gelatin-gmia.com/images/GMIA_Gelatin_Manual_2012.pdf

Goh YF, I Shakir, and R Hussain. 2013. "Electrospun fibers for tissue engineering, drug delivery, and wound dressing." *Journal of Materials Science* 48 (8):3027–3054.

Hatano K, H Inoue, T Kojo, T Matsunaga, T Tsujisawa, C Uchiyama, and Y Uchida. 1999. "Effect of surface roughness on proliferation and alkaline phosphatase expression of rat calvarial cells cultured on polystyrene." *Bone* 25 (4):439–445.

Huang ZM, YZ Zhang, M Kotaki, and S Ramakrishna. 2003. "A review on polymer nanofibers by electrospinning and their applications in nanocomposites." *Composites Science and Technology* 63 (15):2223–2253.

Jamadi ES, L Ghasemi-Mobarakeh, M Morshed, M Sadeghi, MP Prabhakaran, and S Ramakrishna. 2016. "Synthesis of polyester urethane urea and fabrication of elastomeric nanofibrous scaffolds for myocardial regeneration." *Materials Science and Engineering: C* 63: 106–116.

Jin RM, N Sultana, S Baba, S Hamdan, and AF Ismail. 2015. "Porous PCL/Chitosan and nHA/PCL/Chitosan scaffolds for tissue engineering applications: Fabrication and evaluation." *Journal of Nanomaterials* 2015:1–8.

Kharaziha M, M Nikkhah, SR Shin, N Annabi, N Masoumi, AK Gaharwar, and A Khademhosseini. 2013. "PGS: Gelatin nanofibrous scaffolds with tunable mechanical and structural properties for engineering cardiac tissues." *Biomaterials* 34 (27): 6355–6366.

Khil M-S, SR Bhattarai, H-Y Kim, S-Z Kim, and K-H Lee. 2005. "Novel fabricated matrix via electrospinning for tissue engineering." *Journal of Biomedical Materials Research Part B: Applied Biomaterials* 72B (1):117–124.

Kumbar SG, SP Nukavarapu, R James, LS Nair, and CT Laurencin. 2008. "Electrospun poly (lactic acid-co-glycolic acid) scaffolds for skin tissue engineering." *Biomaterials* 29 (30):4100–4107.

Lakshmanan R, P Kumaraswamy, UM Krishnan, and S Sethuraman. 2016. "Engineering a growth factor embedded nanofiber matrix niche to promote vascularization for functional cardiac regeneration." *Biomaterials* 97:176–195.

Langer R, and JP Vacanti. 1993. "Tissue engineering." *Science* 260 (5110):920–926.

Li J, I Minami, M Shiozaki, L Yu, S Yajima, S Miyagawa, and S Li. 2017. "Human pluripotent stem cell-derived cardiac tissue-like constructs for repairing the infarcted myocardium." *Stem Cell Reports* 9 (5):1546–1559.

Lim MM, and N Sultana. 2014. Drug loading, drug release and in vitro degradation of poly(caprolactone) electrospun fibers. Paper read at *IEEE Conference on Biomedical Engineering and Sciences (IECBES)*, December 8–10, Miri, Malaysia.

Lim MM, and N Sultana. 2016. "In vitro cytotoxicity and antibacterial activity of silver-coated electrospun polycaprolactone/gelatine nanofibrous scaffolds." *3 Biotech* 6 (2):211.

Lim MM, and N Sultana. 2017. "Comparison on in vitro degradation of polycaprolactone and polycaprolactone/gelatin nanofibrous scaffold." *Malaysian Journal of Analytical Sciences* 21 (3):627–632.

Lim MM, T Sun, and N Sultana. 2015. "In vitro biological evaluation of electrospun polycaprolactone/gelatine nanofibrous scaffold for tissue engineering." *Journal of Nanomaterials* 2015:10.

Linh NTB, and BT Lee. 2012. "Electrospinning of polyvinyl alcohol/gelatin nanofiber composites and cross-linking for bone tissue engineering application." *Journal of Biomaterials Applications* 27 (3):255–266.

Lowery JL, N Datta, and GC Rutledge. 2010."Effect of fiber diameter, pore size and seeding method on growth of human dermal fibroblasts in electrospun poly (ε-caprolactone) fibrous mats." *Biomaterials* 31 (3):491–504.

Lubasová D, L Martinová, D Mareková, and P Kostecká. 2010. Cell growth on porous and non-porous polycaprolactone nanofibers. Paper read at Nanocon, October 12–14, at Olomouc, Czech Republic.

Ma Z, Z Mao, and C Gao. 2007. "Surface modification and property analysis of biomedical polymers used for tissue engineering." *Colloids and Surfaces B: Biointerfaces* 60 (2):137–157.

Ndreu A. 2007. *Electrospun Nanofibrous Scaffolds for Tissue Engineering*, Graduate School of Natural and Applied Sciences, Middle East Technical University, Ankara, Turkey.

Pörtner R, and C Giese. 2007. An overview on bioreactor design, prototyping and process control for reproducible three-dimensional tissue culture, *Drug Testing In Vitro* (U Marx and V Sandig Eds.). Wiley-VCH Verlag, Weinheim, Germany.

Rajput M. 2012. *Optimization of Electrospinning Parameters to Fabricate Aligned Nanofibers for Neural Tissue Engineering*, National Institute of Technology Rourkela, Rourkela.

Safaeijavan R, M Soleimani, A Divsalar, A Eidi, and A Ardeshirylajimi. 2014. "Comparison of random and aligned PCL nanofibrous electrospun scaffolds on cardiomyocyte differentiation of human adipose-derived stem cells." *Iranian Journal of Basic Medical Sciences* 17 (11):903.

Schnell E, K Klinkhammer, S Balzer, G Brook, D Klee, P Dalton, and J Mey. 2007. "Guidance of glial cell migration and axonal growth on electrospun nanofibers of poly-ε-caprolactone and a collagen/poly-ε-caprolactone blend." *Biomaterials* 28 (19):3012–3025.

Shkadov VY, and AA Shutov. 2001. "Disintegration of a charged viscous jet in a high electric field." *Fluid Dynamics Research* 28 (1):23–39.

Sultana N, and TH Khan. 2012. "In vitro degradation of PHBV scaffolds and nHA/PHBV composite scaffolds containing hydroxyapatite nanoparticles for bone tissue engineering." *Journal of Nanomaterials* 2012:1–12.

Sultana N, and M Wang. 2012. "PHBV/PLLA-based composite scaffolds fabricated using an emulsion freezing/freeze-drying technique for bone tissue engineering: Surface modification and in vitro biological evaluation." *Biofabrication* 4 (1):015003.

Sun T, TH Khan, and N Sultana. 2014. "Fabrication and in vitro evaluation of nanosized hydroxyapatite/chitosan-based tissue engineering scaffolds." *Journal of Nanomaterials* 2014:1–8.

Taylor G. 1969. "Electrically driven jets." *Proceedings of the Royal Society of London. A. Mathematical and Physical Sciences* 313 (1515):453–475.

Toth JM, HS An, T-H Lim, Y Ran, NG Weiss, WR Lundberg, R-M Xu, and KL Lynch. 1995. "Evaluation of porous biphasic calcium phosphate ceramics for anterior cervical interbody fusion in a caprine model." *Spine* 20 (20):2203–2210.

Woodruff MA, and DW Hutmacher. 2010. "The return of a forgotten polymer—polycaprolactone in the 21st century." *Progress in Polymer Science* 35 (10):1217–1256.

Xu T, JM Miszuk, Y Zhao, H Sun, and H Fong. 2015. "Electrospun polycaprolactone 3D nanofibrous scaffold with interconnected and hierarchically structured pores for bone tissue engineering." *Advanced Healthcare Materials* 4 (15):2238–2246.

Yang J, G Shi, J Bei, S Wang, Y Cao, Q Shang, G Yang, and W Wang. 2002. "Fabrication and surface modification of macroporous poly (L-lactic acid) and poly (L-lactic-co-glycolic acid) (70/30) cell scaffolds for human skin fibroblast cell culture." *Journal of Biomedical Materials Research* 62 (3):438–446.

Yildirim ED, R Besunder, D Pappas, F Allen, S Güçeri, and W Sun. 2010. "Accelerated differentiation of osteoblast cells on polycaprolactone scaffolds driven by a combined effect of protein coating and plasma modification." *Biofabrication* 2 (1):014109.

Zhang Y, CT Lim, S Ramakrishna, and ZM Huang. 2005a. "Recent development of polymer nanofibers for biomedical and biotechnological applications." *Journal of Materials Science: Materials in Medicine* 16 (10):933–946.

Zhang Y, H Ouyang, CT Lim, S Ramakrishna, and ZM Huang. 2005b. "Electrospinning of gelatin fibers and gelatin/PCL composite fibrous scaffolds." *Journal of Biomedical Materials Research Part B: Applied Biomaterials* 72B (1):156–165.

4 Electrospun Collagen Nanofiber Membranes for Regenerative Medicine

Jonathan P. Widdowson, Nidal Hilal, and Chris J. Wright

CONTENTS

4.1 COLLAGEN FIBER MOLECULAR STRUCTURE

Collagen is a highly conserved protein that can be found right across the animal spectrum, from ancient sponges to mammals. The primary structure of collagen has Glycine -X-Y repeats, where X can be any amino acid, and Y is often proline or hydroxyproline (Shoulders and Raines 2010), The individual chains form a left-handed helix, and the three chains wind around one another in a right-handed superhelix (Myllyharju and Kivirikko 2001). This forms the characteristic tertiary structure of a triple helix (Figure 4.1) and makes collagen fibers insoluble with high tensile strength (Gelse et al. 2003). Collagen is produced and secreted predominantly

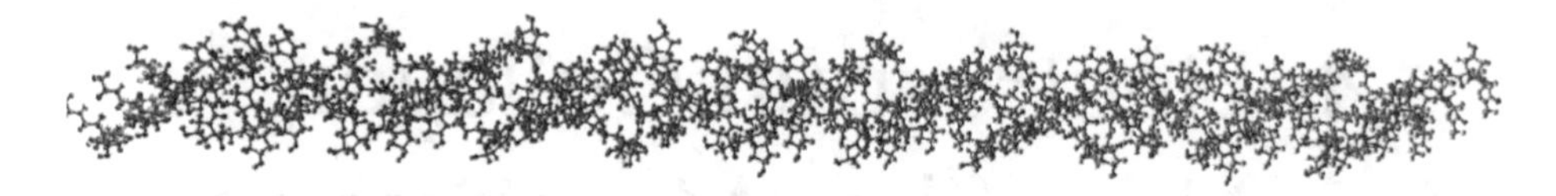

FIGURE 4.1 Triple helix peptide of Gly-Pro-Hyp repeats arranged with a hybrid Lui Storey and Conjugate Descent optimization (LS-CD) using Abalone software to build amino acid structure.

by fibroblast cells. The process follows the standard pathway for a secreted protein, where the precursor collagen chains are synthesized as a longer chain termed procollagen. These chains are transported to the lumen of the rough endoplasmic reticulum where the procollagen undergoes modifications such as glycosylation. Specific proline and lysine residues in the central region of the chains undergo hydroxylation before disulfide bonds are formed between N and C terminal propeptide sequences to align the three chains and begin the formation of the triple helix from C to N terminus (Lodish et al. 2000). The final assembly of collagen within the ECM is a discontinuous super twisted right-handed microfibril. The simplest depiction of multicellular animals requires a combination of macromolecules to bind cells together, provide structural integrity in tissues and allow the overall stability seen in animals. This is achieved by the ECM a highly organized network of glycoproteins, proteoglycans and glycosaminoglycans. The ECM provides physical stability while promoting cell growth, including migration, differentiation, and adhesion by acting as a substrate for the cells. The composition of the ECM varies depending on the tissue, in direct governance from the surrounding cell types. While cells are responsible for the production of ECM, the interactions with ECM components and membrane receptors cause the activation and upregulation of specific cellular signaling pathways associated with controlling gene expression. Within this ECM, collagens are the main group of structural proteins. They are in fact the most abundant protein found in animals, providing between 25% and 35% of the whole body protein. To date, there are 28 forms of collagen as described in Table 4.1, all of which share the characteristic triple helix structure of Gly-X-Y repeats.

Variance in the amino acid composition in collagen can be observed to have distinct influence in the host organism, for example, the amino acid repeat Gly-Gly-Y that is found in abundance in the type I collagen of rainbow trout (11% of total triple helical region for the $\alpha 3$ (1) chain) is shown to loosen the helix in the trout's skin sufficiently to lower denaturation temperature by 0.5% compared with muscle collagen of the same fish (Saito et al. 2001). Furthermore, a reduction in the overall abundance of proline in position II and proline hydroxylation in position III, as can be seen in marine animals shows a reduction in denaturation temperature when compared with mammals (Nagai et al. 2000; Engel and Bächinger 2005; Nalinanon et al. 2007; Barzideh et al. 2014). It is this reduced temperature which is a hindrance in the use of marine collagens for tissue engineering in humans. The current overall consensus suggests that the thermal stability of collagen for a given species is correlated with the surrounding environmental temperature, and body temperature (Rigby 1968). The detailed analysis of the molecular folding and nucleation steps in the formation of collagen triple helix peptides has been examined in detail by (Xu et al. 2002).

TABLE 4.1
Collagen Proteins Classified by Their Form with Additional Details on Proteins Which Are Poorly Classified

Type	Form	Additional Details
I, II, III, V, XI	Fibrillar	Full Triple Helix Fibril Formation
IV	Non-Fibrillar	Basement Membrane Protein
VI, VII	Non-Fibrillar	Cell Anchoring Proteins
VIII, X	Non-Fibrillar	Short Chain
IX, XII, XIV, XVI, XIX, XX, XXI, XXII	Non-Fibrillar	Fibril Associated Collagens with Interrupted Triple Helices (FACIT)
XIII, XVII, XXIII	Non-Fibrillar	Membrane Associated Collagens with Interrupted Triple Helices (MACIT)
XV	Non-Fibrillar	Multiple Triple Helix Domains with Interruptions (Multiplexin)
XVIII	Non-Fibrillar	Basement Membrane—Cleaved to form Endostatin Peptide for Proteolytic Cleavage
XXIV	Fibrillar	May Regulate Type I Fibrillogenesis
XXV	Non-Fibrillar	Brain-Specific membrane bound collagen. Proteolytic processing releases CLAC which interacts with senile plaques in Alzheimer's disease
XXVI	Non-Fibrillar	May play an important role in adult reproductive organs as well as their development
XXVII	Non-Fibrillar	May play a role in calcification of cartilage and the transition of cartilage to bone
XXVIII	Non-Fibrillar	Present in neuronal cells and their surrounding basement membranes

The origins in the use of collagen are rooted to the use of gelatine in food products. Gelatine is a denatured form of collagen that was produced and used in the late medieval period by boiling animal products, while the first patent awarded for gelatine production occurred in England in 1754 (Gelatin 2018). Collagen has been used for many decades in different applications within healthcare and cosmetics. The repeating pattern structure, while difficult to examine at the time, was first partially examined within the tendon of kangaroo and catgut during the 1930s using x-ray diffraction (Clark et al. 1935; Wyckoff et al. 1935). This first insight into the morphology of collagen at a molecular level led to research spanning the last nine decades. Notable developments following this research focused mainly on describing the monomers which form collagen; the quaternary packing structure of collagen was described by the work of Ramachandran, which elucidated the triple helical packing structure of fibrillar collagens with three chains entangled. This was further described as one chain in helix form, coiling in the opposing direction to the other two (Ramachandran et al. 1968). This model was modified by groups around the world to gradually understand the complex structures that underpin collagen (Venkatraman 2017).

Other work detailing the packing structure of collagen monomers led to many misconceptions regarding how the secondary structure of collagen is described, including whether collagen forms microfibrils or staggered sheets (Wess et al. 1998); The microfibrillar structure of collagen within major structures including cornea and tendon has been imaged using electron microscopy (Raspanti et al. 1990; Holmes et al. 2001; Holmes and Kadler 2006), which has led to the accepted description of collagen packing structure as a discontinuous super-twisted right-handed microfibril; the subunit for tropocollagen within tissues (Petruska and Hodge 1964).

Today, collagens are used across the scientific spectrum, from life science research as a basic 3D scaffold for cell culture, work in tissue engineering advances, to composite biomaterials for structural materials research. There are several medical uses for collagen already available, as well as in the cosmetic industry, where variants of collagen are advertised for their appearance enhancing and rejuvenating effects. There has been a large debate surrounding the need for collagen to become more versatile, unfortunately at present harsh solvents are often required for usable concentrations above 10% to be achieved (Dong et al. 2009); these solvents often causes denaturation of the collagen into gelatine derivatives (Zeugolis et al. 2008a).

4.2 COLLAGEN NANOFIBERS AND WOUND HEALING

Following injury of the skin, the natural processes of wound healing begin to repair the skin, beginning with the formation of blood clots. These clots become a scab at the surface, sealing the wound; fibroblasts migrate into the lower regions of the blood clot from the edges of the surrounding tissue. The fibroblasts begin breaking down the clot and replacing it with scar tissue formed of collagen fibers (Clark 1996, 2012). The deposited collagen differs from healthy tissue by its alignment structure; where normal skin typically contains a "basket weave" fiber structure, the scar tissue develops single directional aligned fibers of collagen as remodeling occurs by continual interaction with fibroblasts (Dallon 1998; Dallon et al. 1999; Dallon and Sherratt 2000).

Collagen fibers are deposited and arranged in a linear format during wound closure, and this causes a weakening of the tissue once restored, referred to as scarring, the degree of scar severity is dependent on a wide array of factors, including age (Ashcroft et al. 1998), skin color (Murray et al. 1981), and location (Bayat et al. 2003). This scarring occurs due to the repair, or incomplete regeneration of the tissue (Min et al. 2006b), rather than the regeneration seen in non-injured skin which acts as an exact copy of morphology and functionality (Min et al. 2006a).

Collagen plays a role in aiding regeneration at all stages of wound healing during the repair and regeneration of the tissues (Brett 2008). Thus, the extraction and fabrication of collagen fibers through processes such as electrospinning has great potential in terms of bioactivity and control. Indeed, the ultimate wound dressing or tissue engineering device must be a highly accurate analogue of the natural tissue and the ECM. Modern techniques are striving to achieve this through improved fabrication of 3D fibrous materials in parallel with stem cell research. The efficient electrospinning of collagen, maintaining its integrity and functionality, is arguably the optimum technique and choice of polymer for achieving this; other polymers will not have the

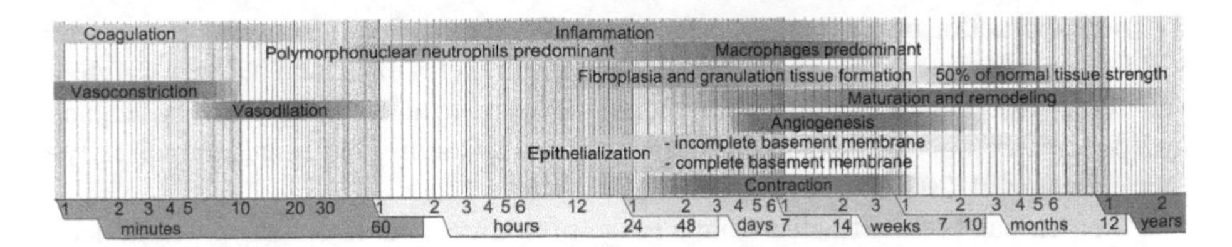

FIGURE 4.2 Wound regeneration timeline. (From Häggström, M., *WikiJournal Med.*, 1, 8, 2014. With permission.)

innate functionality and biocompatibility. Collagen has been shown to be an essential part of wound regeneration, beginning when platelets aggregate around exposed collagen as part of the body's own process for physically blocking the wound and ceasing blood flow. During the inflammation phase, the peptide fragments of collagen have a chemotactic effect in the recruitment of cells essential to wound healing, while the collagen-derived peptides stimulate fibroblast proliferation during the proliferation phase (Schultz and Mast 1999; Brett 2008). Collagen has also been shown to aid in vessel reformation with endothelial cells (Montesano et al. 1983).

When migrating cells such as keratinocytes encounter type 1 collagen in a wound, the cells secrete matrix metalloproteinases (MMPs) which denature the collagen to gelatine in order to expose the active RGD site (Arg-Gly-Asp) sequences, which are responsible for the creation of granulation tissue (Brett 2008). Previous research has also shown that when the collagen triple helix is unwound through partial denaturation, elements of the molecule are exposed that up regulate many pathways important in wound healing (Castillo-Briceño Patricia et al. 2011; Banerjee et al. 2015).

Despite this, collagen accumulation does not occur until the third day of wound healing, during the as termed "late phase" (Clark 1998). Collagen is an essential part of the later stages of wound healing, where the collagen matrix is compacted by fibroblasts using their actin-rich microfilament structures (Ehrlich and Moyer 2013). Collagen in combination with the $\alpha 1\beta 1$ integrin aids the proliferation of fibroblasts (Pozzi et al. 1998), which is a vital process to restore the structure of the extra cellular matrix in a wound. The maturation phase of tissue repair begins when levels of collagen production and degradation equalize (Greenhalgh 1998). Throughout maturation, type III collagen, produced during proliferation, is replaced by type I collagen (Dealey 1999). The timeline for these events are seen in Figure 4.2.

4.3 COLLAGEN EXTRACTION

Collagen is traditionally isolated from mammalian animals using solubilization in acetic acid. The use of pepsins to digest collagen which would traditionally be insoluble, is a modern addition to increase the yield of extractions. This allowed collagen to be extracted from sources such as cartilage. The use of pepsins causes a significant increase in the removal of the non-helical telomeric regions of collagen, which should not be disruptive to the α-helical structure of collagen (Miller 1972). The use of a base, usually sodium hydroxide, was added to disrupt the cells and strips them from the collagen matrix.

These extraction processes have been adapted and modified in many ways, though yield has never been shown to rise above 2% from wet weight or 20% dry weight, and some groups have modified the technique used to determine yield, based on hydroxyproline content to demonstrate small increases in yield more dramatically (Nalinanon et al. 2007).

Collagen has been extracted from a range of sources, but is commonly obtained from bovine, porcine and equine sources for *in vivo* use (Silvipriya et al. 2015). However, these sources have potential problems with the risk of Bovine Spongiform Encephalopathy (BSE), other Transmittable Spongiform Encephalopathies (TSEs) and potential viral vectors that could be transmissible to humans (Asher 1999; Lupi 2002). More recently, jellyfish and other marine animals have emerged as a source of collagen that is an attractive alternative to existing sources due to a plentiful supply (Williams 2015) and a safer source through lack of BSE risk and potential viral vectors (Song et al. 2006). The major drawback of these sources, particularly their use for *in vitro* or *in vivo* testing, is due to the low denaturation temperatures seen with marine collagens (Sadowska et al. 2003; Bae et al. 2008; Kittiphattanabawon et al. 2010; Addad et al. 2011; Widdowson et al. 2017). There has been significant research into cross-linking these materials to improve these characteristics (Addad et al. 2011; Widdowson et al. 2017), but studies suggest that an achievable native collagen scaffold is unlikely when using marine collagens due to their inherent molecular structure (Li et al. 2013) as described above in further detail.

Once extracted, collagen can be used to form tissue scaffolds based upon various methods. Often a solution of collagen is lyophilized using freeze drying to form an open architecture based scaffold which is ideal for cell migration. This collagen can also be integrated into a solution of HFP or TFE on its own or as a copolymer and electrospun to form nanofiber scaffolds (Chen et al. 2007). In both instances the resulting scaffold must be cross-linked using various chemical (Barnes et al. 2007), enzymatic (Torres-Giner et al. 2009) or photoreactive methods (Cherfan et al. 2013) due to the extraction processing of the collagen that renders the material soluble to physiological and acidic conditions.

4.4 ELECTROSPINNING OF COLLAGEN NANOFIBER MEMBRANES

Electrospinning of soluble collagens provides a suitable way of producing scaffolds that closely mimic the high porosity and surface area often seen in the ECM of tissues, and thus presents great potential in the fabrication of 3D constructs for tissue engineering and wound healing (Figure 4.3).

Electrospinning was developed in the first half of the twentieth century (Formhals 1934), and further investigated with added detail (Reneker and Chun 1996; Reneker et al. 2000); though the origins of the process were initially explored by Boys (1887), and arguably by Gilbert (1628), where electro-spraying occurred through the addition of electrically charged amber near water. Patents related to collagen electrospinning were first filed in the late 1990s (Simpson et al. 2003), with journal articles in the early 2000s describing the use of HFP to electrospin collagen (Matthews et al. 2002).

FIGURE 4.3 Collagen solution being electrospun with constant droplet being replenished and a single Taylor cone producing a jet of collagen solution which lands on the grounding target as collagen nanofibers.

The process of electrospinning involves a simple process detailed in Figure 4.4, which briefly consists of a high-voltage power supply, which supplies positive charge to an electrospinning needle and a grounding electrode coupled to the collector plate. The electrospinning needle is supplied by a syringe loaded with polymer and solvent, which is fed at a controlled rate using a syringe pump. Electrospinning setups can be either horizontal or vertical depending on solution properties such as low viscosity liquids using gravity to prevent spraying of material onto the mat (Figure 4.5).

The optimized conditions for collagen electrospinning are dependent on the quality of collagen, as described later in this chapter, as well as being down to individual electrospinning setups in different research groups. The most commonly used solvent, HFP is generally optimum for dissolving an 8% (w/v) concentration of collagen, though this has been as high as 20% (Li et al. 2015). A separation distance in the range of 8–20 cm and voltage of 15–25 kV are normally used for

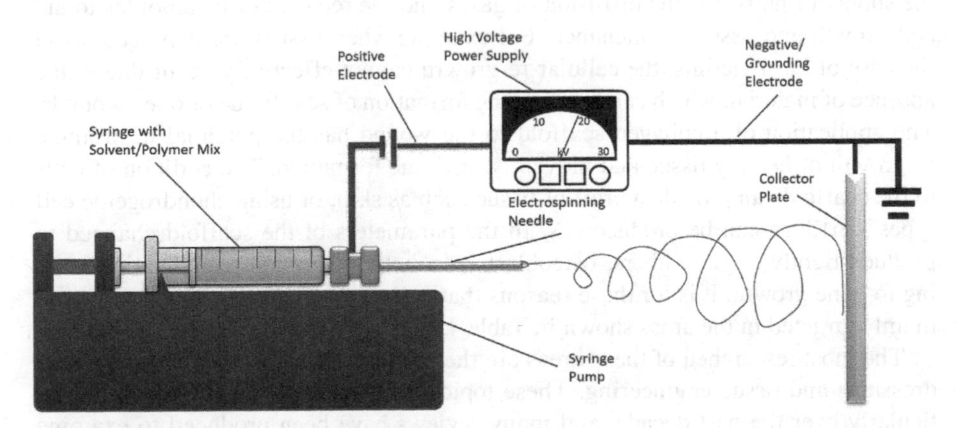

FIGURE 4.4 A basic electrospinning set up using a syringe pump and high voltage power supply connected with a grounded collector plate.

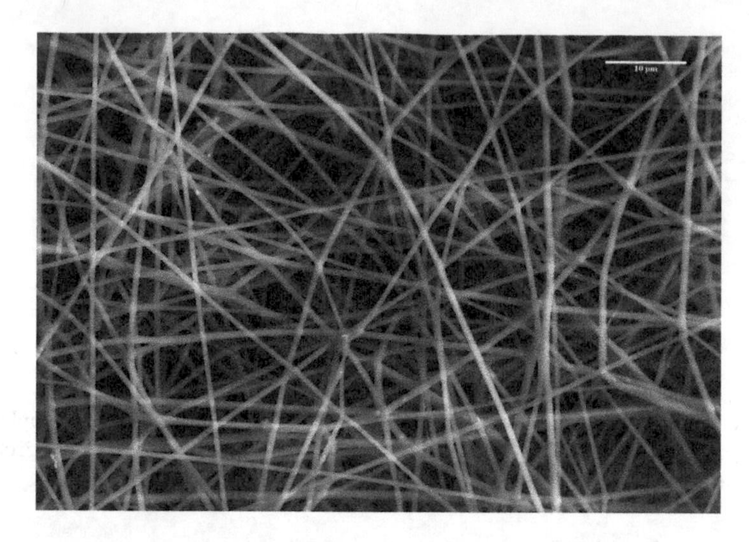

FIGURE 4.5 SEM micrograph of needle electrospun collagen fibers. Magnification of 1500X. Fiber diameter is shown to be 646 ± 121 nm. Scale bar = 10 μm.

the electric field, while flow rate varies widely between groups but is usually about 1 mL/h (Matthews et al. 2002; Rho et al. 2006; Zhou et al. 2016). Most groups have carried out electrospinning on bench scale standard setups (Figure 4.4), with the replacement of the static grounding target with a rotating mandrel collector in some articles (Zhong et al. 2006; Dong et al. 2009) or other designs to collect the fibers (Buttafoco et al. 2006).

The applications of collagen electrospun mats utilize the native functionality of collagen as part of the ECM, namely adherence, migration, proliferation and differentiation of cells alongside the establishment of morphological structure that allows the supply of nutrients, the diffusion of gases and the removal of metabolites to aid cell growth and tissue maintenance. For example when tissues are damaged, as in the case of burn victims, the cellular re-growth cannot effectively occur due to the absence of material, which can result in the formation of scar tissue or open wounds. The application of a collagen scaffold to the wound has the potential to promote re-growth of healthy tissue and prevent scar tissue formation. The addition of cells to the scaffold can provide a graft of tissue such as skin, or using chondrogenic cell types, cartilage can be produced. With the parameters of the scaffold changed to produce tightly aligned fibers, osteoblasts can help to promote calcification leading to bone growth. It is for these reasons that collagen nanofibrous membranes are mainly targeted in the areas shown in Table 4.2.

The most researched of these areas are the use of collagen nanofibers for wound dressings and tissue engineering. These topics have garnered much attention, particularly over the past decade, and many reviews have been produced to examine them (Ma et al. 2005; Pham et al. 2006; Vasita and Katti 2006; Lu and Guo 2018; Mortimer et al. 2018). Despite the apparent interest in collagen as an effective biomaterial, there are large issues that currently surround its use in regenerative medicine,

TABLE 4.2
Example Fields and Uses for Collagen Electrospun Nanofiber Membranes

Field	Use	Examples
Tissue engineering	Artificial skin, blood vessels, tendon	Law et al. (2017); He et al. (2005); Chainani et al. (2013)
Regenerative medicine	Bioresorbable grafts	Sell et al. (2009)
Wound dressings	Bioactive bandage	Zahedi et al. (2010)
Drug delivery	Controlled release	Hall Barrientos et al. (2017)
3D cell culture	Model skin	Nisbet et al. (2009)
Drug screening/trials	Model system	Hartman et al. (2009)

focused mainly on the current use of fluoroalcohols and their denaturing effects on collagen and other proteins when used for electrospinning, but also on the need to cross-link the collagen electrospun scaffolds and the effects of the use of some of these cross-linkers, particularly glutaraldehyde.

4.4.1 The Nativity of Electrospun Collagen Nanofiber Membranes

There is an ongoing debate in the use of collagen in the field of electrospinning, particularly around the potential denaturation of the protein during this process, possibly resulting in the denatured form of collagen, known as gelatin. The effect of such a finding would result in a reduced similarity to the natural ECM, in which collagen is a major constituent. This would make electrospun collagen a less appealing prospect to those trying to create an analogous *in vitro* version of the ECM for cell-based applications.

One major paper on the topic (Zeugolis et al. 2008b) utilizes an array of techniques to visualize the presence or absence of the native collagen α-helical structure, which can be identified by the presence of the 67 nm banding pattern (during gel electrophoresis), as well as using the Fourier-transform infrared spectroscopy (FTIR) footprint of collagen when compared to a gelatin standard. The group attempted to answer the question of whether the electrospinning process denatures the protein, removing its ability to form the triple helical arrangement, which results in the quarter staggered banding pattern seen in native collagen, also known as the D-banding pattern. To this end, the research group used 1,1,1,3,3,3-hexafluoro-2-propanol (HFP) and 2,2,2-trifluoroethanol (TFE) as solvents to dissolve the collagen; these are the commonly used solvents for the electrospinning of collagen and remain the favored choice at the time of writing. The re-dissolved fibers were analyzed using a variety of techniques including transmission electron microscopy (TEM) and circular dichroism (CD) to detect the presence of the α-chain structures present in native collagen. Their findings suggest initially that the electrospun fibers are denatured, where 99.5% of collagen is converted to gelatin. It should be noted that denaturation of collagens which have been dissolved in fluoroalcohol (HFP or TFE) without electrospinning taking place remains high (93%), suggesting that it is the reactions with these solvents that have caused denaturation to occur before electrospinning has taken place, and as such it is impossible to tell from the findings whether the electrospinning of these fibers causes any further denaturation. It would have been useful as a further experiment of this study to carry out self-assembly testing on redissolved samples of electrospun collagen to give an indication if renaturation of the collagen fibers is possible following fluoroalcohol based electrospinning.

In 2008, the aforementioned solvents (HFP and TFE) were the only solvents found to be viable for electrospinning collagen, due to the lack of quality assurance, with some collagens being less soluble in acetic acid than others (Zeugolis et al. 2008b). In the following year, Dong et al. (2009) published research using a benign solvent mixture of phosphate buffered saline (PBS) and ethanol to dissolve and electrospin their collagen. The paper suggests the mixture allows for the electrospinning of collagen without denaturation prior to electrospinning as seen with HFP and TFE. This is backed up using FTIR data to compare the electrospun scaffolds to native collagen. This group however fails to investigate the presence or absence of banding present in the collagen fibers produced, and their findings of water soluble collagen both before and post electrospinning provide little reassurance, as collagen should remain relatively insoluble in water, but rather unable to hold form when moistened.

Other groups have since attempted different solvent systems to replace fluoroalcohols in the electrospinning of collagen, such as Liu et al. (2010), who use 40% acetic acid to electrospin a 25% collagen solution and examine the degree of degradation compared to HFP using CD. The group found a helical fraction of 28.89% for acetic acid and 12.51% for HFP but failed to further examine other qualities such as the presence of periodic D-banding or FTIR amide 1–2 peak ratios. Barrientos et al produced co-electrospun nanofibers of PEO and collagen blends with a 1% collagen in 8% total polymer concentration in HFP which when examined with atomic force microscopy (AFM) clearly displayed the periodic D-banding pattern of collagen within the fibers (Hall Barrientos et al. 2017). DSC measurements are not sufficiently clear due to PEO being included as a copolymer. No further examination of the fibers displayed data to suggest denaturation of the collagen had significantly occurred or been prevented, despite acknowledgement from the group that HFP is known to denature collagen (Figure 4.6).

In order to fully assess the true nature of collagen post electrospinning, the collagen must be extensively tested to characterize several parameters. There is little literature into electrospinning of collagen nanofibers dissolved in a benign solvent which would not significantly alter the ability of collagen to form its native triple helical structure and this will require thorough experimentation to be carried out. The parameters to effectively determine the retention of the collagen's native structure include those tested by Dong et al. (2009) and Zeugolis et al. (2008b), namely to include TEM analysis, SEM analysis, CD, FTIR, and finally differential scanning calorimetry (DSC) experimentation. Together, these techniques can show the definitive presence of collagen or gelatin in the product, by examining the α-helix formation and assess any changes in denaturation temperature.

There are however disparities in the argument proposed by Zeugolis et al., in particular when groups have observed the characteristic 67 nm banding pattern (which is typically only seen in native collagen fibers) on electrospun collagen fibers. Groups such as Matthews et al. (2002) have shown this to be present in collagen electrospun fibers produced using HFP as the solvent system as seen in Figure 4.7.

FIGURE 4.6 (a) AFM image of a PLA-collagen electrospun fiber displaying periodic D-banding pattern typical of collagen. (b) SEM image of PLA-collagen electrospun fibers. (Modified to renumber from Hall Barrientos, I.J. et al., *Int. J. Pharm.*, 531, 67–79, 2017.)

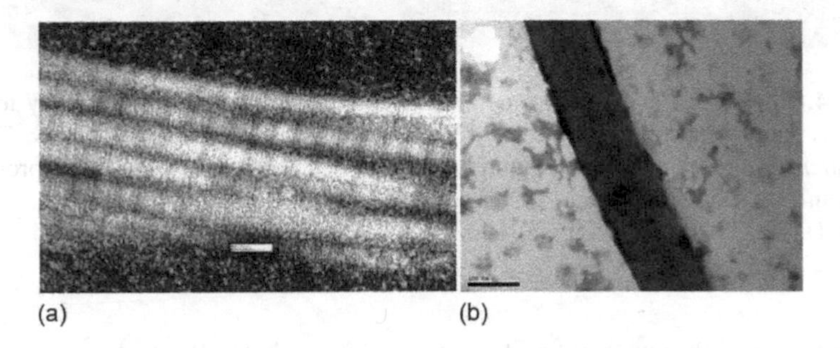

FIGURE 4.7 Comparison of TEM images of electrospun collagen fibers from: (a) electrospun calfskin collagen displaying 67 nm banding pattern within fiber. (Scale bar = 100 nm.) (Reprinted with permission from Matthews, J. A. et al., *Biomacromolecules*, 3, 232–238, 2002. Copyright 2002 American Chemical Society.) (b) Electrospun collagen produced in house that has no banding present. (Scale bar = 100 nm.) (Reprinted with permission from (Dimitrios I. Zeugolis et al. 2008a. Copyright 2008 Elsevier.)

The main issue is the lack of comprehensive research findings, particularly when techniques that examine only small areas of a sample are used such as TEM, which can lead to picking of features present in a sample where it represents a minority; Zeugolis did after all show a 99.5% denaturation, meaning 0.5% presence of triple helix could be present. Therefore, more research is required to determine the exact reasons that collagen may be denatured when being electrospun and exactly how much is caused by each influencing factor of the process. As well as the use of solvent being of concern, the quality of the extract of collagen being used also affects how well the solution can be electrospun (Zeugolis et al. 2008a). This relates to many factors from the extraction conditions, with critical points that must be addressed. Figure 4.8 shows how collagen production can be influenced by several extraction conditions.

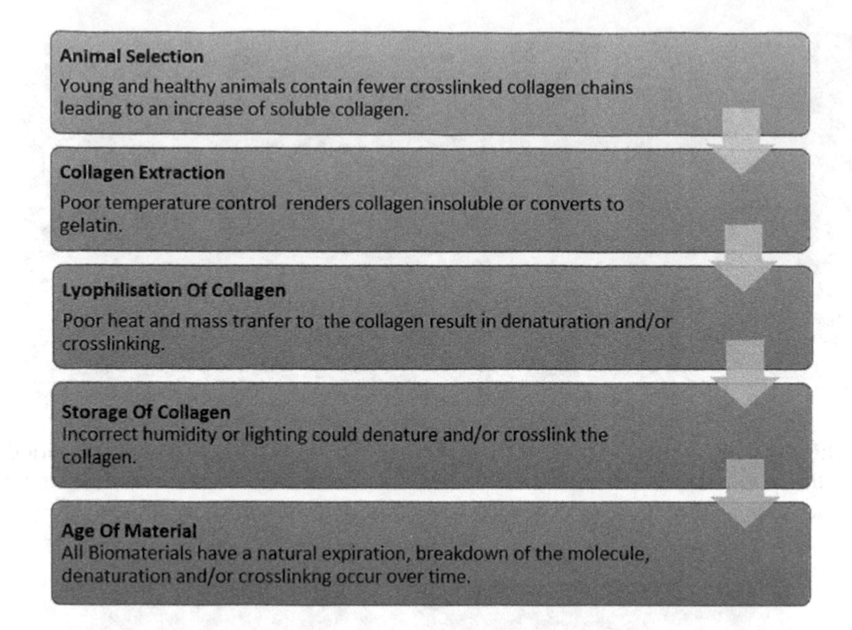

Animal Selection
Young and healthy animals contain fewer crosslinked collagen chains leading to an increase of soluble collagen.
Collagen Extraction
Poor temperature control renders collagen insoluble or converts to gelatin.
Lyophilisation Of Collagen
Poor heat and mass tranfer to the collagen result in denaturation and/or crosslinking.
Storage Of Collagen
Incorrect humidity or lighting could denature and/or crosslink the collagen.
Age Of Material
All Biomaterials have a natural expiration, breakdown of the molecule, denaturation and/or crosslinkng occur over time.

FIGURE 4.8 Chart of extraction stages which contain critical points necessary to avoid solubility issues with collagen extracts. (With kind permission from Taylor & Francis: *J. Biomater. Sci.*, Collagen solubility testing, a quality assurance step for reproducible electrospun nano-fiber fabrication, a technical note, *polymer edition* 19, 1307–1317, 2008b, Zeugolis, D.I., Li, B., Lareu, R.R., Chan, C.K., and Raghunath, M.)

It is because of these solvent and extraction issues that collagen has remained an undesirable polymer for electrospinning and some research has moved onto gelatin electrospinning instead (Sajkiewicz and Kołbuk 2014). This denatured form of collagen is a more affordable alternative to collagen, which is also more easily handled using benign solvents such as acetic acid and phosphate buffered saline. Research on the electrospinning of native collagen has stalled in recent years, due to the condemning findings of the Zeugolis group, and papers published in the post 2008 years have either proceeded in the use of fluoroalcohols, knowing their denaturing effects, or attempted solvent systems that reduce the denaturing aspects of the process. This has diverted attention away from applied scaffolds produced by electrospinning collagen and stunted progression of collagen fiber application in the fields of tissue engineering and regenerative medicine.

4.4.2 Cross-linking of Electrospun Collagen Nanofiber Membranes

For physiologically soluble polymers it is usually required that the scaffold is cross-linked. Glutaraldehyde has for many years been regarded as the standard method of cross-linking (Niu et al. 2013), particularly with proteins due to its efficiency and high degree of cross-linking across different polymers (Barbosa et al. 2014). This chemical has many setbacks with regard to tissue engineering scaffolds,

TABLE 4.3
Examples of Cross-Linking Agents Used for Physiologically Soluble Polymers in Electrospinning Applications to Increase Cytocompatibility in Comparison with Glutaraldehyde and Their Corresponding Polymers

Cross-Linking Agent	Example Polymers for Use	Type
1-ethyl-3-(3-dimethylaminopropyl)carbodiimide hydrochloride (EDC)	Any protein	Chemical
Genipin	Any protein	Chemical
Rose bengal and 532 nm excitation	Collagen	Photochemical
Citric acid	Zein, collagen	Chemical
Thermal cycling induced crystallization	PVA	Thermal
Lysyl oxidase	Collagen	Native enzymatic

foremost is the calcification of scaffolds and surrounding tissues, which are exposed to residual cross-linking agents that may lead to device failure (Golomb et al. 1987). In recent years, alternatives to glutaraldehyde have arisen with the desire to increase cytocompatability *in vivo,* examples of these can be seen in Table 4.3.

Niu et al. (2013) compared PCL/Collagen scaffolds that were cross-linked with either glutaraldehyde vapor or genipin, and examined their efficacy with cell infiltration, survival and proliferation. They found that nanofiber-based scaffolds increased cell proliferation while microfiber scaffolds showed better infiltration of cells. They also showed that genipin showed higher levels of cytocompatability than scaffolds cross-linked with glutaraldehyde. The increased use of less cytotoxic cross-linking agents shows a shift in the field away from the use of compounds that undo the beneficial effects of electrospun scaffolds.

EDC (1-ethyl-3-(3-dimethylaminopropyl) carbodiimide hydrochloride) has arisen as the favored cross-linking agent for electrospun collagen scaffolds but is not itself without issues. First described in relation to electrospun collagen by Barnes et al. (2007), the technique uses EDC alone or in combination with N-hydroxysuccinimide (NHS). The group showed that the cross-linker was successfully able to create an insoluble electrospun collagen scaffold by linking the carboxylic functional groups of aspartic and glutamic amino acids to form O-isoacylurea, which then undergoes nucleophilic attack by the amine functional groups of lysine and hydroxylysine amino acids on adjacent collagen fibrils, creating an iso-peptide bond. The group found that when EDC cross-linking is used, the fibrous structure of collagen is lost as seen in Figure 4.9. This has been observed by other groups (Dong et al. 2009) and techniques such as those seen in Figure 4.10, which stretch or compress the scaffold that have successfully prevented the loss of fibers in the scaffolds.

Work carried out in house by our research group has utilized a method that injects EDC over collagen samples that are being compressed by a frame (Figure 4.10b), reducing the potential points which can shrink and lose their fibrous state as in other methods that use a complete Teflon frame plate to compress the collagen

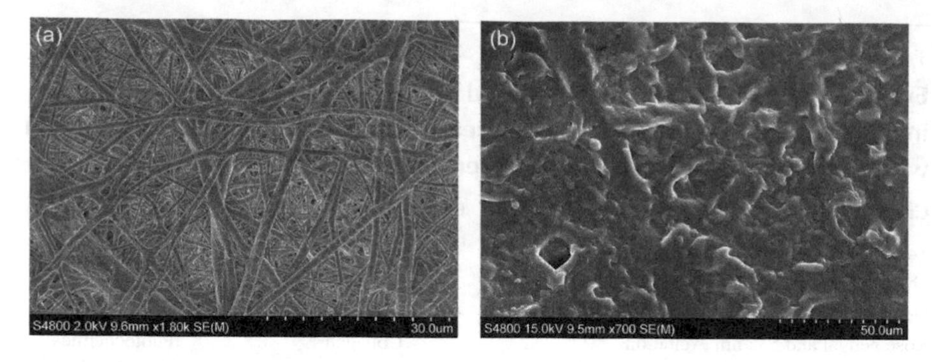

FIGURE 4.9 SEM images of electrospun collagen fibers (a) before EDC cross-linking (b) after EDC cross-linking.

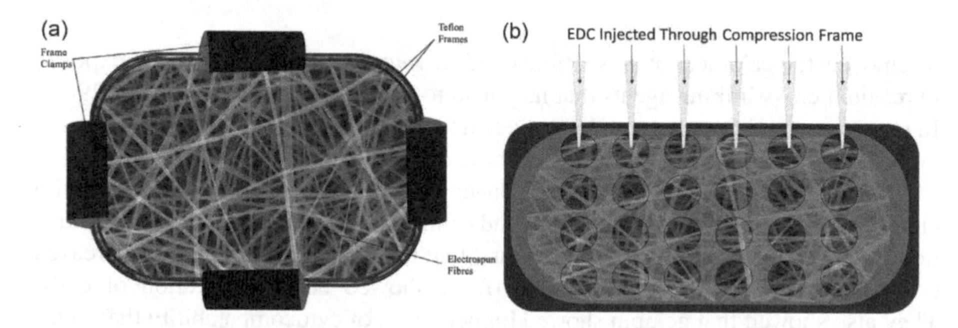

FIGURE 4.10 Different compression methods for preventing the loss of fibers seen when electrospun collagen scaffolds are cross-linked with EDC. (a) Teflon frame method used by Dong et al. (From Dong, B. et al., *Macromol. Rapid Commun.*, 30, 539–542, 2009.) (b) Compression frame with EDC injection ports used in house.

(Figure 4.10a) (Dong et al. 2009). The fibers shown in Figure 4.10 are cross-linked collagen fibers produced in house, cross-linked by injecting EDC over the collagen fibers while they are being compressed.

Liu et al. (2010) cross-linked scaffolds using rose bengal as a photo-initiator (532 nm) and excitation with an argon laser, showing scaffold integrity after 21 days, with a scaffold weight loss of 47.7% on day 7 and 68.9% on day 15. The scaffolds retained their nanofiber structure, despite the weight loss, with apparent reduction in fiber diameters. The group note, however, that in order for the cross-linking to occur, the rose bengal-treated scaffolds need to be immersed in either water or ethanol prior to laser excitation, the water would immediately degrade the scaffold, the ethanol can remain and if not removed may be cytotoxic to cells.

Enzymatic cross-linking may offer a biologically acceptable method to stabilize electrospun collagen chains, and the use of transglutaminase (TG) for this purpose has been examined (Torres-Giner et al. 2009). TG acts to catalyze the formation of amide cross-links between glutamine and lysine residues (O'Halloran et al. 2006). This has been shown to be non-cytotoxic and can be used on scaffolds, which are seeded with cells. The technique has the downside that the increase in mechanical integrity is low, reducing its effectiveness (Orban et al. 2004; Halloran et al. 2008). Lysyl oxidase (LO) is an alternative to TG, which can be used for the enzymatic cross-linking of electrospun collagen (Hapach et al. 2015). LO facilitates the formation of intra- and inter-molecular covalent cross-links between collagen fibers that are essential for collagen fibril formation (Prockop et al. 1979). This is hindered by the limited access to isolated LO for use in a cross-linking solution, with many studies using transfection (Lau et al. 2006) or hypoxia induction (Makris et al. 2014) in cells to produce endogenous LO instead, which would not suit the application for electrospun collagen sufficiently due to the immediate dissolution of collagen in the culture medium.

Physical methods have also been examined in the cross-linking of electrospun collagen scaffolds. Drexler and Powell assessed the well-documented process of dehydrothermal (DHT) cross-linking for its effectiveness when applied to electrospun collagen scaffolds (Drexler and Powell 2011). The group found potential for its effectiveness, with degradation by collagenase being significantly lower than the uncross-linked control samples, and almost equal to EDC cross-linked samples. The DHT scaffolds were not as resistant as DHT and EDC samples and displayed a lower increase in strength than with EDC or DHT and EDC combined. Finally, the group found significantly lower fibroblast cell viability than the EDC cross-linked scaffolds when assessed by a metabolic activity (MTT) assay. The results suggest that the use of DTT may allow for the reduced use of chemical cross-linkers, for applications where accelerated cell regeneration is less important, such as dressings and intubations.

There is a positive move away from the use of glutaraldehyde for cross-linking of electrospun collagen scaffolds, and a growing selection of alternative cross-linkers, which have lower cytotoxic effects than glutaraldehyde, benefitting the surrounding tissues upon implantation. The issue remains that collagen requires cross-linking after electrospinning in order to prevent solubilization, and this issue needs to be addressed as the currently accepted hypothesis is that this is due to the denaturation of the collagen which has been electrospun.

4.5 MODIFICATION OF ELECTROSPUN COLLAGEN NANOFIBER MEMBRANES FOR DIFFERENT TISSUES

One of the largest challenges for electrospun scaffolds is not simply the degree of porosity, which is usually in the region of 91.6% (Li et al. 2002), but how accessible the pores are for cell infiltration. Wang et al. (2010) propose the use of the Darcy permeability coefficients to determine how cells will respond to surfaces and the architecture of tissue engineering scaffolds through scaffold porosity. They also describe three types of pore shown in Figure 4.11. A blind end pore leads to a section of scaffold where material cannot infiltrate further. A closed pore is isolated inside the scaffold and is therefore inaccessible to cells. The desirable architecture is an open pore network where the pores branch between each other and allow cell migration

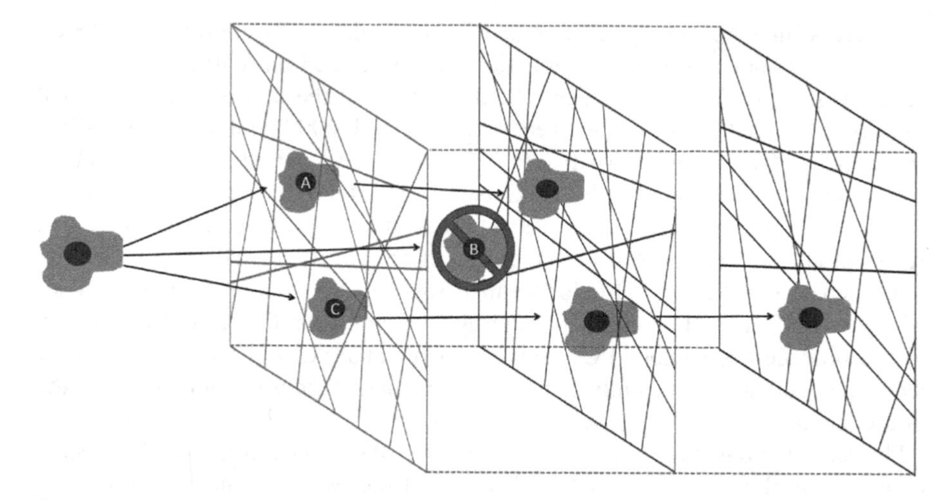

FIGURE 4.11 Cell migration is influenced by three types of pore. A: Blind end pore where cells cannot completely infiltrate. B: A closed pore inside the scaffold cannot be accessed by cells. C: Open pore network allowing complete cellular infiltration. (Reprinted from Mortimer, C.J. et al., "Electrospinning of Functional Nanofibers for Regenerative Medicine: From Bench to Commercial Scale." In *Novel Aspects of Nanofibers*, edited by Tong Lin. Rijeka, InTech, 2018.)

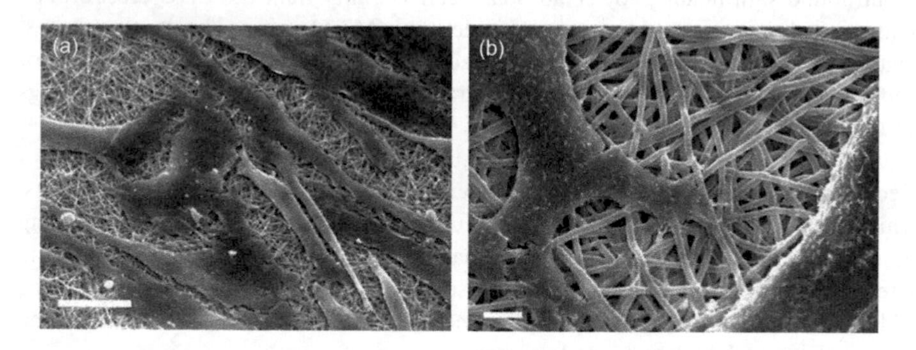

FIGURE 4.12 SEM images of vascular wall mesenchymal stem cells grown on genipin (5%) cross-linked mats for 7 days. Bars: (a) 50 μm and (b) 5 μm. (Reprinted with from *Acta Biomaterialia*, Panzavolta, S. et al., electrospun gelatin nanofibers: optimization of genipin cross-linking to preserve fiber morphology after exposure to water, 7, Copyright 2010, with permission from Elsevier.)

to occur. Blind end pores are often found in electrospun scaffolds which can be seen as cells becoming confluent on the outside of the scaffold without sufficient inward infiltration, as seen in Figure 4.12.

Li et al. (2006) assessed the state of nanofibrous scaffolds for tissue engineering, with great focus on the technology of electrospinning. This review investigated the ideal characteristics for a tissue engineered scaffold, and the questions which should be answered for a particular tissue niche. These begin with architecture; are the fibers of aligned or random orientation? The porosity of a scaffold must allow for cell migration and nutrient diffusion, and if a scaffold is to recreate blood vessels, are red blood cells contained? This is one of the potential downfalls for an electrospun scaffold, as pore size cannot be easily controlled during production. To avoid this, various methods such as salt leaching have been employed to increase and control pore size (Lee et al. 2005). The mechanical properties of a scaffold must also be assessed to match the tissue niche it will be integrated into as these can have drastic effects on cell morphology and cell proliferation and differentiation (Hubbell 1995). Chang and Wang (2011) reviewed how scaffold topography and chemistry influence the successfulness of cell growth on a given scaffold in a given niche.

4.5.1 Ideal Characteristics in Cell Interactions

The ideal conditions for cellular growth of a particular cell type rely on a very well controlled niche, and in order for tissue engineering scaffolds to best suit the growth of a specific cell type, it must be optimized to meet the unique requirements of each cellular environment. Many of the current market offerings were reviewed by Zhong et al. (2010) who noted a lack of electrospun collagen dressings being used currently in wound healing applications. The following will detail some of the most commonly tested cell types on electrospun scaffolds and how control has been exerted to meet the unique requirements necessary for that type of cells expansion and infiltration.

4.5.2 Mesenchymal Stem Cells

Mesenchymal stem cells (MSCs) have been used extensively to assess the basic attributes of electrospun collagen scaffolds. Shih et al. (2006) demonstrated that culturing of MSCs on an electrospun collagen scaffold had no effect on the osteogenic differentiation pathway when compared to the standard polystyrene cell culture flasks. Further work was carried out by Li et al. (2005), who compared various ECM proteins and found that cellular growth was faster on these than the standard tissue culture treated polystyrene (TCPS). This showed maintenance of typical fibroblastoid morphologies seen in Figure 4.13, and that human embryonic palatal mesenchymal (HEPM) cells attach, spread, and form oriented monolayers on protein fiber matrices.

4.5.3 Fibroblasts

As a major cell component in the formation and renewal of the dermis, in particular the excretion of collagen, fibroblasts are of particular interest in collagen-based tissue engineering of skin grafting materials. Due to the interest in collagen as a bioactive wound dressing material, it is important that the interactions between electrospun collagen scaffolds and fibroblasts are examined in great detail. Zhong et al. (2006) produced aligned nanofibers of collagen to compare the growth of rabbit conjunctiva fibroblasts on randomly orientated scaffolds. They found that despite experiencing lower initial cell adhesion in the aligned samples, higher cell proliferation and confluence was seen and the shape and directional growth of the cells was seen to follow

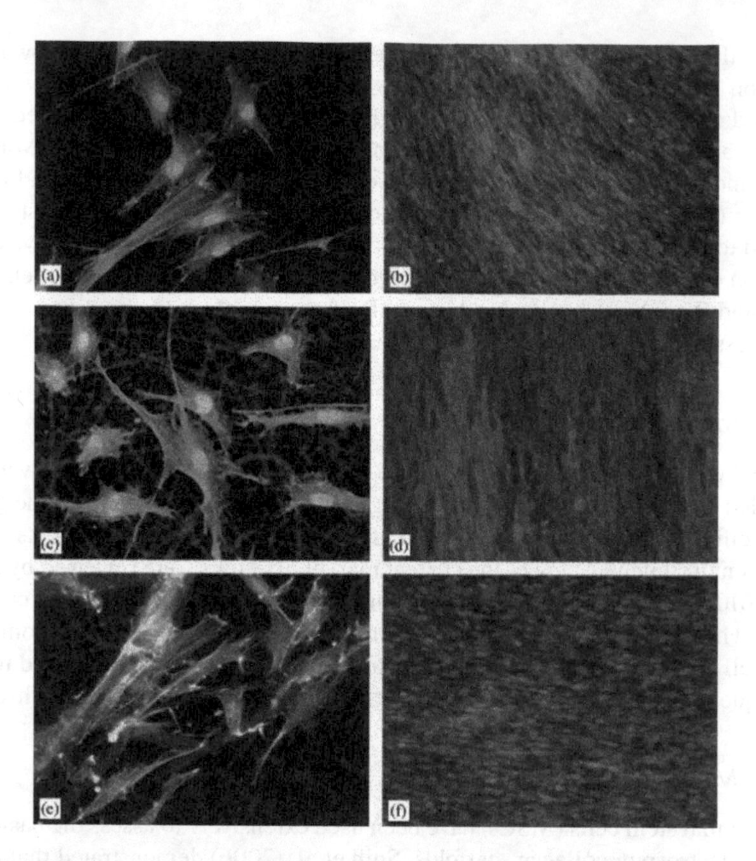

FIGURE 4.13 Morphology of HEPM cells on protein fiber matrices. Staining for nuclei-bisbenzimide, actin cytoskeleton-phalloidin, fibers-autofluorescence. (a, b) TCPS; (c, d) Gelatin; (e, f) Elastin; (a, c, e) HEPM after 48 h in culture (original magnification 400×); (b, d, f) HEPM monolayer at confluence after 72 h in culture (original magnification 200×). (Reprinted from *Biomaterials*, Li, M. et al., electrospun protein fibers as matrices for tissue engineering, 26, 5999–6008, Copyright 2005, with permission from Elsevier.)

the alignment of the collagen fibers. The findings of this study failed to point out that linear regeneration of tissue forming a wound closure results in a weakened, incomplete repair; described above to form scar tissue, this is an undesirable quality in a tissue engineered scaffold. This may have good applications in other tissues however, particularly those which naturally require alignment, such as muscle and bone.

The use of coaxial electrospinning by Zhang et al. (2005) allowed the production of collagen/poly(ϵ-caprolactone) (PCL) (core/shell) coaxial nanofibers that could be compared to PCL nanofibers coated in collagen. The group compared proliferation and cell morphology for human dermal fibroblasts (HDF) between the groups of coaxial, coated, pure collagen and PCL fibers and found that the three groups containing collagen experienced significantly higher proliferation than virgin PCL. The coaxial collagen/PCL outperformed the coated PCL in all areas, demonstrating cell infiltration into the scaffold (Figure 4.14c), while pure collagen scaffolds

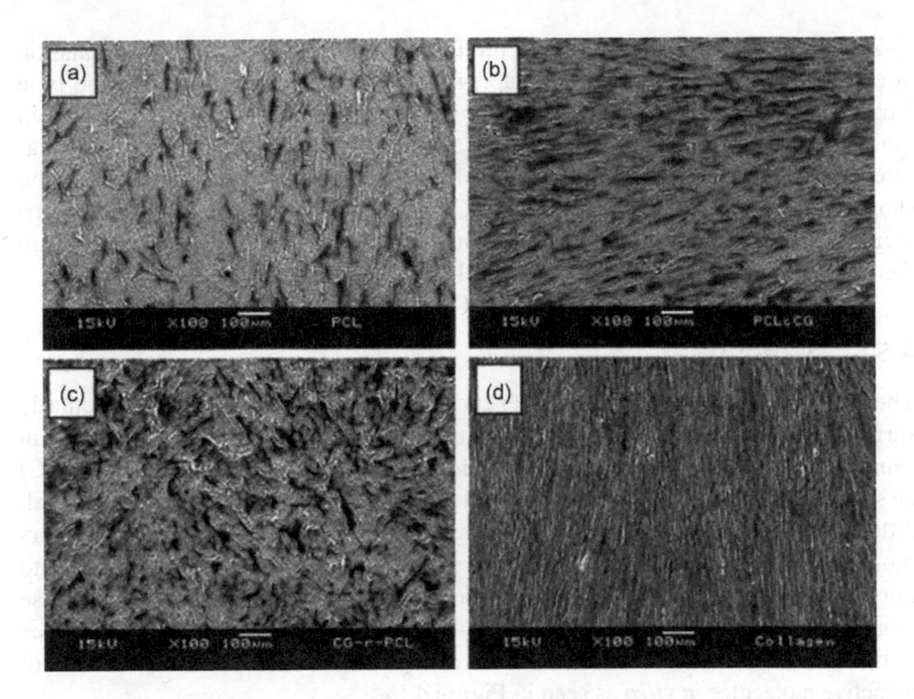

FIGURE 4.14 Cell morphology of HDF at low magnification on different fibrous scaffolds: (a) pure PCL; (b) PCL with surface roughly collagen coated; (c) individually collagen-coated PCL (Coaxial Collagen/PCL); and (d) pure collagen nanofibers. (Reprinted with permission from Zhang, Y. Z. et al., *Biomacromolecules* 6, 2583–2589, 2005. Copyright 2005 American Chemical Society.)

experienced complete confluence across the scaffold (Figure 4.14d). The results of this *in vitro* study demonstrate the importance of collagen in the interaction between tissue engineering material and the body, with translational implications in areas such as coatings for bone replacements (hip, knee) suggesting that integrating electrospun collagen onto the surface of such devices could reduce rejection and quicken recovery periods significantly.

4.5.4 Keratinocytes

As the primary cell component of the epidermis, keratinocytes have also been studied in detail. Studies on cell attachment and spreading of human oral and epidermal keratinocytes (Rho et al. 2006), immortal human keratinocytes (HaCaTs) (Zhou et al. 2016), and human dermal keratinocytes (Sadeghi-Avalshahr et al. 2017) have shown success in using electrospun collagen to mimic the native environment of keratinocytes *in vitro* and carried forward to *in vivo* research.

Rho et al. (2006) found that cells seeded onto collagen electrospun fibers formed from HFP and cross-linked with glutaraldehyde vapor showed poor adhesion from keratinocytes, even when compared with tissue culture polystyrene, suggesting the structure of the collagen had been altered sufficiently to retard cell growth significantly.

Zhou et al. (2016) found significant proliferation of HaCaTs as well as stimulation of epidermal differentiation with the upregulation of involucrin, filaggrin, and type I transglutaminase on their electrospun nanofibers using collagen extracted from tilapia fish. The study didn't demonstrate a sufficiently high denaturation temperature for the marine derived collagen, despite evidence from Chen et al. (2011) who showed a denaturation temperature of 48°C (62°C with EDC cross-linking) and furthermore did not take into consideration the damaging and harmful effects that are well documented with HFP and glutaraldehyde.

4.5.5 Neuronal

The use of collagen in the assistance of neural regrowth has been studied, particularly surrounding recovery from spinal cord injury (SCI). Collagen has been electrospun using HFP in combination with PCL to aid the growth of neural stem cells (NSC) on scaffolds that have been manipulated to produce tubular structures (Hackett et al. 2010). This allowed the group to also assess the controlled release of growth factors (fibroblast growth factor 2 and nerve growth factor) to stimulate differentiation into oligodendrocyte lineages and large primary neurosphere proliferation. The release of growth factors in a controllable manner was observed and provides good evidence of the useful application of co-electrospun collagen/PCL in the controlled release of bioactive molecules *in vitro* as seen in Figure 4.15.

Research surrounding the use of neuronal cells in electrospun scaffolds containing collagen has demonstrated the importance of alignment of fibers in the

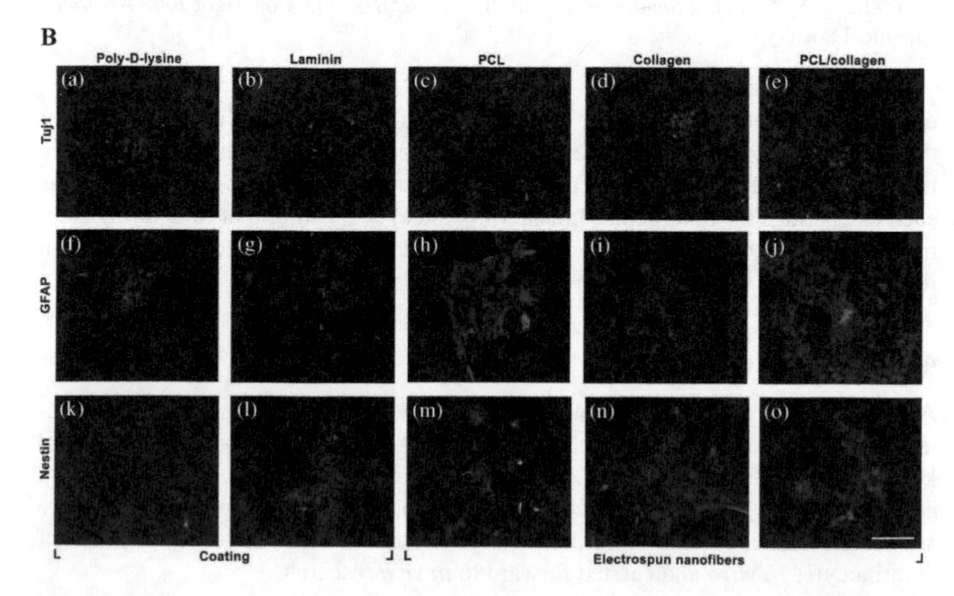

FIGURE 4.15 Immunofluorescence analysis of NSCs cultured on various treatments. Representative images of cells on each substrate were stained for Tuj1 (A–E), GFAP (F–J), and Nestin (K–O). Cell nuclei were counterstained using DAPI. All cells were imaged at 20X; scale bar = 100 μm. (Reprinted from Hackett, J. M. et al., *Materials*, 3, 3714–3728, 2010.)

future regeneration of nerve tissues. The growth of pheochromocytoma-derived cell line (PC12) cells on electrospun scaffolds of Poly (3-hydroxybutyrate-co-3-hydroxyvalerate) (PHBV) alone or in combination with 25% or 50% collagen was assessed for cell phenotype and neurite extension on the electrospun scaffolds (Prabhakaran et al. 2013). The results clearly display the advantage of the inclusion of collagen within the scaffolds, significantly enhancing cell proliferation, and alignment of the nanofibers led to neurite extension in the direction of scaffold alignment.

Finally, by creating a gradient of stromal cell-derived factor-1α (SDF1α), which is bound to electrospun collagen scaffolds through a collagen binding domain, it is possible to direct cell migration and polarize morphology toward the area of higher SDF1α concentration (Li et al. 2015). This experiment displays another useful quality of collagen when influencing cell interactions, where collagen's native presence in the ECM has evolutionary and secondary attributes which influence cell behaviors, rather than simply acting as structural support, with collagen binding domains serving as but one example.

4.6 FUTURE OUTLOOK FOR COLLAGEN ELECTROSPINNING

The future outlook for the electrospinning of collagen membranes revolves around the improved confidence that will be gained as methods continue to develop away from the use of fluoroalcohols and glutaraldehyde toward more benign solvent and cross-linking systems. This will include an added focus on ensuring a clear distinction between native collagens in products and the avoidance of denatured and hydrolyzed collagen products, improving the usefulness of the membranes, in particular in relation to their effects on cellular systems.

A greater control of mechanical properties will be gained alongside the improved monitoring, where composites will be used to best emulate the niche environment required, and modifications to the electrospinning apparatus to enable an increased control on the scaffold orientation, composition and characteristics, which will benefit the recipients of such membranes greatly. Such composite membranes, given current medical trends, may look to control or facilitate the electric potentials of the wound bed, a hot topic in current wound and tissue engineering research (Huttenlocher and Horwitz 2007). This may further boost re-epithelialization of wounds and the combination of such therapies with scaffolds is inevitable.

As has already been displayed with neuronal stem cells above, the controlled release of drugs and growth factors is already underway, and a large increase in the research into this sector is expected as immobilized molecules can be released from the collagen electrospun membranes in a controlled manner, and directional and differential segments can be established within a fabricated construct, achieving multiple clinical objectives within a single scaffold.

The inclusion of non-electrospun products into the scaffold is not limited to molecules, but can also include the incorporation of sensing technologies, such as diagnostic and monitoring equipment which can monitor and improve patient prognosis in real time. As these technologies improve, their size and cost to produce decreases, as has been shown with microneedle technologies. The opportunities for their integration into "smart bandages" gain traction and present yet another opportunity for electrospun collagen scaffolds.

The use of electrospinning has traditionally been held back by a lack of scalable equipment to bring the technology from bench to commercial scale, but given the rise of companies such as Elmarco, The Electrospinning Company, SNC, and others, it has become possible to produce commercial quantities of electrospun membranes, using multi-needle and needle-less based technologies (Mortimer et al. 2018).

Other technology developments, such as on body electrospinning, may lead to the *in vivo* production of fibers to accompany the developments currently being seen in cell-based therapies, which if successful would lower the need for surgery to implant the scaffolding devices. The improved biological compatibility of cross-linking agents and solvent systems in the electrospinning of collagen scaffolds would greatly support the potential for this achievement.

4.7 CONCLUSIONS

The preceding decade has been a turbulent time in the electrospinning of collagen membranes, with many of the traditionally accepted methods for the fabrication and stabilization of these membranes called into question. Their efficacy has been shown to be greatly reduced by the use of fluorocarbon solvents, and the use of glutaraldehyde as the "gold standard" cross-linking agent has caused calcification of surrounding tissues, lowering the efficacy of the scaffold in medically relevant settings.

The move away from these aggressive methods in recent years, though slow, has further emphasized the need for collagen in medical scaffolds, and the desire to see the body's most abundant protein utilized for its structural and chemical properties. There remains a great deal of interest, and this has led to several research groups exploring novel methods to preserve the collagen during electrospinning and to use a cross-linker which is much more biocompatible.

As the techniques for electrospinning continue to improve and the fundamental understandings of control parameters advance, the technology will allow for increased control and variety of collagen-based electrospun scaffolds to be produced, both at bench and commercial scale. Even with the ongoing debate over the denaturation of collagen when electrospun, the biomaterial has shown itself to be extremely useful when interacting with cellular systems, and with slight modification, the increased control gained from modern electrospinning setups has shown that collagen has many desirable qualities, which current synthetic polymers are unable to mimic. As new techniques enable a more native collagen to be electrospun, the potential improved clinical results may cement electrospun collagen's place as the top polymer of choice for clinical use, including wound, vascular, nervous, bone, and dental applications.

REFERENCES

Addad S, JY Exposito, C Faye, S Ricard-Blum, and C Lethias. 2011. "Isolation, Characterization and Biological Evaluation of Jellyfish Collagen for Use in Biomedical Applications." *Marine Drugs* 9 (6): 967–983. doi:10.3390/md9060967.

Ashcroft GS, MA Horan, and MW Ferguson. 1998. "Aging Alters the Inflammatory and Endothelial Cell Adhesion Molecule Profiles during Human Cutaneous Wound Healing." *Laboratory Investigation; a Journal of Technical Methods and Pathology* 78 (1): 47–58. http://www.ncbi.nlm.nih.gov/pubmed/9461121.

Asher DM. 1999. "The Transmissible Spongiform Encephalopathy Agents: Concerns and Responses of United States Regulatory Agencies in Maintaining the Safety of Biologics." *Developments in Biological Standardization* 100: 103–118. http://www.ncbi.nlm.nih.gov/pubmed/10616181.

Bae I, K Osatomi, A Yoshida, K Osako, A Yamaguchi, and K Hara. 2008. "Biochemical Properties of Acid-Soluble Collagens Extracted from the Skins of Underutilised Fishes." *Food Chemistry* 108 (1): 49–54. doi:10.1016/j.foodchem.2007.10.039.

Banerjee P, L Suguna, and C Shanthi. 2015. "Wound Healing Activity of a Collagen-Derived Cryptic Peptide." *Amino Acids* 47 (2): 317–328. doi:10.1007/s00726-014-1860-6.

Barbosa O, C Ortiz, Á Berenguer-Murcia, R Torres, RC Rodrigues, and R Fernandez-Lafuente. 2014. "Glutaraldehyde in Bio-catalysts Design: A Useful Crosslinker and a Versatile Tool in Enzyme Immobilization." *RSC Advances.* 4 (4): 1583–1600. doi:10.1039/C3RA45991H.

Barnes CP, CW Pemble, DD Brand, DG Simpson, and GL Bowlin. 2007. "Cross-Linking Electrospun Type II Collagen Tissue Engineering Scaffolds with Carbodiimide in Ethanol." *Tissue Engineering* 13 (7): 1593–1605. doi:10.1089/ten.2006.0292.

Barzideh Z, AA Latiff, CY Gan, S Benjakul, and AA Karim. 2014. "Isolation and Characterisation of Collagen from the Ribbon Jellyfish (*Chrysaora* sp.)." *International Journal of Food Science and Technology* 49 (6): 1490–1499. doi:10.1111/ijfs.12464.

Bayat A, DA McGrouther, and MWJ Ferguson. 2003. "Skin Scarring." *BMJ (Clinical Research Ed.)* 326 (7380): 88–92. http://www.ncbi.nlm.nih.gov/pubmed/12521975.

Boys CV. 1887. "On the Production, Properties, and Some Suggested Uses of the Finest Threads." *Proceedings of the Physical Society of London* 9 (1): 8.

Brett D. 2008. "A Review of Collagen and Collagen-Based Wound Dressings." *Wounds* 20 (12): 1–11.

Buttafoco L, NG Kolkman, P Engbers-Buijtenhuijs, AA. Poot, PJ Dijkstra, I Vermes, and J Feijen. 2006. "Electrospinning of Collagen and Elastin for Tissue Engineering Applications." *Biomaterials* 27 (5): 724–734. doi:10.1016/j.biomaterials.2005.06.024.

Castillo-Briceño Patricia P, D Bihan, M Nilges, S Hamaia, J Meseguer, A García-Ayala, R W Farndale, and V Mulero. 2011. "A Role for Specific Collagen Motifs during Wound Healing and Inflammatory Response of Fibroblasts in the Teleost Fish Gilthead Seabream." *Molecular Immunology* 48 (6–7): 826–834. doi:10.1016/j.molimm.2010.12.004.

Chainani A, KJ Hippensteel, A Kishan, NW Garrigues, DS Ruch, F Guilak, and D Little. 2013. "Multilayered Electrospun Scaffolds for Tendon Tissue Engineering." *Tissue Engineering Part A* 19 (23–24): 2594–2604. doi:10.1089/ten.tea.2013.0165.

Chang H-I, and Y Wang. 2011. "Cell Responses to Surface and Architecture of Tissue Engineering Scaffolds." *Regenerative Medicine and Tissue Engineering—Cells and Biomaterials*, 569–588. doi:10.5772/21983.

Chen S, N. Hirota, M Okuda, M Takeguchi, H Kobayashi, N Hanagata, and T Ikoma. 2011. "Microstructures and Rheological Properties of Tilapia Fish-Scale Collagen Hydrogels with Aligned Fibrils Fabricated under Magnetic Fields." *Acta Biomaterialia* 7 (2): 644–652. doi:10.1016/J.ACTBIO.2010.09.014.

Chen Z, X Mo, and F Qing. 2007. "Electrospinning of Collagen-Chitosan Complex." *Materials Letters* 61 (16): 3490–3494. doi:10.1016/j.matlet.2006.11.104.

Cherfan D, EE Verter, S Melki, TE Gisel, FJ Doyle, G Scarcelli, SH Yun, RW Redmond, and IE Kochevar. 2013. "Collagen Cross-Linking Using Rose Bengal and Green Light to Increase Corneal Stiffness." *Investigative Ophthalmology and Visual Science* 54 (5): 3426–3433. doi:10.1167/iovs.12-11509.

Clark GL, EA Parker, JA Schaad, and WJ Warren. 1935. "New Measurements of Previously Unknown Large Interplanar Spacings in Natural Materials." *Journal of the American Chemical Society* 57 (8): 1509–1509. doi:10.1021/ja01311a504.

Clark R. 2012. *The Molecular and Cellular Biology of Wound Repair.* Springer US, New York City.

Clark RAF. 1996. *The Molecular and Cellular Biology of Wound Repair.* Plenum Press, New York.

Clark RAF. 1998. "Overview and General Considerations of Wound Repair." In *The Molecular and Cellular Biology of Wound Repair*, pp. 3–33. Boston, MA: Springer US. doi:10.1007/978-1-4615-1795-5_1.

Dallon JC. 1998. "A Mathematical Model for Fibroblast and Collagen Orientation." *Bulletin of Mathematical Biology* 60: 101–129. http://www.macs.hw.ac.uk/~jas/paperpdfs/dallon1.pdf.

Dallon JC, and JA Sherratt. 2000. "A Mathematical Model for Spatially Varying Extracellular Matrix Alignment." *SIAM Journal on Applied Mathematics* 61 (2): 506–527. doi:10.1137/S0036139999359343.

Dallon JC, JA Sherratt, and PK Maini. 1999. "Mathematical Modelling of Extracellular Matrix Dynamics Using Discrete Cells: Fiber Orientation and Tissue Regeneration." *Journal of Theoretical Biology* 199. http://www.idealibrary.com.

Dealey C. 1999. *The Care of Wounds: A Guide for Nurses.* Blackwell Science, Chichester, UK. https://capitadiscovery.co.uk/edgehill/items/115403.

Dong B, O Arnoult, ME Smith, and GE Wnek. 2009. "Electrospinning of Collagen Nanofiber Scaffolds from Benign Solvents." *Macromolecular Rapid Communications* 30 (7): 539–542. doi:10.1002/marc.200800634.

Drexler JW, and HM Powell. 2011. "Dehydrothermal Crosslinking of Electrospun Collagen." *Tissue Engineering Part C: Methods* 17 (1): 9–17. doi:10.1089/ten.tec.2009.0754.

Ehrlich HP, Moyer KE. 2013. "Cell-Populated Collagen Lattice Contraction Model for the Investigation of Fibroblast Collagen Interactions." In: Gourdie R, Myers T. *Wound Regeneration and Repair: Methods in Molecular Biology (Methods and Protocols)*, Vol. 1037. Humana Press, Totowa, NJ. doi:10.1007/978-1-62703-505-7_3.

Engel J, and HP Bächinger. 2005. "Structure, Stability and Folding of the Collagen Triple Helix." *Topics in Current Chemistry* 247: 7–33. doi:10.1007/b103818.

Formhals A. 1934. Process and Apparatus for Preparing Artificial Threads: US, 1975504.

Gelatin. 2018. Encyclopedia of Food and Culture. Accessed August 3, 2018. https://www.encyclopedia.com/science-and-technology/biochemistry/biochemistry/gelatin.

Gelse K, E Pöschl, and T Aigner. 2003. "Collagens—Structure, Function, and Biosynthesis." *Advanced Drug Delivery Reviews* 55 (12): 1531–1546. doi:10.1016/j.addr.2003.08.002.

Gilbert W. 1628. *De Magnete.* London, UK: Peter Short.

Golomb G, FJ Schoen, MS Smith, J Linden, M Dixon, and RJ Levy. 1987. "The Role of Glutaraldehyde-Induced Cross-Links in Calcification of Bovine Pericardium Used in Cardiac Valve Bioprostheses." *The American Journal of Pathology* 127 (1): 122–130. https://www.ncbi.nlm.nih.gov/pmc/articles/PMC1899585/pdf/amjpathol00145-0127.pdf.

Greenhalgh DG. 1998. "The Role of Apoptosis in Wound Healing." *The International Journal of Biochemistry & Cell Biology* 30 (9): 1019–1030. doi:10.1016/S1357-2725(98)00058-2.

Hackett JM, TNT Dang, EC Tsai, and X Cao. 2010. "Electrospun Biocomposite Polycaprolactone/Collagen Tubes as Scaffolds for Neural Stem Cell Differentiation." *Materials* 3 (6): 3714–3728. doi:10.3390/ma3063714.

Häggström M. 2014. "Medical Gallery of Mikael Häggström 2014." *WikiJournal of Medicine* 1 (2). doi:10.15347/wjm/2014.008.

Hall Barrientos IJ, E Paladino, P Szabó, S Brozio, PJ Hall, CI Oseghale, MK Passarelli et al. 2017. "Electrospun Collagen-Based Nanofibres: A Sustainable Material for Improved Antibiotic Utilisation in Tissue Engineering Applications." *International Journal of Pharmaceutics* 531 (1): 67–79. doi:10.1016/j.ijpharm.2017.08.071.

Halloran DO, S Grad, M Stoddart, P Dockery, M Alini, and AS Pandit. 2008. "An Injectable Cross-Linked Scaffold for Nucleus Pulposus Regeneration." *Biomaterials* 29 (4): 438–447. doi:10.1016/J.BIOMATERIALS.2007.10.009.

Hapach LA, JA Vanderburgh, JP Miller, and CA Reinhart-King. 2015. "Manipulation of In Vitro Collagen Matrix Architecture for Scaffolds of Improved Physiological Relevance." *Physical Biology*. doi:10.1088/1478-3975/12/6/061002.

Hartman O, C Zhang, EL Adams, MC Farach-Carson, NJ Petrelli, BD Chase, and JF Rabolt. 2009. "Microfabricated Electrospun Collagen Membranes for 3-D Cancer Models and Drug Screening Applications." *Biomacromolecules* 10 (8): 2019–2032. doi:10.1021/bm8012764.

He W, T Yong, WE Teo, Z Ma, and S Ramakrishna. 2005. "Fabrication and Endothelialization of Collagen-Blended Biodegradable Polymer Nanofibers: Potential Vascular Graft for Blood Vessel Tissue Engineering." *Tissue Engineering* 11 (9–10): 1574–1588. doi:10.1089/ten.2005.11.1574.

Holmes DF, and KE Kadler. 2006. "The 10 + 4 Microfibril Structure of Thin Cartilage Fibrils." *PNAS* 103 (46): 17249–17254. doi:doi10.1073pnas.0608417103.

Holmes DF, CJ Gilpin, C Baldock, U Ziese, AJ Koster, and KE Kadler. 2001. "Corneal Collagen Fibril Structure in Three Dimensions: Structural Insights into Fibril Assembly, Mechanical Properties, and Tissue Organization." *Proceedings of the National Academy of Sciences of the United States of America* 98 (13): 7307–7312. doi:10.1073/pnas.111150598.

Hubbell JA. 1995. "Biomaterials in Tissue Engineering." *Bio/Technology (Nature Publishing Company)* 13 (6): 565–576. doi:10.1038/nbt0695-565.

Huttenlocher A, and AR Horwitz. 2007. "Wound Healing with Electric Potential." *New England Journal of Medicine* 356 (3): 303–304. doi:10.1056/NEJMcibr066496.

Kittiphattanabawon P, S Benjakul, W Visessanguan, H Kishimura, and F Shahidi. 2010. "Isolation and Characterisation of Collagen from the Skin of Brownbanded Bamboo Shark (*Chiloscyllium punctatum*)." *Food Chemistry* 119 (4): 1519–1526. doi:10.1016/j.foodchem.2009.09.037.

Lau Y-KI, AM Gobin, and JL West. 2006. "Overexpression of Lysyl Oxidase to Increase Matrix Crosslinking and Improve Tissue Strength in Dermal Wound Healing." *Annals of Biomedical Engineering* 34 (8): 1239–1246. doi:10.1007/s10439-006-9130-8.

Law JX, LL Liau, A Saim, Y Yang, and R Idrus. 2017. "Electrospun Collagen Nanofibers and Their Applications in Skin Tissue Engineering." *Tissue Engineering and Regenerative Medicine*. Springer Science + Business Media. doi:10.1007/s13770-017-0075-9.

Lee YH, JH Lee, IG An, C Kim, DS Lee, YK Lee, and JD Nam. 2005. "Electrospun Dual-Porosity Structure and Biodegradation Morphology of Montmorillonite Reinforced PLLA Nanocomposite Scaffolds." *Biomaterials* 26 (16): 3165–3172. doi:10.1016/j.biomaterials.2004.08.018.

Li WJ, CT Laurencin, EJ Caterson, RS Tuan, and FK Ko. 2002. "Electrospun Nanofibrous Structure: A Novel Scaffold for Tissue Engineering." *Journal of Biomedical Materials Research* 60 (4): 613–621. doi:10.1002/jbm.10167.

Li X, H Liang, J Sun, Y Zhuang, B Xu, and J Dai. 2015. "Electrospun Collagen Fibers with Spatial Patterning of SDF1α for the Guidance of Neural Stem Cells." *Advanced Healthcare Materials* 4 (12): 1869–1876. doi:10.1002/adhm.201500271.

Li M, MJ Mondrinos, MR Gandhi, FK Ko, AS Weiss, and PI Lelkes. 2005. "Electrospun Protein Fibers as Matrices for Tissue Engineering." *Biomaterials* 26 (30): 5999–6008. doi:10.1016/j.biomaterials.2005.03.030.

Li W-J, RM Shanti, and RS Tuan. 2006. "Electrospinning Technology for Nanofibrous Scaffolds in Tissue Engineering." *Nanotechnologies for the Life Sciences* 9. doi:10.1002/9783527610419.ntls0097.

Li ZR, B Wang, CF Chi, QH Zhang, YD Gong, JJ Tang, HY Luo, and GF Ding. 2013. "Isolation and Characterization of Acid Soluble Collagens and Pepsin Soluble Collagens from the Skin and Bone of Spanish Mackerel (*Scomberomorous niphonius*)." *Food Hydrocolloids* 31 (1): 103–113. doi:10.1016/j.foodhyd.2012.10.001.

Liu T, WK Teng, BP Chan, and SY Chew. 2010. "Photochemical Crosslinked Electrospun Collagen Nanofibers: Synthesis, Characterization and Neural Stem Cell Interactions." *Journal of Biomedical Materials Research—Part A* 95 (1): 276–282. doi:10.1002/jbm.a.32831.

Lodish H, A Berk, SL Zipursky, P Matsudaira, D Baltimore, and J Darnell. 2000. *Molecular Cell Biology,* 4th ed. W. H. Freeman: New York.

Lu P, and Y Guo. 2018. Electrospinning of Collagen and Its Derivatives for Biomedical Applications. In *Novel Aspects of Nanofibers*. doi:10.5772/intechopen.73581.

Lupi O. 2002. "Prions in Dermatology." *Journal of the American Academy of Dermatology* 46 (5): 790–793. doi:10.1067/mjd.2002.120624.

Ma Z, M Kotaki, R Inai, and S Ramakrishna. 2005. "Potential of Nanofiber Matrix as Tissue-Engineering Scaffolds." *Tissue Engineering* 11 (1–2): 101–109. doi:10.1089/ten.2005.11.101.

Makris EA, DJ Responte, NK Paschos, JC Hu, and KA Athanasiou. 2014. "Developing Functional Musculoskeletal Tissues through Hypoxia and Lysyl Oxidase-Induced Collagen Cross-Linking." *Proceedings of the National Academy of Sciences of the United States of America* 111 (45): E4832-E4841. doi:10.1073/pnas.1414271111.

Matthews JA, GE Wnek, DG Simpson, and GL Bowlin. 2002. "Electrospinning of Collagen Nanofibers." *Biomacromolecules* 3 (2): 232–238. doi:10.1021/bm015533u.

Miller EJ. 1972. "Structural Studies on Cartilage Collagen Employing Limited Cleavage and Solubilization with Pepsin." *Biochemistry* 11 (26): 4903–4909. doi:10.1021/bi00776a005.

Min S, SW Wang, and W Orr. 2006a. Graphic General Pathology: 2.2 Complete Regeneration. *Pathology*. https://web.archive.org/web/20160313151452/http://pathol.med.stu.edu.cn/pathol/listengcontent2.aspx?contentid=492.

Min S, SW Wang, and W Orr. 2006b. "Graphic General Pathology: 2.3 Incomplete Regeneration." *Pathology*. Accessed April 27, 2018. https://web.archive.org/web/20160311182730/http://pathol.med.stu.edu.cn/pathol/listEngContent2.aspx?contentID=493.

Montesano R, L Orci, and P Vassalli. 1983. "In Vitro Rapid Organization of Endothelial Cells into Capillary-like Networks Is Promoted by Collagen Matrices." *Journal of Cell Biology* 97 (5 I): 1648–1652. doi:10.1083/jcb.97.5.1648.

Mortimer CJ, JP Widdowson, and CJ Wright. 2018. "Electrospinning of Functional Nanofibers for Regenerative Medicine: From Bench to Commercial Scale." In *Novel Aspects of Nanofibers*, edited by Tong Lin. Rijeka: InTech. doi:10.5772/intechopen.73677.

Murray JC, SV Pollack, and SR Pinnell. 1981. "Keloids: A Review." *Journal of the American Academy of Dermatology* 4 (4): 461–470. doi:10.1016/S0190-9622(81)70048-3.

Myllyharju J, and KI Kivirikko. 2001. "Collagens and Collagen-Related Diseases." *Annals of Medicine* 33 (1): 7–21. doi:10.3109/07853890109002055.

Nagai T, W Worawattanamateekul, N Suzuki, T Nakamura, T Ito, K Fujiki, M Nakao, and T Yano. 2000. "Isolation and Characterization of Collagen from Rhizostomous Jellyfish (*Rhopilema asamushi*)." *Food Chemistry* 70 (2): 205–208. doi:10.1016/S0308-8146(00)00081-9.

Nalinanon S, S Benjakul, W Visessanguan, and H Kishimura. 2007. "Use of Pepsin for Collagen Extraction from the Skin of Bigeye Snapper (*Priacanthus tayenus*)." *Food Chemistry* 104 (2): 593–601. doi:10.1016/j.foodchem.2006.12.035.

Nisbet DR, JS Forsythe, W Shen, DI Finkelstein, and MK Horne. 2009. "Review Paper: A Review of the Cellular Response on Electrospun Nanofibers for Tissue Engineering." *Journal of Biomaterials Applications* 24 (1): 7–29. doi:10.1177/0885328208099086.

Niu G, T Criswell, E Sapoznik, S-J Lee, and S Soker. 2013. "The Influence of Cross-Linking Methods on the Mechanical and Biocompatible Properties of Vascular Scaffold." *Journal of Science and Applications: Biomedicine* 1 (1): 1–7.

O'Halloran DM, RJ Collighan, M Griffin, and AS Pandit. 2006. "Characterization of a Microbial Transglutaminase Cross-Linked Type II Collagen Scaffold." *Tissue Engineering* 12 (6): 1467–1474. doi:10.1089/ten.2006.12.1467.

Orban JM, LB Wilson, JA Kofroth, MS El-Kurdi, TM Maul, and DA Vorp. 2004. "Crosslinking of Collagen Gels by Transglutaminase." *Journal of Biomedical Materials Research—Part A* 68 (4): 756–762. doi:10.1002/jbm.a.20110.

Panzavolta S, M Gioffrè, ML Focarete, C Gualandi, L Foroni, and A Bigi. 2011. "Electrospun Gelatin Nanofibers: Optimization of Genipin Cross-Linking to Preserve Fiber Morphology after Exposure to Water." *Acta Biomaterialia* 7: 1702–1709. doi:10.1016/j.actbio.2010.11.021.

Petruska JA, and AJ Hodge. 1964. "A Subunit Model for the Tropocollagen Macromolecule." *Proceedings of the National Academy of Sciences* 51 (5): 871–876. doi:10.1073/pnas.51.5.871.

Pham QP, U Sharma, and AG Mikos. 2006. "Electrospinning of Polymeric Nanofibers for Tissue Engineering Applications: A Review." *Tissue Engineering* 12 (5): 1197–1211. doi:10.1089/ten.2006.12.1197.

Pozzi A, KK Wary, FG Giancotti, and HA Gardner. 1998. "Integrin A1β1 Mediates a Unique Collagen-Dependent Proliferation Pathway In Vivo." *Journal of Cell Biology* 142 (2): 587–594. doi:10.1083/jcb.142.2.587.

Prabhakaran MP, E Vatankhah, and S Ramakrishna. 2013. "Electrospun Aligned PHBV/Collagen Nanofibers as Substrates for Nerve Tissue Engineering." *Biotechnology and Bioengineering* 110: 2775–2784. doi:10.1002/bit.24937/abstract.

Prockop DJ, KI Kivirikko, L Tuderman, and NA Guzman. 1979. "The Biosynthesis of Collagen and Its Disorders." *New England Journal of Medicine* 301 (2): 77–85. doi:10.1056/NEJM197907123010204.

Ramachandran GN, BB Doyle, and ER Bloot. 1968. "Single-Chain Triple Helical Structure." *Biopolymers* 6 (12): 1771–1775. doi:10.1002/bip.1968.360061213.

Raspanti M, V Ottani, and A Ruggeri. 1990. "Subfibrillar Architecture and Functional Properties of Collagen: A Comparative Study in Rat Tendons." *Journal of Anatomy* 172 (October): 157–164. http://www.ncbi.nlm.nih.gov/pubmed/2272900.

Reneker DH, and I Chun. 1996. "Nanometre Diameter Fibres of Polymer, Produced by Electrospinning." *Nanotechnology* 7 (3): 216–223. doi:10.1088/0957-4484/7/3/009.

Reneker DH, AL Yarin, H Fong, and S Koombhongse. 2000. "Bending Instability of Electrically Charged Liquid Jets of Polymer Solutions in Electrospinning." *Journal of Applied Physics* 87 (9): 4531–4547. doi:10.1063/1.373532.

Rho KS, L Jeong, G Lee, BM Seo, YJ Park, SD Hong, S Roh, JJ Cho, WH Park, and BM Min. 2006. "Electrospinning of Collagen Nanofibers: Effects on the Behavior of Normal Human Keratinocytes and Early-Stage Wound Healing." *Biomaterials* 27 (8): 1452–1461. doi:10.1016/j.biomaterials.2005.08.004.

Rigby BJ. 1968. "Amino-Acid Composition and Thermal Stability of the Skin Collagen of the Antarctic Ice-Fish [19]." *Nature*. doi:10.1038/219166a0.

Sadeghi-Avalshahr A, S Nokhasteh, AM Molavi, M Khorsand-Ghayeni, and M Mahdavi-Shahri. 2017. "Synthesis and Characterization of Collagen/PLGA Biodegradable Skin Scaffold Fibers." *Regenerative Biomaterials* 4 (5).: 309–314. doi:10.1093/rb/rbx026.

Sadowska M, I Kołodziejska, and C Niecikowska. 2003. "Isolation of Collagen from the Skins of Baltic Cod (*Gadus morhua*)." *Food Chemistry* 81 (2): 257–262. doi:10.1016/S0308-8146(02)00420-X.

Saito M, Y Takenouchi, N Kunisaki, and S Kimura. 2001. "Complete Primary Structure of Rainbow Trout Type I Collagen Consisting of Alpha1(I) Alpha2(I) Alpha3(I) Heterotrimers." *European Journal of Biochemistry/FEBS* 268: 2817–2827.

Sajkiewicz P, and D Kołbuk. 2014. "Electrospinning of Gelatin for Tissue Engineering—Molecular Conformation as One of the Overlooked Problems." *Journal of Biomaterials Science. Polymer Edition* 25 (18): 2009–2022. doi:10.1080/09205063.2014.975392.

Schultz GS, and BA Mast. 1999. "Molecular Analysis of the Environments of Healing and Chronic Wounds: Cytokines, Proteases and Growth Factors." *Wounds* 2: 7–14. http://www.awma.com.au/journal/0701_01.pdf.

Sell SA, MJ McClure, K Garg, PS Wolfe, and GL Bowlin. 2009. "Electrospinning of Collagen/Biopolymers for Regenerative Medicine and Cardiovascular Tissue Engineering." *Advanced Drug Delivery Reviews* 61 (12): 1007–1019. doi:10.1016/j.addr.2009.07.012.

Shih Y-RV, C-N Chen, S-W Tsai, YJ Wang, and OK Lee. 2006. "Growth of Mesenchymal Stem Cells on Electrospun Type I Collagen Nanofibers." *Stem Cells* 24 (11): 2391–2397. doi:10.1634/stemcells.2006-0253.

Shoulders MD, and RT Raines. 2010. "Collagen Structure and Stability." *Annual Review of Biochemistry*, 929–958. doi:10.1146/annurev.biochem.77.032207.120833.COLLAGEN.

Silvipriya KS, KK Kumar, AR Bhat, BD Kumar, A John, and P Lakshmanan. 2015. "Collagen: Animal Sources and Biomedical Application." *Journal of Applied Pharmaceutical Science* 5 (3): 123–127. doi:10.7324/JAPS.2015.50322.

Simpson DG, GL Bowlin, GE Wnek, PJ Stevens, ME Carr, JA Matthews, and S Rajendran. 2003. Electroprocessed collagen and tissue engineering. *PCT Int. Appl.*, issued May 28, 2003. https://patents.google.com/patent/US20040037813A1/en.

Song E, SY Kim, T Chun, HJ Byun, and YM Lee. 2006. "Collagen Scaffolds Derived from a Marine Source and Their Biocompatibility." *Biomaterials* 27 (15): 2951–2961. doi:10.1016/j.biomaterials.2006.01.015.

Torres-Giner S, JV Gimeno-Alcañiz, MJ Ocio, and JM Lagaron. 2009. "Comparative Performance of Electrospun Collagen Nanofibers Cross-Linked by Means of Different Methods." *ACS Applied Materials and Interfaces* 1 (1): 218–223. doi:10.1021/am800063x.

Vasita R, and DS Katti. 2006. "Nanofibers and Their Applications in Tissue Engineering." *International Journal of Nanomedicine*. doi:10.2147/nano.2006.1.1.15.

Venkatraman V. 2017. Madras Triple Helix: The World Has Nearly Forgotten the Indian Scientist Who Cracked the Structure of Collagen—Quartz. https://qz.com/983439/madras-triple-helix-the-world-has-nearly-forgotten-the-indian-scientist-who-cracked-the-structure-of-collagen/.

Wang Y, PE Tomlins, AGA Coombes, and M Rides. 2010. "On the Determination of Darcy Permeability Coefficients for a Microporous Tissue Scaffold." *Tissue Engineering. Part C, Methods* 16 (2): 281–289. doi:10.1089/ten.tec.2009.0116.

Wess TJ, AP Hammersley, L Wess, and A Miller. 1998. "Molecular Packing of Type I Collagen in Tendon." *Journal of Molecular Biology* 275 (2): 255–267. doi:10.1006/JMBI.1997.1449.

Widdowson JP, AJ Picton, V Vince, CJ Wright, and A Mearns-Spragg. 2017 "*In vivo* Comparison of Jellyfish and Bovine Collagen Sponges as Prototype Medical Devices." *Journal of Biomedical Materials Research Part B: Applied Biomaterials* 106 (4): 1524–1533. doi:10.1002/jbm.b.33959.

Williams J. 2015. "Are Jellyfish Taking Over The World?" *Journal of Aquaculture & Marine Biology* 2 (3): 1–13. doi:10.15406/jamb.2015.02.00026.

Wyckoff RWG, RB Corey, and J Biscoe. 1935. "X-Ray Reflections of Long Spacing From Tendon." *Science* 82 (2121): 175–176. doi:10.1126/science.82.2121.175.

Xu Y, M Bhate, and B Brodsky. 2002. "Characterization of the Nucleation Step and Folding of a Collagen Triple-Helix Peptide." *Biochemistry* 41 (25): 8143–8151. doi:10.1021/bi015952b.

Zahedi P, I Rezaeian, SO Ranaei-Siadat, SH Jafari, and P Supaphol. 2010. "A Review on Wound Dressings with an Emphasis on Electrospun Nanofibrous Polymeric Bandages." *Polymers for Advanced Technologies*. doi:10.1002/pat.1625.

Zeugolis DI, ST Khew, ESY Yew, AK Ekaputra, YW Tong, LYL Yung, DW Hutmacher, C Sheppard, and M Raghunath. 2008a. "Electro-Spinning of Pure Collagen Nano-Fibres—Just an Expensive Way to Make Gelatin?" *Biomaterials* 29 (15): 2293–2305. doi:10.1016/j.biomaterials.2008.02.009.

Zeugolis DI, B Li, RR Lareu, CK Chan, and M Raghunath. 2008b. "Collagen Solubility Testing, a Quality Assurance Step for Reproducible Electro-Spun Nano-Fibre Fabrication. A Technical Note." *Journal of Biomaterials Science. Polymer Edition* 19 (10): 1307–1317. doi:10.1163/156856208786052344.

Zhang YZ, J Venugopal, ZM Huang, CT Lim, and S Ramakrishna. 2005. "Characterization of the Surface Biocompatibility of the Electrospun PCL-Collagen Nanofibers Using Fibroblasts." *Biomacromolecules* 6 (5): 2583–2589. doi:10.1021/bm050314k.

Zhong SP, YZ Zhang, and CT Lim. 2010. "Tissue Scaffolds for Skin Wound Healing and Dermal Reconstruction." *Wiley Interdisciplinary Reviews: Nanomedicine and Nanobiotechnology* 2 (5): 510–525. doi:10.1002/wnan.100.

Zhong S, WE Teo, X Zhu, RW Beuerman, S Ramakrishna, L Yue, and L Yung. 2006. "An Aligned Nanofibrous Collagen Scaffold by Electrospinning and Its Effects on In Vitro Fibroblast Culture." *Journal of Biomedical Materials Research Part A: An Official Journal of The Society for Biomaterials, The Japanese Society for Biomaterials, and The Australian Society for Biomaterials and the Korean Society for Biomaterials* 79 (3): 456–463.

Zhou T, N Wang, Y Xue, T Ding, X Liu, X Mo, and J Sun. 2016. "Electrospun Tilapia Collagen Nanofibers Accelerating Wound Healing via Inducing Keratinocytes Proliferation and Differentiation." *Colloids and Surfaces B: Biointerfaces* 143 (July): 415–422. doi:10.1016/J.COLSURFB.2016.03.052.

Williams L 2015. "Are Fish Turning [illegible] Wound Dressings [illegible]" [illegible] 7(5):1-[illegible]. doi:10.[illegible].
W[illegible] RW, RB [illegible] and [illegible] 1988. "[illegible] Long Spacing Fibrils [illegible] Structure [illegible]" [illegible] (12): 173-[illegible]. doi:10.1016/[illegible].
[illegible] Y, [illegible] and [illegible] 20[illegible]. "[illegible] from [illegible] Triple-Helix [illegible]" [illegible].
Zahedi [illegible] Rezaeian, [illegible] Ramakrishna [illegible]. [illegible] Wound Dressings [illegible].
[illegible] D[illegible], [illegible], [illegible] Y, [illegible] Zhang [illegible]. [illegible] of [illegible] Based [illegible] Wound [illegible] (11): [illegible].
[illegible] 2008. [illegible] Design [illegible] Nano-Fibro[illegible] [illegible] Journal of [illegible] Engineering [illegible].
Zhang YZ, [illegible] ZM Huang, CT Lim and [illegible] 2005. "Characterization of [illegible] Surface [illegible] of the [illegible] Collagen [illegible] Using [illegible]." [illegible] 6 (5): 2583-[illegible]. doi:10.1021/bm050314k.
Zhong, [illegible] Zhang and CT Lim. 2010. "[illegible] for Skin Wound Healing and [illegible] Tissue [illegible]." [illegible] Nanomedicine and Nanobiotechnology 2 [illegible] 510-525. doi:10.1002/wnan.100.
Zhong S, [illegible] Teo, [illegible] RW Beuerman, S Ramakrishna [illegible] Yung. 2006. "An Aligned Nanofibrous Collagen Scaffold by Electrospinning [illegible] In Vitro [illegible]." [illegible] Official Journal of the Society for Biomaterials, The Japanese Society for Biomaterials, and the Australian Society for Biomaterials and the Korean Society for Biomaterials [illegible].
Zhou T, N Wang, Y Xue, [illegible] Ding, X Liu, X Mo and J Sun. 2016. "[illegible] Collagen Nanofibers Accelerating Wound Healing [illegible] Proliferation and Differentiation." Colloids and Surfaces B: Biointerfaces [illegible]. doi:10.1016/j.colsurfb.2016.03.052.

5 Nanofiber Membranes for Oily Wastewater Treatment

Issa Sulaiman Al-Husaini, Abdull Rahim Mohd Yusof, Woei-Jye Lau, Ahmad Fauzi Ismail, and Mohammed Al-Abri

CONTENTS

5.1 INTRODUCTION

Industrially, oil and gas operations are among the sectors that produce large amounts of wastewater containing not only high concentrations of oil and grease but also many hazardous hydrocarbon mixtures and heavy metals. The presence of oil and grease in wastewater occurs in different forms such as free, dispersed or emulsified, which predominantly have different particle sizes. Typically, the oil droplets are dispersed extremely well as small droplets (<10 mm) and the conventional techniques such as gravity separation, centrifugation and air flotation are not effective enough in separating them from the wastewater. Typical oily wastewaters may contain between 50 and 1000 ppm (equivalent to mg/L) total oil and grease (TOG), depending on the type of application. The maximum content of TOG in discharge water however is limited to 5–40 ppm according to regulation, with a typical requirement of 10–15 ppm (Gohari et al., 2014; Ma et al., 2016).

With increasing environmental awareness and tighter regulations, it is necessary to develop new technologies and materials that are capable of addressing these challenges. Although various oil/water separation methods have already been reported in the literature and could offer great promise, most of them are associated with their own limitations. Thus, further development of cost-effective, environmentally friendly, and recyclable/reusable materials for effective oil/water separation is in high demand. Based on the statistics from the ISI Web of Science, a skyrocketing increase in the number of research articles related to oil/water separation has occurred between 2013 and 2016. There were close to 800 papers published for the period of 1998–2016 and more than half of the papers were recorded between 2013 and 2016.

Recent progress in fabricating electrospun nanofiber membranes with superwetting surfaces provides a new future to dramatically improve the membrane performance, in particular water permeability and antifouling resistance. Generally, the design of electrospun nanofiber membranes for oil/water separation requires the surfaces to be either superhydrophobic (low water affinity)-superoleophilic (high oil affinity) or superhydrophilic (high water affinity)-superoleophobic (low oil affinity) to achieve excellent oil-water separation efficiency (Gupta et al., 2017).

Considering the fast development of this area, this chapter will provide a review on the development of nanofiber membranes for oily wastewater treatment. Specifically, it will highlight the technique of electrospinning in fabricating nanofiber membranes with "oil-removing" or "water-removing" types of superwetting surfaces. A comprehensive review on the progress of nanofiber membranes incorporated with advanced materials (both organic and inorganic) for oil/water separation will also be provided. Finally, conclusions and perspectives concerning the future development of superwetting surface nanofiber membranes are provided.

5.2 ELECTROSPINNING PROCESS FOR NANOFIBER MEMBRANES PREPARATION

Although the first patent related to electrospinning was filed about one century ago, it did not gain substantial attention in industrial and academic settings until the activities of the Reneker's group in the 1990s (Reneker and Chun, 1996). After more than two decades of research and development, electrospun nanofibers have been found useful and practical for many applications. These include filtration, protective textiles, drug delivery, tissue engineering, electronic and photonic devices, sensor technology and catalysis (Pham et al., 2006; Homaeigohar, 2011; Choong, 2015; Ma et al., 2016). Up to now, the electrospinning method remains a promising and intriguing method for preparation of nanofibers with different diameters and morphologies (Anwar et al., 2013). The electrospinning apparatus for nanofiber fabrication is composed of three main components, i.e., a high voltage supplier (typically 10–40 kV), a polymer solution (usually contained in a syringe) and a metallic collecting plate (grounded collector).

Ultrafine fibers from micro to nanoscale can be produced with ease by forcing a polymer melt or solution through a spinneret with an electrical driving force. Polymer solutions are more commonly used because the spinning process can be done at room temperature and pressure. Electrospinning of polymer melts

TABLE 5.1
Influences of Electrospinning Parameters on the Nanofiber Morphology

Electrospinning Parameters	Effect on Nanofiber Morphology
High viscosity of polymer solution	Larger fibers diameter
Increase in surface tension	Higher number of beaded fibers and beads
Lower solution conductivity	Smaller fibers diameter
Higher applied voltage	Smaller fibers diameter followed by larger diameter
Greater distance between spinneret and collector	Smaller fiber diameter (beaded morphologies, however, occur if the distance between the capillary and collector is too short)
Higher humidity	Smaller fibers diameter followed by larger diameter
Increase in flow rate	Larger fibers diameter (beaded morphologies occur if the flow rate is too high)

typically requires high temperature and vacuum condition. The polymer solution contained in a syringe is pumped at a low flow rate to a tube connected to a spinneret. In this process, a high voltage is applied between two electrodes connected to the spinning solution/melt and to the collector to create an electrically charged jet of polymer solution from a syringe. The electric field is subjected to the tip of the needle containing a droplet of the polymer solution. The surface of the droplet is electrified by the electric field. Increase of the intensity of the electric field changes the hemispherical surface of the fluid at the tip of the needle to a conical shape known as the Taylor cone (Anwar et al., 2013). The Taylor cone spins down as a fiber to reach the collector plate, forming a fiber mat. It is possible to create nanofibers with different morphology, pore size and thickness by varying electrospinning parameters. Table 5.1 summarizes briefly the effect of these parameters on the properties of fibers.

5.3 SUPERWETTING MATERIALS FOR OIL/WATER SEPARATION

Fabrication of nanofiber membranes with selective superwetting properties can be achieved through manipulating chemical composition and/or surface morphological structure (Liang and Guo, 2013; Raza et al., 2014). In recent years, research into the role of wettability in oil/water separation has attracted more and more attention. Wettability is an intrinsic property of a solid surface that determines the wetting phenomenon when liquid comes into contact with the solid surface. Super-wetting surfaces with superhydrophobicity, superhydrophilicity, superoleophobicity or superoleophilicity can be designed and fabricated with appropriate surface structure and composition design. Materials with superhydrophobicity and superoleophilicity can selectively filter or absorb oil from oil/water mixtures, and are usually called "oil-removing" types of materials. Similarly, materials with the opposite wettability, superhydrophilicity and superoleophobicity are "water-removing" types of materials (Xue et al., 2014). The following two subsections will further describe the property differences of these two materials.

5.3.1 "Oil-Removing" Types of Superwetting Materials

The properties of this material, which can effectively absorb oil molecules from oil/water mixtures, can be attributed to its unique superhydrophobic and superoleophilic properties. As the wettability of a solid surface is determined by its chemical composition and morphology, the surface energy of the chosen material should be between those of oil (20–30 mN/m) and water (~72 mN/m). To achieve such properties, the material needs to be hydrophobic and oleophilic. The effective combination of wettability properties and size exclusion can create a material that repels water but allows oil to flow freely. Unlike ultrafiltration membranes, which are used in the separation of a small amount of emulsions, this material can separate oil from the mixture containing extremely high oil concentration. Such material was developed by Feng et al. (2004) for the first time in 2004 to coat a polytetrafluoroethylene (PTFE) layer on top of a stainless steel mesh support. PTFE was selected as it is both hydrophobic (water contact angle (WCA) of 100–110°) and oleophilic (hexadecane contact angle of 50°–60°). Once treated, the PTFE-coated mesh exhibited hierarchical micro/nanostructures and special wettability properties. Figure 5.1a shows water droplet forms on the surface, while diesel oil would spread and permeate through the mesh in only 240 ms. Figure 5.1b demonstrates the quick separation of diesel oil/water mixture by gravity (Feng et al., 2004).

5.3.2 "Water-Removing" Types of Superwetting Materials

Compared with the "oil-removing" types of super-wetting material, the "water-removing" type with superhydrophilic and superoleophobic properties, which can selectively separate water from oil/water mixtures, has some unique advantages and has been the focus of recent research. Typical superoleophobic surfaces possess an oil contact angle (OCA) larger than 150° and extraordinarily low

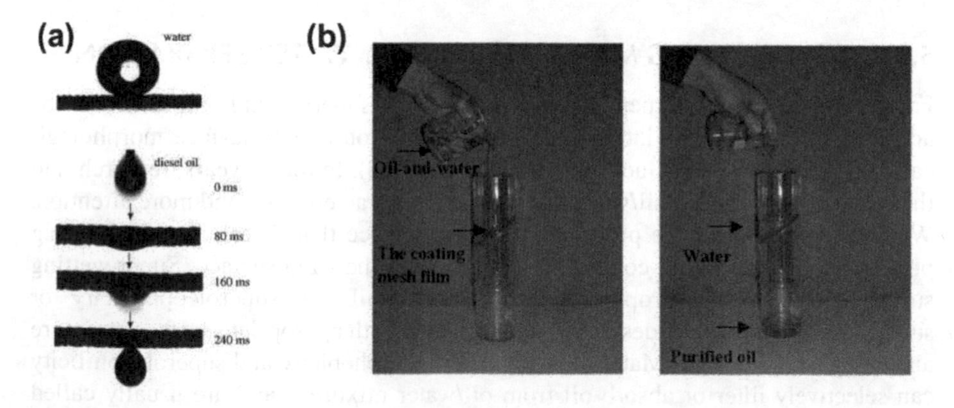

FIGURE 5.1 Oil/water separation using the PTFE-coated mesh film, (a) superhydrophobic and superoleophilic properties of the as-prepared mesh and (b) simple demonstration of oil/water separation by gravity force. (From Feng, L. et al., *Angewandte Chemie International Edition*, 43, 2012–2014, 2004; Xue, Z. et al., *Journal of Materials Chemistry A*, 2, 2445–2460, 2014.)

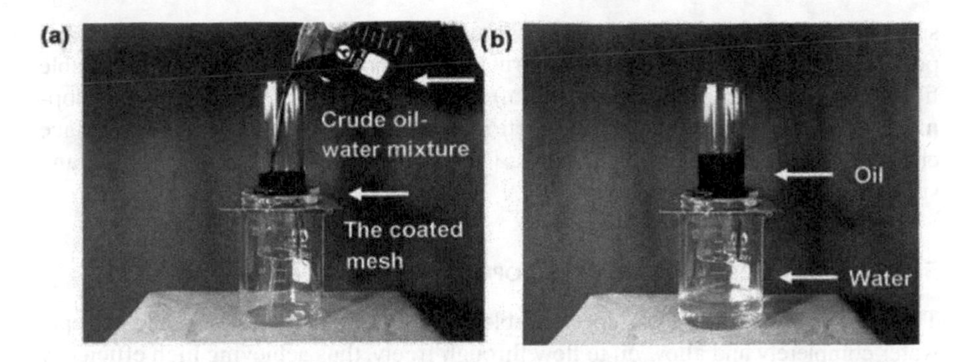

FIGURE 5.2 The use of "water-removing" type of polyacrylamide (PAM)-hydrogel coated mesh for oil/water separation, (a) before and (b) after crude oil/water mixture separation. (From Xue, Z. et al., *Adv. Mater.*, 23, 4270–4273, 2011.)

WCA typically less than 10°. With "water-removing" material, the superoleophobicity prohibits fouling by oils, and overcomes the limitations of "oil-removing" material. Durability, reuse and recycling are all improved. As the separation process is gravity-driven, it has been shown that "oil-removing" type filtration films are not suitable for most oils and organic solvents. It is because the higher-density water settles below the oil preventing oil permeation through the membrane. "Water–removing" materials do not have this problem. Their application is also better in uses such as fuel purification, oil retention barriers of industrial outlet sewer pipes, and in the separation of high viscosity oils. These benefits of "water-removing" materials have driven their development, however, when compared to materials with superhydrophobic and superoleophilic properties, these "water-removing" materials are much harder to create as the surface energy of the material has to be larger than that of water and, at the same time, smaller than that of oil (Xue et al., 2014; Wang et al., 2016).

Xue et al. (2011) developed this concept and created a novel superhydrophilic and underwater superoleophobic hydrogel-coated mesh in an oil/water/solid three-phase system. This developed material could effectively and selectively separate water from various oil/water mixtures (e.g., vegetable oil, gasoline, diesel, and crude oil/water) without any extra power. Figure 5.2a and b shows the oil/water separation using the coated mesh in which water could easily pass through the mesh while oil molecules are prevented from passing. In this case, water is trapped in this hierarchically rough structure and becomes a repulsive liquid phase for oil. As such, it prevents the penetration of oil by creating superoleophobic and low-adhesive surfaces in water.

5.4 NANOFIBER MEMBRANES FOR OIL/WATER SEPARATION

Compared to the microporous asymmetric membranes made by the phase inversion technique, the inherent properties of electrospun nanofiber membranes offer better performance with respect to separation efficiency and energy consumption. Furthermore, the electrospun nanofiber membranes exhibit significantly higher specific surface area to volume ratios compared to the asymmetric membranes and possess highly interconnected pore structures, nanoscale pore sizes, controllable fiber diameter (ranging from mm to nm) and chemistry. Given the rapid developments in this area, the following subsections are organized according to the surface chemistry of the nanofiber membranes, i.e., superhydrophobic/superoleophilic and superhydrophilic/superoleophobic.

5.4.1 Superhydrophobic/Superoleophilic Nanofiber Membranes

The "oil-removing" type of superwettable nanofiber membranes is aimed to repel water completely and allow oil to flow through freely, thus achieving high efficiency and selectivity. Fang et al. (2016) fabricated electrospun *N*-perfluorooctyl-substituted polyurethane (PU) membranes with self-healing ability for oil/water separation. The self-healing was realized by a low-surface-energy fluorine containing polymer migrating to the membrane surface. The controllable wettability was achieved by tuning membrane composition (composed of *N*-perfluorooctyl-substituted PU (PU-C_8F_{17}) and *N*-alkyl-substituted PU (PU-$C_{18}H_{37}$)), so that membranes could separate both immiscible oil/water mixtures and water-in-oil emulsions with high separation efficiency and high flux. The composite membranes also showed excellent durability against abrasion and maintained high permeate flux and high separation efficiency after self-healing cycles.

Zhu et al. (2011) fabricated a novel, high-capacity oil sorbent consisting of polyvinyl chloride (PVC)/polystyrene (PS) fibers by an electrospinning process. The sorption capacity, oil/water selectivity and sorption mechanism of the PVC/PS sorbent were studied. The results showed that the sorption capacities of the PVC/PS sorbent for motor oil, peanut oil, diesel and ethylene glycol were 146, 119, 38 and 81 g/g, respectively. It was about five to nine times higher than that of a commercial polypropylene (PP) sorbent. The PVC/PS sorbent also had an excellent oil/water selectivity (about 1,000 times) and high buoyancy in the clean-up of oily water. Microscopy indicated that voids among fibers were the key for the high capacity. To further improve the elasticity and recoverability of the PS-based fiber sorbents, Lin et al. (2012) added PU fiber into the PS fiber via multi-nozzle electrospinning. It was shown that the properties of the as-prepared fibers were significantly improved upon incorporation with the PU fibers, which enabled reuse of the fiber sorbents.

To enhance the mechanical properties and practical applications of nanofiber sorbents, an alternative approach—core–shell electrospinning was proposed by Lin et al. (2013). The authors reported the composite PS/PU fibers with a high specific surface area of 19.57 m^2/g could be prepared via co-axial electrospinning for potential use as an oil sorbent (Figure 5.3a). The composite fiber mats also showed good oil sorption capacities for motor oil (64.40 g/g) and sunflower seed oil (47.48 g/g). With respect to reusability, the composite fibers maintained a high oil sorption capacity even after five sorption cycles (Figure 5.3b).

Using PDMS-block-poly(4-vinylpyridine) (PDMS-b-P4VP), Li et al. (2016) developed nanofiber membranes that could show excellent pH switchability for oil/water separation. As shown in Figure 5.4a, the WCA of the nanofiber membrane was recorded at 155° at pH 7. However, when the water pH was reduced to 4, the water

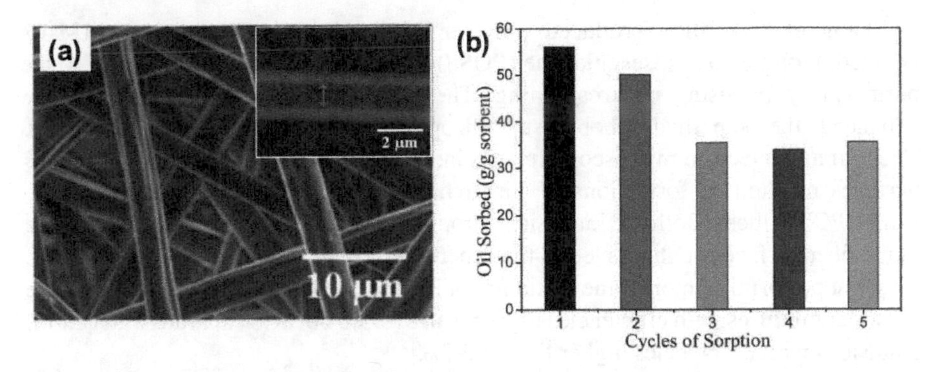

FIGURE 5.3 (a) FESEM and TEM (inset) images of composite PS/PU fibers prepared via co-axial electrospinning and (b) Reusability of the composite PS/PU fibers for motor oil. (From Lin, J. et al., Nanoscale, 5, 2745–2755, 2013.)

droplet gradually wetted and spread over the membrane surface, showing an improved hydrophilic nature (Figure 5.4b). Figure 5.4c further compares the switchability of the membranes for up to five cycles by varying the water pH between 7 and 4. When an oil (n-hexane) droplet was applied on the fiber mat, it was absorbed as shown in Figure 5.4d. The authors also demonstrated effective oil/water separation from the layered mixtures by varying the pH of the solution, driven by gravity. The PDMS-b-P4VP nanofiber membranes showed controlled separations for oil (n-hexane) and water with high fluxes of approximately 9,000 and 27,000 L/m^2·h, respectively.

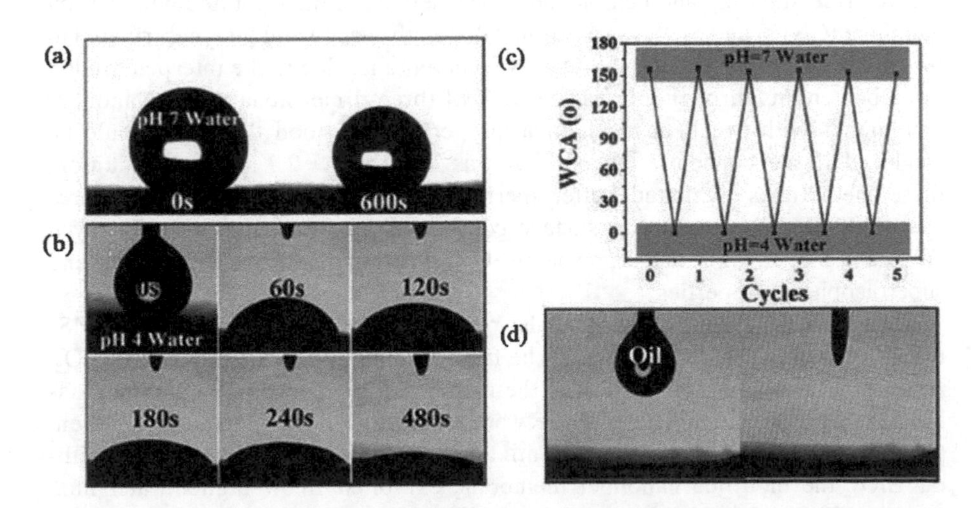

FIGURE 5.4 CA measurements of PDMS-b-P4VP nanofiber membranes in air, (a) images of a neutral water droplet on the mat surface at 0 s (left) and 600 s (right), (b) dynamic images of a pH 4 water droplet spreading over the mat within 480 s, (c) pH-switchable water wettability of the membrane between superhydrophobicity and superhydrophilicity and (d) absorption of n-hexane droplet as soon as it contacts the membrane surface. (From Li, J. et al., *Langmuir*, 32, 13358–13366, 2016.)

Zhang et al. (2015) produced polymer of intrinsic microporosity (PIM-1)/ polyhedral oligomeric silsesquioxane (POSS) fiber membranes with complete water permeability by using electrospinning. The addition of POSS particles greatly enhanced the superhydrophobic–superoleophilic characteristics of membrane. The scanning electron microscopy images indicated that the incorporation of POSS particles resulted in formations of hierarchical structures on the surface of the PIM-1/POSS fibers. Both the intrinsic hydrophobic nature of POSS and the increase in the fiber surface roughness led to the superhydrophobicity and superoleophilicity. The best performing membrane could not only separate a wide range of immiscible oil/water mixtures with efficiencies higher than 99.95% but also separate water-in-oil emulsions with efficiencies higher than 99.97%.

With the advancement of nanoscience and nanotechnology over the past decade, the use of nanomaterials (especially inorganic nanoparticles) has become a hot topic in the research of nanofiber membrane development. Unlike conventional polymeric membranes, nanomaterial-incorporated membranes showed better performance with respect to flux and high rejection due to the improved surface characteristics (Xue et al., 2014). Celik et al. (2011) also found that the presence of finely dispersed inorganic nanoparticles in the polymeric membrane was very useful in the improvement of membrane antifouling performance.

A variety of metal oxide nanoparticles have been previously used to blend with host polymers to enhance their performances. These include magnesium oxide (MgO), iron oxide (Fe_3O_4), aluminum oxide (Al_2O_3), titanium dioxide (TiO_2), silica dioxide (SiO_2), zinc oxide (ZnO), etc. A simple and commonly used approach is to directly blend nanoparticles with polymeric solution. Tai et al. (2014) fabricated a novel, free-standing and flexible nanofiber membrane incorporated with carbon nanofiber (CNFs) and SiO_2. It was argued that both carbon and SiO_2 contributed to the improved properties of the SiO_2-CNFs membranes due to the interpenetration of carbon chains through a linear network of three-dimensional SiO_2. Nanofiber membranes with excellent mechanical properties and good flexibility could be produced if the concentration of SiO_2 was kept below 2.7 wt.%. In addition, these membranes exhibited better thermal stability at elevated temperatures up to 300°C. With respect to surface contact angle, the membranes displayed WCA and OCA of 144° and 0°, respectively, indicating their ultrahydrophobic and superoleophilic properties.

Li et al. (2016) on the other hand also reported that the performance of PDMS-b-P4VP nanofiber membranes could be further improved by incorporating SiO_2. As demonstrated, the properties of the membrane with respect to thermal stability and pH switchability response were enhanced, along with their excellent performance in separating oil/water mixtures. Compared to the membrane without SiO_2, the modified nanofiber membrane exhibited 18.5% higher water flux, reaching 32,000 $L/m^2 \cdot h$ while maintaining oil flux at 9,000 $L/m^2 \cdot h$. Shang et al. (2012) modified the surface of nanofiber membranes via an *in-situ* polymerization technique as illustrated in Figure 5.5. With the presence of polybenzoxazine (F-PBZ) functional layer that incorporated SiO_2 on the outer surface of fiber, it was reported that the resultant membrane could achieve superhydrophobicity with WCA of ~161° and superoleophilicity with OCA of <38°. When tested with

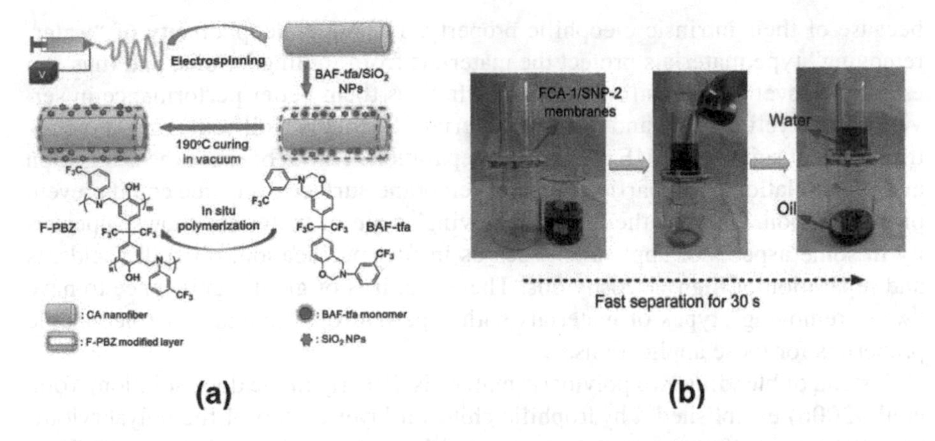

FIGURE 5.5 (a) Schematic for the strategy using *in-situ* polymerization approach to synthesize of F-PBZ/SiO_2-modified CA nanofiber membranes and (b) experimental separation process of oil/water mixture. (From Shang, Y. et al., *Nanoscale*, 4, 7847–7854, 2012.)

oil/water mixtures, the modified membranes exhibited fast and efficient separation solely by a gravity-driven process. Oil quickly permeated through the F-PBZ/SiO_2-modified cellulose acetate (CA) nanofiber membranes while water was retained above the membranes.

Jiang et al. (2015) evaluated the effect of iron oxide ((Fe_3O_4) nanoparticles on the properties of nanofiber membrane composed of polyvinylidene fluoride (PVDF)/polystyrene (PS) and the findings showed that the nanomaterials-incorporated membranes had good oil sorption capacity coupled with improved mechanical properties. The addition of Fe_3O_4 nanoparticles provided magnetic properties that help for easier pick-up. Oil sorption test results showed adequate sorption capacity of 35–45 g/g for four different types of oils and these values were higher than that of conventional polypropylene (PP) fibers.

Li et al. (2014) fabricated superhydrophobic and superoleophilic electrospun nanofibrous membranes by a facile route combining the amination of electrospun polyacrylonitrile (APAN) nanofibers and immobilization of silver (Ag) nanoclusters with an electroless plating technique followed by n-hexadecyl mercaptan surface modification. By introducing the hierarchically rough structures and low surface energy, the pristine superhydrophilic APAN membranes could be endowed with a superhydrophobic (WCA: 171.1°) and a superoleophilic surface (OCA: 0°). Furthermore, the membranes displayed self-cleaning surface properties arising from the extremely low WCA hysteresis (3.0°) and a low water-adhesion property. More importantly, the extremely high liquid entry pressure of water (175 kPa) and the robust fiber morphology of the APAN membranes immobilized with Ag nanoclusters could achieve excellent separation and stability for oil/water separation.

5.4.2 Superhydrophilic/Superoleophobic Nanofiber Membranes

Compared with the "oil-removing" type, this kind of material has some unique advantages. Firstly, the "oil-removing" type of materials is easily fouled by oils because of their intrinsic oleophilic property. The superoleophobicity of "water-removing" type materials protect the materials from fouling by oils, and thus can essentially overcome this limitation, which gives them better performance in service life, recycling of oil and reuse of materials. Secondly, "oil-removing" type filtration is not suitable for UF of oil/water separation. This is because the absorption and accumulation of oil particles on the membrane surface over time could prevent oil permeation. Thirdly, the "water-removing" type of materials shows superiority in some aspects of application such as in fuel purification, oil spill accidents and separation of high viscosity oils. Therefore, it is of great significance to have "water-removing" types of materials with superhydrophilic and superoleophobic properties for these applications.

Instead of blending two polymeric materials directly in the dope solution, Yoon et al. (2006) established a hydrophilic chitosan layer on top of the polyacrylonitrile (PAN) nanofiber membrane supported by a non-woven substrate. As illustrated in Figure 5.6a and b, a PAN nanofiber was first placed onto a non-woven poly(ethylene terephathalate) (PET) substrate followed by forming a hydrophilic layer onto the PAN via spin-casting to form a three-tier nanofiber membrane. Because of the high porosity, thin and smooth barrier layer of the membrane, outstanding performances including high flux were reported as shown in Figure 5.6c.

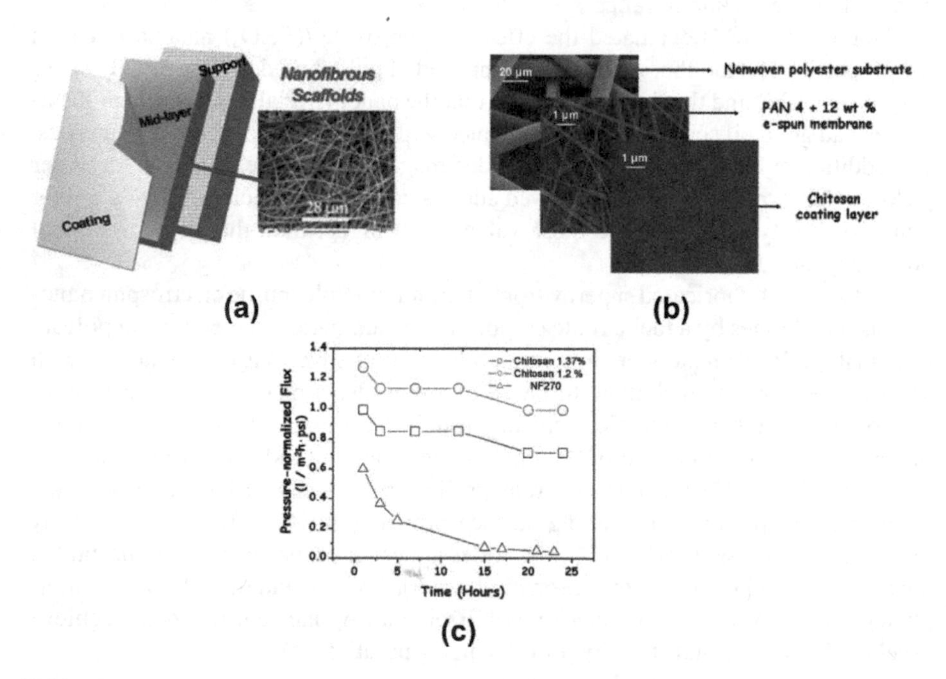

FIGURE 5.6 (a) Schematic structure of nanofiber membrane composed of three layers, (b) SEM surface image of each layer and (c) normalized flux of three-layer membranes (coated with 1.37% and 1.2% chitosan) and commercial membrane (NF270). (From Yoon, K. et al., *Polymer*, 47, 2434–2441, 2006.)

During the prolonged filtration experiment using an oil-in-water emulsion as the feed, the water flux of chitosan coated membranes was maintained at a much higher value than the commercial membrane with no obvious loss of oil rejection.

Organic materials such as fluorinated materials could also be utilized to improve superoleophobic properties of nanofiber membrane for oil/water separation. Ahmed et al. (2014) impregnated cellulose ionic liquid into a nanofiber membrane to produce cellulose/polyvinylidene fluoride-co-hexafluoropropylene (PVDF–HFP) nanofiber membranes that displayed superhydrophilicity (WCA: 0°) and superoleophobicity (OCA: 169°). The modified nanofiber membranes also showed smaller pore size and narrower pore size distribution coupled with increased mechanical properties in comparison to the unmodified nanofiber membranes as shown in Figure 5.7.

Raza and coworkers (2013) demonstrated a facile approach for fabricating superhydrophilic and prewetted oleophobic nanofiber membranes that allowed effective separation of both water rich immiscible and monodispersed oil/water mixtures with extremely high separation efficiency and separation capacity. The *in-situ* crosslinked polyacrylonitrile/polyethylene glycol nanofibrous (x-PEGDA@PG NF) membranes have shown superhydrophilicity with ultralow time of wetting and promising oleophobicity (OCA: 105°). Solely driven by gravity, the x-PEGDA@PG membranes with mean pore size of 1.5–2.6 μm possessed very high water flux of 10,975 L/m^2·h. More importantly, the membranes exhibited good separation efficiency (residual oil content in the filtrate was 26 ppm) as well as high separation capacity that could separate 10 L of an oil/water mixture continuously without experiencing flux decline. This made the membranes an important candidate for treating real emulsified wastewater on a mass scale.

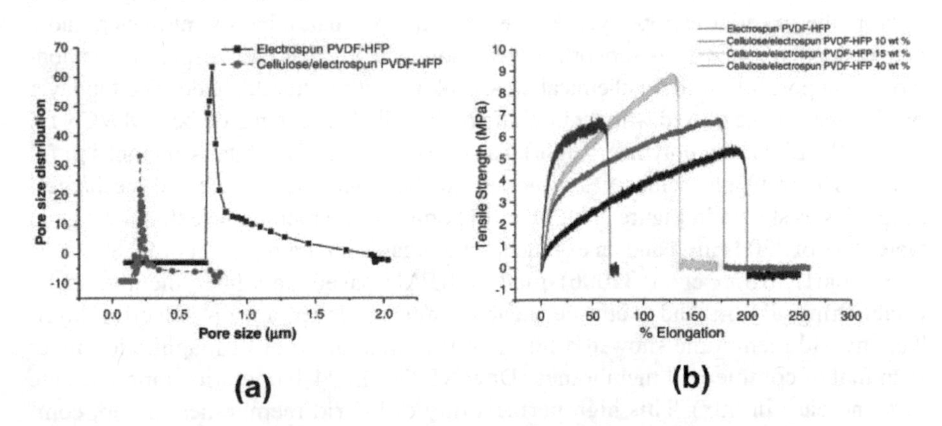

FIGURE 5.7 Comparison between the PVDF-HFP nanofiber membranes with and without cellulose ionic liquid modification, (a) pore size distribution and (b) tensile strength. (From Ahmed, F.E. et al., *Desalination*, 344, 48–54, 2014.)

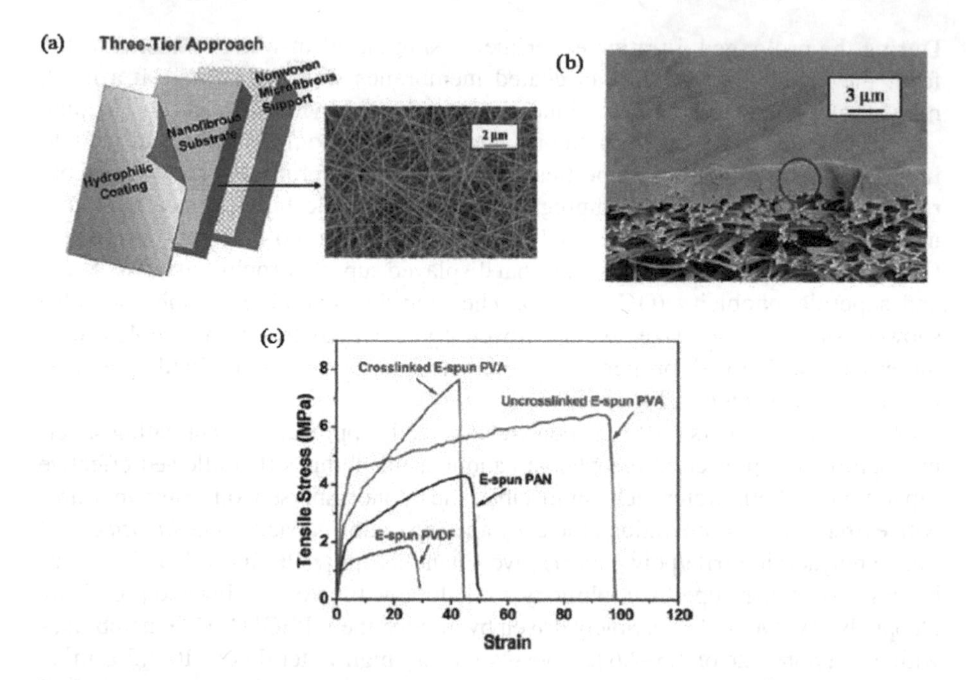

FIGURE 5.8 (a) Illustration of the structure of the 3-tier nanofiber membrane (left) and representative SEM image of electrospun PVA mid-layer (right), (b) cross-sectional structure of top selective layer and (c) tensile stress and strain curve of cross-linked PVA nanofiber. (From Wang, X. et al., *Environ. Sci. Technol.*, 39, 7684–7691, 2005.)

A novel three-tier arrangement of composite nanofiber membranes was also reported by Wang and his coauthors (2005). As illustrated in Figure 5.8a and b, the three-tier composite structure consisted of a nonporous hydrophilic polyether-b-polyamide hydrophilic material as top layer, an electrospun PVA nanofiber as mid-layer, and a conventional nonwoven as support. A water soluble PVA was electrospun on the nonwoven support followed by chemical cross-linking with glutaraldehyde. The top layer of PVA membrane coated with hydrophilic multiwalled carbon nanotubes (MWCNTs) (up to 10 wt.% of the polymer weight) was tested for selective water–oil emulsion filtration. The resultant membranes exhibited excellent water resistance and mechanical properties as shown in Figure 5.8c. More importantly, the membrane exhibited a good water flux of 330 $L/m^2 \cdot h$ and an excellent total organic solute rejection of 99.8%.

Similarly, Yoon et al. (2006) modified PAN-based nanofiber membrane by establishing a thin and water-permeable chitosan layer as top selective layer. This hybrid membrane showed a flux rate that was an order of magnitude higher than that of commercial membranes (Dow NF270) in 24-h operation (approximate 30% increase in flux). This high permeability of hybrid membranes did not compromise the separation rate as >99.9% oil rejection was still recorded.

Recent progress in fabricating superhydrophilic and superoleophobic filtration membranes with innovative structures composed of nanomaterials provides a new future to significantly improve nanofiber membrane performance, especially the flux.

With their small diameter, a membrane composed of nanoparticles can possess both an adjustable effective pore size and surface chemistry. Compared with polymer-based membranes, inorganic nanofiber membranes display potential advantages in terms of easy cycling for long-term use, stability under severe conditions and anti-fouling ability.

Yang et al. (2014) designed a novel, flexible, thermally stable and hierarchical porous silica nanofiber membrane (SNM), aiming to separate water from surfactant-stabilized oil-in-water emulsions solely driven by gravity. As shown in Figure 5.9, the pristine electrospun SNM was used as the template for the *in-situ* polymerization and the incorporation of SiO_2. The developed membranes showed superamphiphilicity in air; however, the oleophobicity appeared immediately after being immersed in water, suggesting an ideal membrane candidate for oil/water separation.

In addition to SiO_2, Tai et al. (2015) prepared nanofiber membranes using TiO_2 nanosheets. Microscopy showed that a double diameter of nanofiber could be achieved and roughness of the surface was covered by the TiO_2 nanosheets on the nanofiber membrane after TiO_2 coating. The hierarchical TiO_2 micro/nanostructure that grows on the surface of carbon nanofibers renders the membrane superhydrophilic and when underwater superoleophobic. Coupled with the characteristic of high porosity and micron-scale pore size, the membrane is capable of separating oil from water by gravity. Without compromising the separation efficiency, a permeate flux of 400–700 $L/m^2{\cdot}h$ can be achieved. Good membrane stability was also demonstrated in ultrasonic, thermal and extreme pH conditions. More importantly, such newly developed membrane displayed excellent membrane flexibility for the purpose of oil–water separation.

You et al. (2013) fabricated a new class of membrane by incorporating MWCNTs into the nanofiber membrane. The MWCNTs were introduced onto the top surface of PAN nanofibers by electrospinning in the presence of PVA. The resultant membrane was further post-treated by water/acetone solution

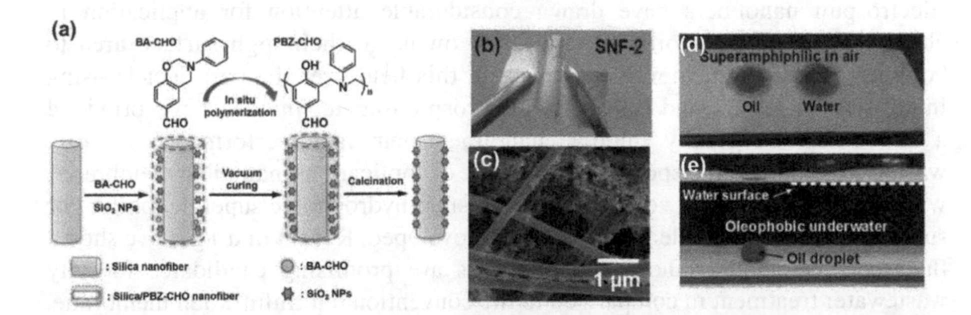

FIGURE 5.9 (a) Schematic illustration of the synthesis of hierarchical SiO_2 incorporated SNMs (SiO_2 at SNMs) by combining the nanofibers with *in-situ* polymerization, (b) optical photograph shows the robust flexibility of SiO_2 at SNMs, (c) FESEM image of SiO_2 at SNMs, (d) droplets of oil and water on the SNMs in air and (e) photograph of an underwater oil droplet on the SNMs. (From Yang, S. et al., Nanoscale, 6, 12445–12449, 2014.)

FIGURE 5.10 Schematic illustration of the fabrication process for PVA–MWCNTs/PAN nanofiber membranes. (From You, H. et al., *Sep. Purif. Technol.*, 108, 143–151, 2013.)

followed by cross-linking using glutaraldehyde to create an integrated barrier film as shown in Figure 5.10. Filtration evaluation confirmed that the incorporation of MWCNTs into the PVA barrier layer could improve the membrane water flux significantly as a result of the existence of more effective water channels. Upon incorporation of 10 wt.% MWCNTs, the resultant membrane showed very good water flux (270 $L/m^2{\cdot}h$) with high rejection rate (99.5%) even at very low feeding pressure (0.1 MPa).

5.5 CONCLUSION AND PERSPECTIVES

Electrospun nanofibers have drawn considerable attention for application in absorption/filtration of oil/water mixtures owing to their high surface area to volume ratio. Great achievements made in this field over the past decade using innovative materials and enhanced electrospinning techniques have provided a new future to greatly improve nanofiber membrane performance for oily wastewater treatment. Depending on the type of applications, nanofiber membranes with different surface characteristics (superhydrophobic/superoleophilic or superhydrophilic/superoleophobic) can be developed. Recent findings have shown that the electrospun nanofiber membranes are promising candidates for oily wastewater treatment in comparison to the conventional ultrafiltration membrane, particularly those incorporating functional nanomaterials. These kinds of nanofiber membranes exhibited not only extremely high water flux/sorption capacity but also achieved substantial degrees of reuse. More research, however, is required in the future, especially on the production scale-up of nanofiber membranes for industrial processes with significantly reduced manufacturing cost.

REFERENCES

Ahmed, F.E., Lalia, B.S., Hilal, N., and Hashaikeh, R. 2014. Underwater superoleophobic cellulose/electrospun PVDF–HFP membranes for efficient oil/water separation. *Desalination. 344*: 48–54.

Anwar, S., Nabeela, A., Nasreen, S., and Sundarrajan, S. 2013. Advancement in electrospun nanofibrous membranes modification and their application in water treatment. *Membranes. 3*: 266–284.

Celik, E., Liu, L., and Choi, H. 2011. Protein fouling behavior of carbon nanotube/polyethersulfone composite membranes during water filtration. *Water Research. 45*: 5287–5294.

Choong, L.T.S. 2015. Application of electrospun fiber membranes in water purification. PhD thesis. Massachusetts Institute of Technology.

Fang, W., Liu, L., Dang, T., Qiao, Z., Xu, C., Wang, Y., Li, T. 2016. Electrospun N-substituted polyurethane membranes with self-healing ability for self-cleaning and oil/water separation. *Chemistry European Journal Communication. 22*: 878–883.

Feng, L., Zhang, Z., Mai, Z., Ma, Y., Liu, B., Jiang, L., and Zhu, D.A. 2004. Super hydrophobic and super-oleophilic coating mesh film for the separation of oil and water. *Angewandte Chemie International Edition. 43*: 2012–2014.

Gohari, R., Halakoo, E., Lau, W.J., Kassim, M.A., Matsuura, T., and Ismail, A.F. 2014. Novel polyethersulfone (PES)/hydrous manganese dioxide (HMO) mixed matrix membranes with improved anti-fouling properties for oily wastewater treatment process. *RSC Advances. 4*: 17587–17596.

Gupta, R.K., Dunderdale, G.J., Englandd, M.W., and Hozumi, A. 2017. Oil/water separation techniques: A review of recent progresses and future directions. *Journal of Materials Chemistry A.5*: 16025–16058.

Homaeigohar, S.S. 2011. Functional Electrospun Nanofibrous Membranes for water filtration. Doctoral Dissertation, Christian-Albrechts Universität Kiel, Germany.

Li, J., Zhou, Y., Jiang, Z., and Luo, Z. 2016. Electrospun fibrous mat with pH-switchable superwettability that can separate layered oil/water mixtures. *Langmuir. 32*: 13358–13366.

Li, X., Wang, M., Wang, C., Cheng, C., and Wang, X. 2014. Facile immobilization of Ag nanocluster on nanofibrous membrane for oil/water separation. *ACS Applied Materials and Interfaces. 6*(17): 15272–15282.

Liang, W., and Guo, Z. 2013. Stable superhydrophobic and superoleophilic soft porous materials for oil/water separation. *RSC Advances.3*: 16469–16474.

Lin, J., Shang, Y., Ding, B., Yang, J., Yu, J., and Al-Deyab, S.S. 2012. Nanoporous polystyrene fibers for oil spill cleanup. *Marine Pollution Bulletin*. 64: 347–352.

Lin, J., Tian, F., Shang, Y., Wang, F., Ding, B., Yu, J., and Guob, Z. 2013. Co-axial electrospun polystyrene/polyurethane fibres for oil collection from water surface. *Nanoscale. 5*: 2745–2755.

Ma, W., Zhang, Q., Hua, D., Xiong, R., Zhao, J., Rao, W., Huang, S., Zhan, X., Chen, F., and Huang, C. 2016. Electrospun fibers for oil-water separation. *RSC Advances. 6*: 12868–12884.

Pham, Q.P., Sharma, U., and Mikos, A. G. 2006. Electrospinning of polymeric nanofibers for tissue engineering applications: A review. *Tissue Engineering. 12*: 1197–1211.

Raza, A., Ding, B., Zainab, G, El-Newehy, M., Al-Deyab, S.S., and Yu, J. 2013. In situ cross-linked superwetting nanofibrous membranes for ultrafast oil/water separation. *Journal of Materials Chemistry A. 0*: 1–3.

Raza, A., Ding, B., Zainab, G., El-Newehy, M., Al-Deyab, S.S., and Yu, J. 2014. In situ cross-linked superwetting nanofibrous membranes for ultrafast oil–water separation. *Journal of Materials Chemistry A*. 2: 10137–10145.

Reneker, D.H., and Chun, I. 1996. Nanometre diameter fibres of polymer, produced by electrospinning. *Nanotechnology. 7*: 216–223.

Shang, Y., Si, Y., Raza, A., Yang, L., Mao, X., Ding, B., and Yu, J. 2012. An in situ polymerization approach for the synthesis of superhydrophobic and superoleophilic nanofibrous membranes for oil-water separation. *Nanoscale. 4*: 7847–7854.

Tai, M.H., Gao, P., Tan, B.Y., Sun, D.D., and Leckie, J.O. 2015. A hierarchical nanostructured TiO2-carbon nanofibrous membrane for concurrent gravity-driven oil-water separation. *Journal of Environmental Science and Development. 6*: 590–595.

Tai, M.H., Gao, P., Tan, B.Y.L., Sun, D.D., and Leckie, J. O. 2014. Highly efficient and flexible electrospun carbon–silica nanofibrous membrane for ultrafast gravity-driven oil–water separation. *ACS Applied Materials and Interfaces. 6*(12): 9393–9401.

Wang, X., Yu, J., Sun, G., and Ding, B. 2016. Electrospun nanofibrous materials: A versatile medium for effective oil/water separation. *Materials Today. 19*(7): 403–414.

Wang, X. Chen, X., Yoon, K., Fang, D., Hsiao, B.S., and Chu, B. 2005. High flux filtration medium based on nanofibrous substrate with hydrophilic nanocomposite coating. *Environmental Science and Technology. 39*: 7684–7691.

Xue, Z., Cao, Y., Liu, N., Feng, L., and Jiang, L. 2014. Special wettable materials for oil/water separation. *Journal of Materials Chemistry A. 2*: 2445–2460.

Xue, Z., Wang, S., Lin, L., Chen, L., Liu, M., Feng, L., and Jiang, L. 2011. A novel superhydrophilic and underwater superoleophobic hydrogel-coated mesh for oil/water separation. *Advanced Materials. 23*: 4270–4273.

Yang, S., Si, Y., Fu, Q., Hong, F. Yu, J., S. Al-Deyab, S.S., El-Newehy, M., and Ding, B. 2014. Superwetting hierarchical porous silica nanofibrous membranes for oil/water microemulsion separation. *Nanoscale. 6*: 12445–12449.

Yoon, K. Kim, K., Wang, X., Fang, D., Hsiao, B.S., and Chu, B. 2006. High flux ultrafiltration membranes based on electrospun nanofibrous PAN scaffolds and chitosan coating. *Polymer. 47*: 2434–2441.

You, H., Li, X., Yang, Y., Wang, B., Li, Z., Wang, X., Zhu, M., and Hsiao, B.S. 2013. High flux low pressure thin film nanocomposite ultrafiltration membranes based on nanofibrous substrates. *Separation and Purification Technology. 108*: 143–151.

Zhang, C., Li, P., and Cao, B. 2015. Electrospun micro-fibrous membranes based on PIM-1/POSS with high oil wettability for separation of oil–water mixtures and cleanup of oil soluble Contaminants. *Industrial & Engineering Chemistry Research. 54*: 8772–8781.

Zhu, H.T., Qiu, S.S., Jiang, W., Wu, D.X., and Zhang, C.Y. 2011. Evaluation of electrospun polyvinyl chloride/polystyrene fibers as sorbent materials for oil spill cleanup. *Environmental Science and Technology. 45*: 4527–4531.

6 Electrospun Nanofibrous Membranes for Membrane Distillation Process

Mohammad Mahdi A. Shirazi and Morteza Asghari

CONTENTS

6.1 INTRODUCTION

Fresh water scarcity is among the major concerns faced by mankind in recent years. The growing population and economic expansion are the main factors affecting the availability of fresh water resources (Sun et al., 2017; Zhang et al., 2018). Changes in production and consumption patterns as well as new marketing strategies have led to an increase in fresh water requirements. Urban population is expected to grow from approximately 2522 million in 1950 to 8909 million in 2050 (Naseri Rad et al., 2016). Forecasts to 2030 indicate an increase in global water uptake at about 40% of the current accessible and reliable supplying sources, meaning an intensification of water consumption (Shirazi and Kargari, 2015).

All these issues have led to a critical and global alarm for the availability of fresh water resources. Therefore, the need for technological innovation to enable novel water management cannot be ignored. A solution for this global challenge is *Desalination*. Desalination is a process for producing fresh water from a saline source such as seawater (Gillam and McCoy, 1967). Seawater is desalinated to produce water suitable for human consumption, irrigation or animal watering. It is worth noting that the modern interest in desalination is mostly focused on cost-effective and highly efficient provision of fresh water for human use (Matsuura, 2001; Fane, 2018). Nanotechnology holds great potential in advancing desalination to improve the efficiency of salt and impurities removal to supply water via safe use of unconventional water resources (Qu et al., 2013; Lu and Astruc, 2018). In this regard, electrospinning technology is an emerging and versatile technique for fabricating micro- and nanofibers, which can be used for desalination and water treatment purposes. This chapter will discuss the application of electrospun nanofibers for desalination and water treatment using membrane distillation processes, by examining the fundamental concepts of this technology, its operating principles and the future perspectives.

6.2 MEMBRANE DISTILLATION (MD) PROCESS

6.2.1 Conventional Desalination Processes

There are different processes for seawater desalination; however, desalination techniques can generally be divided into two main categories: thermal-based desalinations and membrane-based desalinations. The major thermally driven desalinations include multi-effect distillation, multi-stage flash and vapor compression (Brogioli et al., 2018; Zhang et al., 2018). The main membrane-based desalinations include reverse osmosis (RO) and nanofiltration (NF) (Khulbe and Matsuura, 2017; Anand et al., 2018). Both methods are mature and a number of plants either based on distillation or membranes are producing fresh water around the world, especially in the Middle East (Nair and Kumar, 2013).

There are however some important limitations and challenges for these desalination processes. Thermal desalinations are energy sensitive. Thus, due to the price fluctuation of fossil fuels, operation and maintenance expenses can change. Moreover, due to the large equipment size, a large amount of steel is required, which dramatically increases the overall expenses as well as higher corrosion risks. In the case of membrane-based desalinations, the most important challenges are membrane fouling and osmotic-pressure limitations of RO/NF membranes, which consequently increase the input pressure requirement. These limitations have encouraged researcher to look for an alternative for seawater desalination (Ghaffour et al., 2013, 2015; Kaplan et al., 2017).

6.2.2 MD Process

Recently, a new membrane process has been introduced which is the combination of conventional distillation and membrane separation, thus it is called *membrane-distillation* (MD) (Smolders and Franken, 1989).

MD is a non-isothermal separation process, in which only vapor molecules are able to pass through a porous hydrophobic membrane (Alkhudhiri et al., 2012). The driving force of MD is the vapor pressure difference existing between the porous hydrophobic membrane surfaces, i.e., feed channel and permeate channel. MD has a number of attractive advantages including low operating temperatures (40°C–80°C) in comparison to those encountered in conventional distillations, and the solution, which is mainly water in desalination, is not necessarily heated up to the boiling point. Moreover, the hydrostatic pressure encountered in MD is much lower than that used in RO and NF processes (Shirazi and Kargari, 2015). Therefore, MD is expected to be an efficient and cost-effective process, which is also less demanding of membrane characteristics (Bahmanyar et al., 2012). MD has a high rejection factor, theoretically, complete non-volatile solute rejection takes place, as only vapor molecules can pass through the membrane pores. (Lovineh et al., 2013). Moreover, the pore size for an MD membrane is in the range of microfiltration membranes (0.1–0.5 μm). The MD process, therefore, suffers less from fouling, but scaling is still a challenge, especially seawater desalination (Warsinger et al., 2015). MD has the ability to utilize waste thermal energy sources and/or solar energy due to its low operating temperature. Therefore, MD can be investigated as a competitor of RO for desalination of brackish water and seawater (Khayet, 2013).

MD, however, has some drawbacks including low permeate flux compared to RO, high susceptibility of the permeate flux to concentration and temperature polarization, and mass transfer resistance due to trapped air within membrane pores, which also limits MD permeate flux. In addition, the heat lost by membrane heat conduction is another highlighted challenge of MD processes (Martinez-Diez and Vazquez-Gonzalez, 1999; Phattaranawik et al., 2003; Alklaibi and Lior, 2005).

In MD, the hot stream, the liquid feed to be treated, must be in direct contact with the active layer of the membrane and does not penetrate inside the dry pores of the membrane. The hydrophobic nature of the membrane prevents process liquid from entering its pores due to the surface tension forces. As a result, liquid/vapor interfaces are formed at the entrance of the membrane pores (El-Bourawi et al., 2006). The driving force of the MD process is the vapor pressure difference that can be maintained with one of the four following configurations: direct contact membrane distillation (DCMD) where a pure water stream colder than the feed solution is maintained in direct contact with the permeate side of the membrane (Long et al., 2018); sweeping gas membrane distillation (SGMD) where a cold inert gas, usually dried air, sweeps out water vapor molecules in the permeate side of the membrane and the condensation takes place outside the MD module (Moore et al., 2018); air-gap membrane distillation (AGMD) where a stagnant air gap is interposed between the membrane and a cold condensation surface inside the MD module (Ozbey-Unal et al., 2018); and vacuum membrane distillation (VMD) where vacuum pressure is applied in the permeate side of the MD module by means of a vacuum pump, the applied vacuum pressure is lower than the saturation pressure of volatile molecules to be separated from the feed solution, in this case, condensation also occurs outside of

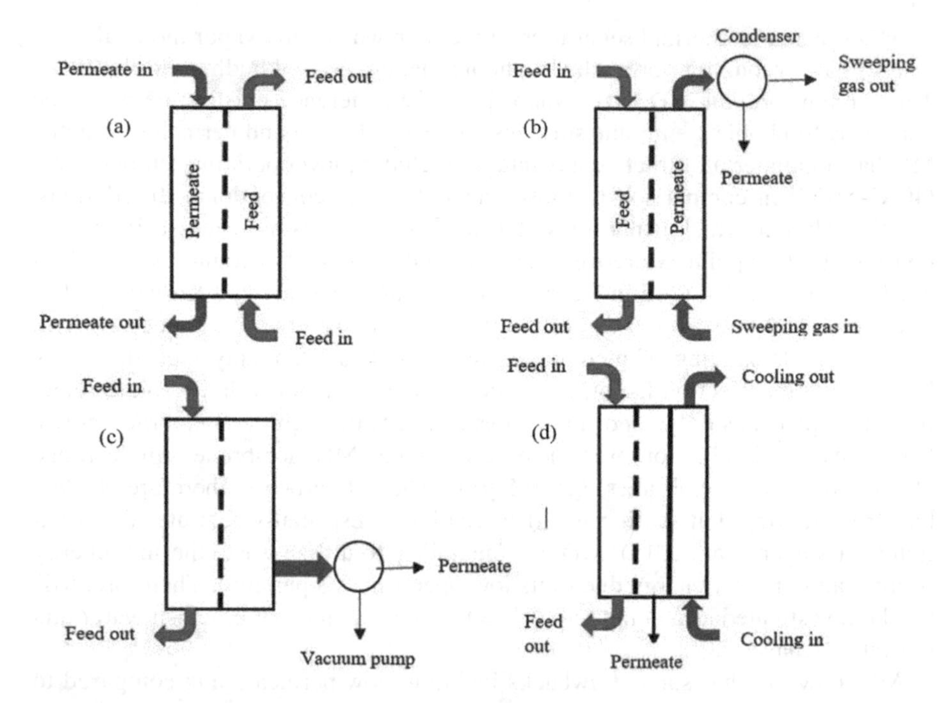

FIGURE 6.1 General scheme of the MD configurations: (a) DCMD, (b) SGMD, (c) VMD and (d) AGMD.

the MD module (He et al., 2018). Figure 6.1 shows the general scheme of each MD configuration. It is worth noting that each one of the mentioned configurations has its own pros and cons for seawater desalination and water treatment. One of the most promising advantages of MD in comparison to the conventional desalinations relies on the lower operating temperature and hydrostatic pressure. Table 6.1 summarizes the summary of the overall concept of the four major MD configurations.

DCMD is the simplest and the most studied configuration, principally for both seawater desalination and brine management (Ashoor et al., 2016); however, the heat lost through membrane conduction is the most important challenge of this configuration. AGMD has the highest energy efficiency, so it is the most promising configuration for use at the commercial scale. However, the air-gap between the membrane and the condensing surface imposes further mass transfer resistance (Duong et al., 2015). Both VMD and SGMD configurations are promising for their higher permeate fluxes and lower heat loss through conduction inside the MD module. However, use of an external condenser can increase the operating cost. The last two configurations, VMD and SGMD, would be more practical for separations where the permeate is not the target product, such as sugar syrup concentration or dewatering of aqueous solutions (Mohammadi and Kazemi, 2014; Shirazi et al., 2014b).

TABLE 6.1
Overview of the Four Major MD Configurations

Mode	Description	Pros	Cons
DCMD	Membrane is in direct contact with both process liquids, i.e., hot and cooling streams	• The simplest configuration to realize, practically • The permeate flux is more stable than most of other configurations • High gained output ratio	• Heat lost through the membrane conduction • Comparing to VMD, the obtained flux is relatively lower • Highest thermal polarization among other configurations • Permeate flux is more sensitive to the salt concentration • The permeate quality is completely sensitive to pore wetting
AGMD	A stagnant air-gap in the permeate side is interposed between the membrane and a condensing plate	• The lowest heat lost among other configurations • No pore wetting risk on the permeate side • Less membrane fouling tendency	• The applied air-gap provides additional resistance to mass transfer • Module design and setup are difficult • Difficult modeling procedure due to the including of several variables • Lowest gained output ratio • Relatively lower permeate flux among other configurations
VMD	Permeate side is vapor or air under vacuum	• High permeate flux • Ability to remove volatile and aroma compounds • No possibility of pore wetting from the permeate side • Lower thermal polarization	• Higher pore wetting risk • Higher fouling tendency in the feed side • Lower selectivity for volatiles • Require external condenser and vacuum in the distillate side
SGMD	Stripping cold inert gas or air is used as carrier for the vapor molecules in the permeate side	• Lowest thermal polarization among other configurations • No pore wetting risk from the permeate side • Promising future perspective, especially for concentrating purposes of aqueous streams	• Additional complexity • Difficult heat recovery • Relatively lower permeate flux • Needing dried and cleaned sweeping gas stream

6.2.3 MD Membranes

MD uses a hydrophobic microporous membrane; however, it must be pointed out that the membrane itself acts only as a barrier to hold the liquid/vapor interfaces at the entrance of the pores; it is not necessary for the MD membrane to be selective as required in other membrane-based separations (Khayet, 2011).

A membrane should satisfy a number of requirements before application in MD for seawater desalination. These requirements are that the membrane must not be wetted and only vapor molecules are present within its pores, so, it should be as hydrophobic as possible (Asghari et al., 2015); the optimal pore size of the membranes used in MD lies between 0.1 and 0.5 μm (Shirazi et al., 2014a); it should have optimal thickness based on the salinity of the feed stream (Eykens et al., 2016b); and should be as porous as possible (Munirasu et al., 2017). Table 6.2 summarizes the required specifications of MD membranes with respect to the optimum range in each case.

TABLE 6.2
Overview of Required Specifications of MD Membranes

Parameter	Optimum Range	Concerns
Pore size	• 0.1–0.5 μm	• Permeate flux • Wetting resistance • Energy efficiency
Hydrophobicity	• Min. value: >90° • Optimum: >140°	Wetting resistance
LEP[a]	• >2.5 bar	Wetting resistance
Porosity	• >85%	• Permeate flux • Mechanical strength • Energy efficiency
Thickness	• For low salinity (brackish water desalination): as thin as possible • For high salinity (RO brine treatment): thick membrane with optimum thickness	• Permeate flux • Mechanical strength • Energy efficiency
Thermal conductivity	• As low as possible (~0.01 $W.m^{-1}.k^{-1}$)	• Energy efficiency • Permeate flux
Mechanical strength	• As high as possible (>60 MPa)	• Long-term performance • Mechanical durability
Compaction	• n/a	• Not compressible
Tortuosity	• As low as possible (~1)	• Permeate flux

[a] LEP: Liquid entry pressure.

Membrane thickness has a great effect on both permeate flux and energy efficiency. In the case of desalination and for low salinity ranges the applied membrane should be as thin as possible. However, for high feed salinities the optimum membrane thickness should be specifically determined for each case (Eykens et al., 2016a). Pore size and its distribution are two other important parameters. While larger pore size can provide slightly higher permeate flux, it also increases the pore wetting risk (Shirazi et al., 2014a); an optimum pore size and narrow pore size distribution are needed to have a balance with reasonable permeate flux and high pore wetting resistance (see Table 6.2 for suggested optimum pore size range). Surface energy is another requisite for MD membrane. This can directly affect the membrane pore wetting risk. It should be noted that not only the material composition, but also the surface morphology and topography of the MD membrane are effective for improving the membrane hydrophobicity (Shirazi et al., 2013b). Porosity, which refers to void volume fraction of the MD membrane, is also important, due to its direct effect on both the permeate flux and energy efficiency. The more porous a membrane, the higher the permeate flux and energy efficiency that can be achieved. However, it should be noted that higher porosity can have negative effect on the mechanical strength (Lalia et al., 2013; Deshmukh and Elimelech, 2017). Tortuosity is the deviation of the sample pore structure from the straight cylindrical structure (Adnan et al., 2012). This parameter is inversely proportional to the MD membrane flux. A higher tortuosity factor can lead to lower permeate flux. However, only a few studies have considered this parameter for MD membranes. The LEP (liquid entry pressure) value is the minimum pressure before saline water wetting the membrane pores. For a safe and long-term MD desalination performance, the LEP value must be lower than the membrane LEP value (see Table 6.2). It is worth noting that even if the inlet pressure in the feed channel is lower than the membrane LEP value, (partial-) pore wetting can occur due to fouling or presence of surfactants/chemicals in the feed (Lin et al., 2015; Han et al., 2017). As the MD process is a thermally driven separation, the membrane thermal conductivity is another crucial parameter. This is more highlighted in DCMD processes (Qtaishat et al., 2008; Khalifa, 2015). Higher porosity reduces the conductive heat and using polymers with lower thermal conductivity should be investigated as a promising alternative for MD membrane fabrication.

The first generation of MD membranes, commercial hydrophobic microporous membranes such as those made from polytetrafluoroethylene (PTFE), polyvinylidene fluoride (PVDF) and polypropylene (PP) meet the requirements discussed above (He et al., 2011a). However, these membranes have specifically been fabricated for microfiltration applications, not MD-based desalination. Therefore, pore wettability may occur and permeate quality may be affected, specifically in long term desalination performance (Gryta, 2005; Zhang et al., 2010). The pore wetting risk may be more serious if the feed stream contains surface active components (e.g. oil/gas produced water treatment) or if the input hydrostatic pressure exceeds the LEP, which is a critical specification for MD membranes (He et al., 2011b; Chew et al., 2017). Table 6.3 summarizes some examples of commercial microfiltration membranes which have been used for MD desalination.

TABLE 6.3
Some Examples of Commercial Microfiltration Membranes for MD Desalination

Membrane Material	Specifications	Supplier	MD Process	References
PTFE	• Pore size: 0.22 μm • Thickness: 178 μm • Porosity: 70% • Contact angle: 132.2°	Millipore	DCMD for real seawater desalination	Shirazi et al. (2012)
PTFE	• Pore size: 0.22 μm • Thickness: 110 μm • Porosity: 83% • Support: Scrim	GE Osmonics	DCMD for desalination	Hwang et al. (2011)
PVDF	• Pore size: 0.45 μm • Thickness: 100 μm • Porosity: 66% • Contact angle: 81°	Membrane-Solutions	AGMD for saline water desalination	He et al. (2011b)
PP	• Pore size: 0.22 μm • Thickness: 200 μm • Contact angle: 116°	Membrane-Solutions	DCMD for desalination	He et al. (2011a)

The second generation of MD membranes has been specifically fabricated for the task using commercial polymers. Different techniques, such as phase inversion, sintering, stretching, track-etching, template leaching, and dip-coating, which enable a membrane to be constructed from a given material and with the desired membrane morphology have been used for fabricating specific membranes for MD desalination. However, among them the phase inversion method is the mostly investigated one (Eykens et al., 2017a). All these membranes can be fabricated both in flat-sheet and hollow-fiber configurations (Wang and Chung, 2015). Khayet (2011) comprehensively reviewed and tabulated a number of examples of fabricated MD membranes, both flat-sheet and hollow-fiber.

All these membranes, however, are faced with some "must-investigate" challenges; a membrane fabricated via the above conventional techniques cannot satisfy all requirements for an MD membrane (see Table 6.3). One of the most important specifications for an MD membrane is the porosity. Membrane porosity not only affects the permeate flux, but also significantly alters the energy efficiency. However, the maximum porosity that can be achieved by the phase inversion method is 60%–75% (Chang et al., 2017; Eykens et al., 2017b; Ragunath et al., 2018). Another critical specification is the membrane surface hydrophobicity. This is crucial for pore wetting prevention in long term desalination operation. However, in order to decrease the membrane surface energy (increase the hydrophobicity) a number of expensive and hard-to-scaleup strategies have been investigated, such as using macromolecules (Shojaie et al., 2017) and carbon nanotubes (Silva et al., 2015), chemical surface modification (Yang et al., 2017) and plasma surface modification (Wu et al., 2018). These issues in addition to the others discussed above reiterate the need for the development of the next generation of MD membranes. Recently, electrospun nanofibrous membranes have been introduced to MD processes (Tijing et al., 2014). In this chapter, this new generation of MD membranes is comprehensively reviewed and discussed.

6.3 ELECTROSPINNING TECHNOLOGY

6.3.1 Overview of Electrospinning

There are different techniques for fabrication of nanofibers. Table 6.4 summarizes the advantages and disadvantages of each technique. Even though each technique has its advantages, the electrospinning has a leading edge over all of them (Sundaramurthy et al., 2014). This is due to the fact that this technology is able to easily control the fabrication, orientation and morphology of the nanofibers (Braghirolli et al., 2014).

Electrospinning relies on repulsive electrostatic forces to draw a viscoelastic solution into nanofibers (Teo and Ramakrishna, 2006). It is mainly suitable for processing polymers, especially *Thermoplastics*. Electrospun fibers are typically in the range of several nanometers to a few microns in diameter (Fridrikh et al., 2003;

TABLE 6.4
Comparison of Some Fabrication Methods for Nanofibers

Method	Pros	Cons
Electrospinning	• Emerging and versatile method • Flexibility of production • Ease of scale-up • High productivity • Controlling of porosity, fiber size and morphology	• Wide range of effective operating parameters • Hard to control 3D structure
Fiber bonding	• High porosity	• Lack required mechanical • Hard to control porosity
Melt blowing	• Under developing method • Ability to process both polymers and resins	• Hard to fabricate nanofibers • Lower porosity
Fiber splitting	• Low processing cost	• Complexity of processing • Time-consuming process
Electro blown	• More feasible nanofiber production	• Sensitive to ambient temperature
Solution blowing	• Coupled method of electrospinning and melt blowing • Higher fiber production rate	• Restricted to viscoelastic materials that can withstand the stresses developed during the drawing process
Force spinning	• Uses centrifugal force, rather than electrostatic force • Processing wide range of polymers	• Under developing

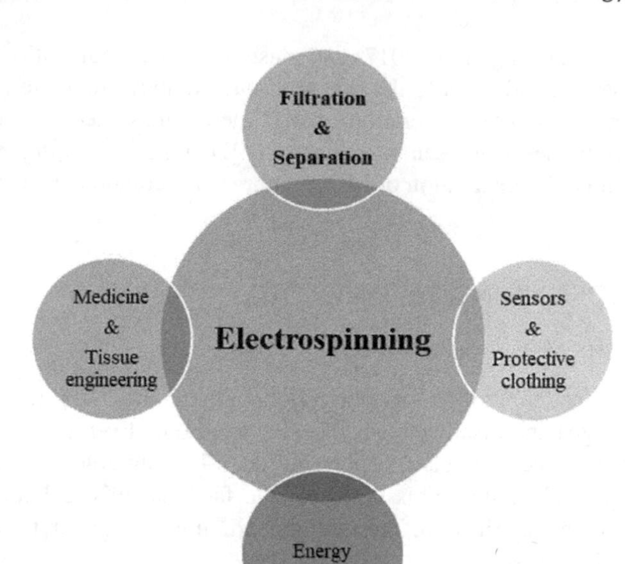

FIGURE 6.2 Major applications of electrospun nanofibers.

Milleret et al., 2012), but the fiber's diameter is controllable. It is to be noted that nanofibers with different morphologies and topographies can also be produced depending on various factors including the polymer used, dope properties, nozzle structure and operating conditions (Persano et al., 2013).

Electrospun nanofibers have many unique characteristics, such as high surface to mass or volume ratio, ability to form a highly porous fibrous membrane with excellent pore interconnectivity, controllability in fiber diameter, surface morphology and fibrous structure, and easiness of being functionalized using various chemicals for electrospinning (Tijing et al., 2014; Wang and Chung, 2015). Figure 6.2 shows an overview of major applications of electrospinning technology.

6.3.2 Operating Parameters in Electrospinning

Different factors can affect the morphology of electrospun nanofibers. This includes electrospinning conditions such as electrostatic potential, electric field strength, electrostatic field shape, tip-to-collector distance, feed injection rate, needle, collector type (which all can directly affect the fiber morphology) (Bhardwaj and Kundu, 2010; SalehHudin et al., 2018) and solvent specifications including solution concentration, viscosity, water content, surface tension, conductivity, dielectric constant, and solvent volatility (which can affect the both electrospinnability and electrospun nanofibers morphology) (Manee-in et al., 2006; Luo et al., 2010, 2012; Saleem et al., 2018). Environmental conditions such as temperature, humidity, local atmosphere flow and pressure also influence the morphology of electrospun fibers (Koski et al., 2004; Gupta et al., 2005; Eda et al., 2007; Hu et al., 2014).

When the concentration of dope solution is low, the viscosity is also low and not enough polymer entanglements occur for fiber formation. A phenomenon known as *Electrospraying* then occurs, due to the instabilities, and micro- or nano-size particles are generated instead of fibers (Yurteri et al., 2010; Scholten et al., 2011). Solvent volatility has also an important influence on the fiber characteristics. The solvent effect can be explained as follows. When a solvent with a low vapor pressure (i.e., low volatility) is used, solvent loss does not occur rapidly, and wet fibers form. On the other hand, if the vapor pressure of the solvent is too high (i.e., high volatility), fibers will not be formed as the polymer jet will solidify upon existing the needle (Son et al., 2004; Yang et al., 2004; Uyar and Besenbacher, 2008). More details on the effect of operating parameters can be found elsewhere (Ahmed et al., 2015; Jung et al., 2016; Sarbatly et al., 2016).

6.3.3 Electrospun Membranes for Conventional Membrane Processes

Microporous polymeric membranes can be fabricated through different methods (Lalia et al., 2013; Karkhanechi et al., 2015). Recently, electrospinning technology has gained considerable attention for the fabrication of microporous polymeric membranes. This is due to the various advantages of electrospinning, mentioned previously. Hence, the development of electrospun membranes with a three-dimensional interconnected pore structure has opened up a new window in membrane separation processes (Liao et al., 2018).

In liquid filtrations membranes separate particulate, bacteria and other suspended contaminants. Electrospun membranes are gaining increased attention as the separative barrier for pressure-driven membrane processes including microfiltration (MF). The main advantages of MF electrospun membranes are the presence of the fibrous network, which provides a high internal surface area and hence enormous loading capacity when compared to phase-inversion membranes with a two-dimensional structure. Having high porosity and an interconnected pore structure are other advantages of MF electrospun membranes. Table 6.5 summarizes some examples of MF electrospun membranes, mostly for water treatment and particulate separation purposes.

In addition to MF application and the use of electrospun membranes as direct filtration media, electrospun polymeric webs can also be used as the support layer for the new generation of thin film composite (TFC) membranes for UF (ultrafiltration), NF (nanofiltration) and RO (reverse osmosis) applications. Typically, TFC membranes comprise of three fundamental layers, which are the top ultrathin selective layer, the middle porous support layer and the bottom nonwoven fabric layer. Recently, there has been a lot of attention on the application of electrospun fibers as the support layer of TFC membranes. Further information in the case of TFC membranes with electrospun nanofibrous supportive webs can be found elsewhere (Subramanian and Seeram, 2013; Shirazi et al., 2017; Liao et al., 2018).

TABLE 6.5
Some Examples of MF Electrospun Membranes

Application	Highlights	Electrospinning	Membrane	References
MBR membrane	• Anti-fouling MF membrane using *para*-aminobenzoate alumoxane (PABA) nanoparticle	• Polymer: Polyacrylonitrile • Solvent: *N*, *N*-Dimethylformamide • Dope injection: 1 mL/h • Voltage: 22 kV • Tip-to-collector distance: 9 cm • Spinning time: 1–6 hours • Humidity: 18% • Temperature: 22°C	• Pore size: 2.31–1.31 μm • Porosity: 22.3%–81.3% • Fiber diameter: 319.1–387.4 μm • Thickness: 12.41–70 μm • Max. pure water flux: 1600 L/m²h • Flux recovery: 65.44%–72.75%	Moradi et al. (2017)
MF of fine particles	• The membranes exhibited high performance as innovative microfiltration media • They could remove 0.2 μm (even 0.1 μm) particles from water with high flux values	• Polymer: Polyacrylonitrile (Mw = 150,000 g·mol⁻¹) • Solvent: *N*, *N*-Dimethylformamide • Needle ID: 0.4 mm • Dope injection: 0.5 mL/h • Voltage: 18 kV • Collector: rotating drum	• Pore size: 0.1–0.35 μm • Porosity: 86%–45% • Operating pressure for MF: 10 psi • Rejection: 42.9%–67% • Flux: 662–8667 L/m²h	Wang et al. (2017)
MF of oily emulsion	• Graphene oxide/PAN hierarchical structured membrane was fabricated • The membrane is superhydrophilic and low-oil-adhesion • The membrane exhibits ultra-fast flux (~10,000 LMH) and excellent efficiency (>98%) • The influence of GO on separation efficiency was systematically investigated	• Polymer: Polyacrylonitrile (Mw = 150,000 g·mol⁻¹) • Additive: Graphene oxide • Solvent: *N*, *N*-Dimethylformamide • Voltage: 15 kV • Tip-to-collector: 15 cm • Dope injection: 1 mL/h • Temperature: 40°C • Humidity: 35%	• Fiber diameter: 0.5–0.9 μm • Surface contact angle: 110–5° • Permeate flux: 5000–13,500 LMH • Rejection ratio: 79.75%–98.1% • Flux recovery: 42.49%–71.53%	Zhang et al. (2017)

(Continued)

TABLE 6.5 (*Continued*)
Some Examples of MF Electrospun Membranes

Application	Highlights	Electrospinning	Membrane	References
MF of oily emulsion	• Novel PVA-coated electrospun nylon 6/Silica composite MF • The MF membrane demonstrated high permeability (4814 LMH/bar) • The MF membrane exhibited improved antifouling properties (85% flux recovery) • The fabricated MF membrane achieved nearly 99% oil rejection	• Polymers: Nylon-6 and Polyvinyl acetate (Mw 140,000) • Solvent: Formic acid and acetic acid • Needle ID: 19.05 mm • Dope injection: 0.18 mL/h • Voltage: 30 kV • Temperature: 25°C • Humidity: 40%	• Mean pore size: 175 nm • Surface contact angle: 21° and 43° • Porosity: 60%–80% • Tensile: ~25 MPa • Oil rejection (0.28 bar): 99%	Islam et al. (2017)

6.4 ELECTROSPUN MEMBRANES FOR MD PROCESSES

The past few years, electrospun nanofibers have been used as promising membrane materials for filtration, desalination and water/wastewater treatment applications. A number of research groups have successfully fabricated and used electrospun membranes as the main filtration media or support material for pressure-driven membrane processes (Shirazi et al., 2017). Recently, research has been carried out on the application of electrospun membranes for osmotic membrane separations such as FO (forward osmosis) and PRO (pressure retarded osmosis) (Liao et al., 2018). With the trend of using nanofibers in membrane-based desalinations, electrospun membranes have also found their way to MD application. Up to 2014, only a few research groups have used electrospun membranes for MD. This research is comprehensively reviewed by Tijing and coworkers (2014). In another review paper, Shirazi and coworkers (2017) also reviewed the published literature from 2014 to 2017. In this chapter, we discuss the recently published research works on MD using electrospun membranes.

6.4.1 DCMD WITH ELECTROSPUN MEMBRANES

Electrospun polymeric membranes have been widely investigated for MD, as mentioned earlier, due to their relatively high porosity, high hydrophobicity and controllable pore size. However, the robustness of such membranes is not guaranteed in long-term desalination performance; mechanical characteristics of electrospun membranes are a major consideration not only for MD process, but also for other filtrations (Shirazi et al., 2013a, 2013c). Various techniques have been used for characterizing and improving the mechanical features of electrospun membranes (Tan and Lim, 2006; Kaur et al., 2014). Among them, heat-press treatment is a simple and effective procedure to improve both morphological and mechanical characteristics of the electrospun membrane (Kaur et al., 2011). Nevertheless, the heat pressing technique is new for MD membranes. Yao and coworkers (2016) comprehensively investigated the effect of heat-press conditions (e.g., temperature, pressure and duration) on the mechanical characteristics and morphological features of electrospun membranes. The heat-pressed membrane samples were then used for DCMD desalination. Results indicated that impressive improvement of mechanical strength and LEP were achieved after using the proposed treatment technique on the electrospun membranes. Moreover, fibers size increased, surface pore size decreased, Young's modulus and stress at break both increased. As expected, heat pressing also decreased the membranes thickness, i.e., ~48–37 μm. The authors concluded that the heat-press treatment with optimal conditions (i.e., temperature at 150°C, pressure at 6.5 kPa, and duration for 8 hours) could successfully improve the characteristics of the membranes for permeate flux and salt rejection in DCMD. Moreover, results indicated that the temperature and duration of heat pressing were more influential than the pressure (Yao et al., 2016).

It is worth noting that the majority of electrospun membranes for MD desalinations have been made of PVDF polymer or PVDF with modified branches (Tijing et al., 2014). The molecular weight and concentration of the base polymer directly affect the fibers diameter. Armand and coworkers (2017) studied the effect of PVDF concentration (20–24.5 wt%) in the dope solution on desalination performance of the electrospun membrane using DCMD. Results indicated that using the 20 wt% concentration of polymer solution led to formation of bead-free nanofibers with the average diameter of 240 nm. Increasing PVDF concentration leads to an increase in fiber diameter, up to 464 nm. Results of DCMD desalination showed that by increasing the PVDF concentration, the permeate flux decreases. For all experiments, the salt rejection was measured at >99%.

Another major challenge for electrospun membranes used in DCMD desalination is the insufficient pore wetting resistance related with hydrophobicity and pore structure of the membrane surface. Some research groups have studied using nanoparticles to improve the characteristics of the PVDF electrospun membranes, such as adding TiO_2 or modified TiO_2 with fluorosilane in order to increase the hydrophobicity (Lee et al., 2017). For example, Ren and coworkers (2017) studied a novel modified electrospun PVDF membrane for DCMD desalination. In this study, a super-hydrophobization technique was used. A virgin PVDF nanofibrous membrane had a first coating of TiO_2 nanoparticles on the surface achieved by a low temperature hydrothermal process. The TiO_2 coated membrane was then fluorosilanized with a low surface energy material such as *1H,1H,2H,2H*-perfluorododecyl trichlorosilane. Results indicated that the proposed modified membrane possessed high hydrophobicity (contact angle: ~157°), high surface roughness (mean roughness parameter: 4.63 μm), considerable wetting resistance (LEP: 158 kPa), well-distributed pore size (mean pore size: 0.81 μm), reasonable porosity (~57%) and modest membrane thickness (~55 μm). The authors concluded that these combined properties made the proposed novel modified membrane an attractive candidate for DCMD desalination. When used for DCMD desalination, during short-term performance using 3.5 wt% NaCl solution, high flux and stable desalination performances were achieved (permeate flux: ~73.4 L/m^{2}h, salt rejection: >99%). The modified membrane in this work was also used for long-term DCMD desalination with actual reverse osmosis brine as feed solution. Results showed high desalination performance with reasonable average permeate flux of ~40.5 L/m^{2}h and salt rejection of >99%.

In addition to PVDF of branch modified PVDF, other polymers have also been investigated for fabricating DCMD membrane. In this regard, Li and coworkers (2014) studied developed a new type of dual-biomimetic hierarchically rough membrane made of polystyrene. The proposed membrane showed superhydrophobic characteristics with a fibrous structure containing both micro- and nano-fibers. The authors used a one-step electrospinning technique with various polymer concentrations (15–30 wt%). Different topographical features including nanopapillose, nanoporous, and microgrooved surface morphologies were observed using scanning electron microscopy. The authors concluded that the superhydrophobicity (contact angle: 150.2°) and high porosity (77.5%) of the as-spun polystyrene nanofibrous membrane make it an attractive candidate for DCMD desalination. Indeed, the membrane samples maintained a high permeate flux (~104.8 kg/m^{2}h) when 20 g/L NaCl aqueous solution was used as feed for a test period of 10 hours without remarkable membrane pore wetting. The results obtained for the electrospun polystyrene membranes in this work were compared with a commercial PVDF membrane and also PVDF nanofibrous membranes reported in the literature. It was concluded that the excellent competency of the polystyrene nanofibrous membrane for MD applications highlight it for further studies (Li et al., 2014). In another work, Ke and coworkers (2016) developed new electrospun polystyrene membranes for DCMD desalination. The proposed membranes were prepared from polystyrene-DMF dope solution with the addition of sodium dodecyl sulfate as a processing aid. Effects of thickness and fiber diameter of nanofibrous membranes on the pore size and its distribution were systematically studied. The authors used a capillary flow porometer, a contact angle meter and the gravimetric method for characterizing the membrane samples. The resulting permeate flux and permeate quality of the electrospun polystyrene membranes were compared with those of commercially available PTFE membranes, when four different feed solutions were used, which were distilled water, simulated brackish water, 35 g/L NaCl solution and real seawater. Effects of flowrates (i.e., 0.2–0.4 GPM) and feed solution temperatures (i.e., 70°C–90°C) on the permeate flux of the optimized as-spun membrane (mean pore size: 0.19 μm) were investigated for 10 hours DCMD operation. The obtained results indicated that the polystyrene membrane has a promising hydrophobicity with an acceptable performance for DCMD desalination.

It is worth noting that not only the hydrophobic electrospun fibers, but also the hydrophilic ones can be used to fabricate an MD membrane. Ray and coworkers (2017) investigated a novel category of improved performance membrane consisting of a hydrophobic mat and a hydrophilic electrospun layer for DCMD desalination. An electrospun layer was fabricated made of polyvinyl alcohol (PVA) and then incorporated with Triton X-100 directly onto a polypropylene membrane. The concept of dual layer membrane is utilized with a hydrophobic membrane on top and a hydrophilic support layer on the bottom in order to enhance the permeate flux and salt rejection. Results indicated that by using the new dual-layer membrane structure the salt rejection was >99% and still maintained a two times higher permeate flux compared to a commercial single layer PP membrane, for a long-term operation of 15 hours.

Table 6.6 summarizes the detailed information for some recently published development of electrospun membranes for DCMD process.

TABLE 6.6
Literature Review for Application of Electrospun Membranes for DCMD Process

Application	Electrospun Membrane	Membrane Distillation	References
DCMD for desalination	**Electrospinning** • Polymer: Hydroxyl copolyimide (HPI) • Solvent: DMAC • Dope solution: 8–21 wt% • Voltage: 19 kV • Needle ID: 0.574 mm • Tip-to-collector distance: 15 cm • Collector: rotating drum (5 rpm) **Membrane** • Pore size: 0.3–1.3 μm • Thickness: 60 μm • Fiber diameter: 180–610 μm • Porosity: 75%–85% • LEP: 0.3–1.0 bar	• Feed: 0.5 M NaCl solution • Feed temperature: 50°C–80°C • Permeate temperature: 20°C • Permeate flux: 80 kg/m^{2}h • Salt rejection: >90%	Park et al. (2018)
DCMD for oily wastewater treatment	**Electrospinning** • Polymer: Cellulose acetate • Solvents: *N*-methyl-2-pyrrolidone (NMP) and acetone • Voltage: 15 and 20 kV • Needle: 18 gauge • Dope injection: 1.5 mL/h • Tip-to-collector distance: 15 cm • Collector: rotating drum (150 rpm) • Temperature: 30°C • Humidity: 30% **Membrane** • Mean pore size: 0.47 μm • Thickness: 303 μm • Porosity: 50.6% • Oil contact angle: 154°	—	Hou et al. (2018)

(Continued)

TABLE 6.6 (*Continued*)
Literature Review for Application of Electrospun Membranes for DCMD Process

Application	Electrospun Membrane	Membrane Distillation	References
DCMD for desalination	**Electrospinning** • Polymer: PVDF-HFP (Kynar Powerflex® LBG, Mw = 450,000 g mol^{-1}) • Solvent: DMAC and acetone • Voltage: 25 kV • Dope solution: 10 wt% • Needle: 18 gauge • Dope injection: 1 mL/h • Tip-to-collector: 15 cm • Temperature: room • Humidity: 55% **Membrane** • Fiber diameter: 0.05–0.45 μm • Contact angle: 153–155° • Mean pore size: 0.57 μm • Max. pore size: 1.93 μm • Porosity: 86.5% • Elastic modulus: 32 MPa • Tensile strength: 12 MPa • Fracture strain: 70%	• Feed temperature: 60°C–70°C • Flux: ~7–10 kg/m^2h • Rejection: >99%	Makanjuola et al. (2018)
DCMD for desalination	**Electrospinning** • Polymer: Linear tri-block SBS copolymer • Solvents: THF and DMF • Dope solution: 15 wt% • Needle: 22 gauge • Voltage: 1.0 kV/cm • Temperature: 21°C • Humidity: 43% **Membrane** • Pore size: 0.58 μm • Porosity: 81% • Thickness: 200 μm • Roughness: 457.4 nm • Contact angle: 132° • Elastic modulus: 9.8 MPa	• Feed temperature: 60°C • Permeate temperature: 25°C • Feed and permeate flowrate: 0.5 L/min • Permeate flux: 11.2 L/m^2h	Duong et al. (2018)

(*Continued*)

TABLE 6.6 (*Continued*)
Literature Review for Application of Electrospun Membranes for DCMD Process

Application	Electrospun Membrane	Membrane Distillation	References
DCMD for desalination	**Electrospinning** • Polymers: PVDF (Mw: 275 kg/mol) and PSf (79 kg/mol) • Solvents: *N, N*-dimethyl acetamide (DMAC) and acetone • Dope solution: 25 wt% • Voltage: 27 kV • Needle ID/OD: 0.6/0.9 mm • Dope injection: 1.23 mL/h • Tip-to-collector distance: 27.5 cm • Spinning time: 0.5–2.5 hours • Temperature: 23°CC • Humidity: 36% **Post treatment** • Drying in an oven: T = 120°C, *t* = 0.5 hours **Membrane** • Pore size: ~500–1885 nm • Porosity: 77%–92% • LEP (H_2O): 2.9–17 MPa • Young's modulus: 20.2–58 MPa	• Feed temperature: 80°C • Permeate temperature: 20°C • Feed sample: saline solutions (12 g/L and 30 g/L) • Permeate flux: ~45–55 kg/m²h • Salt rejection: >99%	Khayet et al. (2018)
DCMD for desalination	**Electrospinning** • Polymer: PVDF (Mw: 410 kDa) • Solvents: DMAC (>99%) and acetone • Dope solution: 13 wt% • Tip-to-collector distance: 150 mm • Collector: rotating drum (140 rpm) • Voltage: 15 kV • Temperature: 25°C • Humidity: 40% **Membrane** • Pore size: 0.24–0.32 µm • Thickness: 46–79 µm • Fiber diameter: 202–445 nm • Porosity: 61.23–66.24% • LEP_w: 68.95–110.32 kPa	• Feed temperature: 28, 38 and 48°C • Permeate temperature: 16°C • Permeate flux: 1.83–2.87 kg/m²h • Salt rejection: >99%	Yang et al. (2018)
DCMD for desalination	**Electrospinning** • Polymer: PVDF (Mw: 146,000 g/mol) • Solvent: DMAC • Additive: LiCl (0.004 wt%) • Dope solution: 18–26 wt% • Dope injection: 0.5 mL/h • Temperature: 25°C • Relative humidity: 45%	• Feed: 42 g/L NaCl solution • Permeate flux: ~30 kg/m²h • Salt rejection: >99%	Ebrahimi et al. (2018)

(*Continued*)

TABLE 6.6 (*Continued*)
Literature Review for Application of Electrospun Membranes for DCMD Process

Application	Electrospun Membrane	Membrane Distillation	References
	Membrane • Fibers diameter: 142–240 nm • Surface contact angle: 108–122° • Thickness: 188–220 μm • Porosity: 80.8–86% • LEP: 100 kPa • Young's modulus: 21.5–40.55 MPa • Tensile strength: 4.57–6.3 MPa		
DCMD for oil wastewater treatment	**Electrospinning** • Polymer: Polyethyleneimine (PEI, Mw = 600) • Solvent: DMAC • Additive: polydopamine/ polyethyleneimine (PDA/PEI) • Additive: SiO_2 nanoparticles • Voltage: 15 kV • Dope injection: 0.25 mL/h • Tip-to-collector distance: 20 cm • Temperature: 23°C • Humidity: 45% **Membrane** • LEP: 42 kPa • Contact angle: 152°	• Feed temperature: 85°C • Feed sample: NaCl, KCl, MgCl, $MgSO_4$, $CaSO_4$ • Permeate temperature: 20°C • Effective membrane area: 16 cm^2 • Permeate flux: 31 L/m^2h • Rejection: 100%	Zhu et al. (2018)
DCMD for desalination	**Electrospinning** • Polymer: Poly(vinylidene fluoride-co-hexafluoropropylene) (PVDF-HFP, MW = 455,000 g/mol) • Solvent: DMF and acetone • Additive: LiCl • Voltage: 14 and 18 kV • Needle: multi-nozzle system • Tip-to-collector distance: 15 cm • Dope injection: 1 mL/h • Collector: rotating drum (700 rpm) • Temperature: 25°C • Humidity: 55% **Membrane** • Pore size: 0.61–0.76 μm • Thickness: 87–100 μm • Porosity: 88.6%–91.6% • LEP: 65.8–96.3 kPa • Fiber diameter: 231–368 nm • Tensile strength: 6–9.1 MPa	• Feed temperature: 60°C • Permeate temperature: 20°C • Effective membrane area: 9.8 cm^2 • Feed flowrate: 450 mL/min • Permeate flux: 5.5–41.1 kg/m^2h	Lee et al. (2017)

(*Continued*)

TABLE 6.6 (*Continued*)
Literature Review for Application of Electrospun Membranes for DCMD Process

Application	Electrospun Membrane	Membrane Distillation	References
DCMD for desalination	**Electrospinning** • Polymers: Polysulfone (Mw: 72,000–82,000 g/mol) and polydimethylsiloxane (PDMS) • Solvent: DMF • Dope solution: 23 wt% • Needle ID: 0.37 mm • Dope injection: 1 mL/h • Tip-to-collector distance: 15 cm • Voltage: 20 kV • Temperature: 25°C • Humidity: 25% **Post-treatment** • Cold press (4 MPa) **Membrane** • Thickness: 56–124 μm • Inter-fiber space: 0.73–3.69 μm • Contact angle: 109–151°	• Feed temperature: 50°C–80°C • Feed flowrate: 0.6 L/min • Permeate flowrate: 0.4 L/min • Effective membrane area: 25 cm^2 • Feed: 30 g/L NaCl aqueous solution • Sampling: every 30 sec • Permeate flux: 21.5 kg/m^2h	Li et al. (2017)
DCMD for dyeing wastewater treatment	**Electrospinning** • Polymer: PVDF-HFP ($Mw = 455{,}000\ g\ mol^{-1}$) and PDMS (Sylgard®) • Additive: LiCl • Solvents: DMF, THF and acetone • Dope solution: 15 wt% • Voltage: 18–30 kV • Tip-to-collector distance: 11–15 cm • Dope injection: 0.7–2 mL/h • Humidity: ~50% **Membrane** • Pore size: 0.49–0.52 μm • Porosity: ~87% • Thickness: 98–102 μm • Surface contact angle: 137–155° • LEP: 0.92–1.26 bar	• Feed: Dye wastewater (100 mg/L) • Feed temperature: 60°C • Permeate temperature: 20°C • Membrane effective area: 9.8 cm^2 • Feed/permeate flowrate: 0.5 L/min • Max. permeate flux: 34 L/m^2h • Rejection: 100%	An et al. (2017)

6.4.2 AGMD with Electrospun Membranes

Application of electrospun membranes for MD desalination was firstly introduced by Feng and coworkers (2008). This was also the first application of nanofibrous membrane for the AGMD process. In this work, the authors used PVDF (Kynar 761) as the base polymer for electrospinning of nanofibers. 18 kV high voltage was used between the needle tip (18 gauge) and the collector plate. The dope solution was injected at 2 mL/h flowrate. The fabricated membrane was then used for simulated seawater desalination via the AGMD process. The fabricated membrane was characterized for morphological features using SEM and AFM microscopes. Based on the obtained results, the authors concluded that the PVDF nanofibrous membrane can be used for producing drinking water via AGMD desalination. The permeate flux of the proposed membrane was comparable to those obtained by commercial microfiltration membranes (5–28 kg/m^{2}h) when 25°C–83°C temperature differences were imposed.

Table 6.7 summarizes the summary of the published work on the application of electrospun membranes for AGMD. As can be observed, there are few works that have used nanofibrous membranes for AGMD, while most of the published literature investigated the DCMD performance of such membranes (see Table 6.6).

Woo and coworkers (2017a) studied o a new PVDF based electrospun membrane for AGMD and the possibility of modifying the membrane with CF_4 plasma surface treatment. In this work, the effect of different durations of plasma treatment on the nanofibrous membrane characteristics was investigated. The proposed membrane was used for treating a real feed solution; RO brine containing surfactant. After plasma treatment, surface energy of the nanofibrous membrane decreased, with the formation of new CF_2-CF_2 and CF_3 bonds. Different low surface tension liquids, including methanol, mineral oil and ethylene glycol, were used for examining the membrane wetting resistance. No significant changes in morphology of the membrane were observed after plasma treatment, while the LEP of the membrane increased from 142 kPa up to 187 kPa for the virgin and the modified membrane, respectively. The authors also observed that using the new plasma-treated membrane, a 100% salt rejection as well as 15.3 L/m^{2}h initial permeate flux were achieved.

As discussed above, nanoparticles such as TiO_2 can be used not only for improving the mechanical features and antifouling properties of the electrospun membranes, but also for enhancing the permeate flux. Various strategies can be used for adding nanoparticles to the nanofibrous mat such as adding to the dope solution or surface coating. Recently, Attia and coworkers (2017) used nanoparticles to improve the performance of AGMD nanofibrous membrane. The authors claimed that this is a novel approach to fabricate superhydrophobic membranes by using an environmentally friendly and cost-effective superhydrophobic nanoparticle (Al_2O_3) to enhance the salt rejection and permeate flux of nanofibrous membrane. In this work, PVDF membranes were fabricated using a new electrospinning technique. Electrospinning parameters (polymer concentration, voltage, solvent ratio, and cationic surfactant) were studied to optimize the membrane. Results indicated that highly porous, bead-less fibers were fabricated. The as-spun membrane samples were then characterized in terms of pore size, LEP, porosity, contact angle, permeate

TABLE 6.7
Some Examples for Application of Electrospun Membranes for AGMD Process

Application	Electrospun Membrane	Membrane Distillation	References
AGMD for desalination	**Electrospinning** • Polymer: PVDF (Kynar 761) • Solvent: DMF • Dope solution: 16 wt% • Dope injection: 0.5 mL/h • Voltage: 24 kV • Needle: 18 gauge • Temperature: 24°C • Humidity: 60% • Collector: rotating drum (550 rpm) **Membrane** • Pore size: • Thickness: 193.7–205.3 μm • Porosity: >90% • Contact angle: 115–131° • LEP: 102–135 kPa • Fiber diameter: 289–327 nm	• Feed: 3.5 wt% NaCl solution • Feed temperature: 60°C • Cooling temperature: 20°C • Feed flow rate: 30 L/h • Salt rejection: >90% • Permeate flux: 10.8–13.6 L/m²h	Jafari et al. (2018)
AGMD for desalination	**Electrospinning** • Polymer: Polyacrylonitrile (PAN, Mw = 85,000) • Solvent: DMF • Dope solution: 11.5 wt% • Voltage: 20 kV • Dope injection: 0.025 mL/min • Tip-to-collector distance: 20 cm • Collector: rotating drum: 2800 rpm • Temperature: 25°C • Humidity: 25% **Membrane** • Pore size: 0.45–0.57 μm • Porosity: 86% • Thickness: 52 μm • Fiber diameter: 0.2 μm • Contact angle: 153°	• Feed temperature: 38°C–65°C • Cooling temperature: 20°C • Permeate flux: 2–11 kg/m²h • Salt rejection: >99%	Cai et al. (2018)

(Continued)

TABLE 6.7 (*Continued*)
Some Examples for Application of Electrospun Membranes for AGMD Process

Application	Electrospun Membrane	Membrane Distillation	References
AGMD for desalination	**Electrospinning** • Polymers: Polyvinylidene fluoride-co-hexafluoropropylene (PH); polyvinyl alcohol (PVA); nylon-6 (N6); and polyacrylonitrile (PAN) • Solvents: DMF • Additive: LiCl • Voltage: 18–27 kV • Dope solution: 20 wt% • Dope injection: 0.6–1.2 mL/h • Tip-to-collector distance: 10–20 cm • Temperature: 20°C–25°C • Humidity: 40%–45% **Membrane** • Mean pore size: 0.13–0.40 μm • Thickness: 92.7–128.9 μm • Porosity: 78%–91% • LEP: 122–184 kPa • Contact angle: 122–148° • Fiber diameter: 239–333 nm • Young's modulus: 18–117 MPa • Tensile strength: 5.9–19.5 MPa	• Feed temperature: 60°C • Feed solution: 3.5 g/L NaCl • Cooling temperature: 20°C • Membrane effective area: 21 cm^2 • Permeate flux: 10.9–15.5 L/m^2h • Salt rejection: >99%	Woo et al. (2017b)
AGMD for heavy metals removal	**Electrospinning** • Polymer: PVDF (Mw = 275,000) • Additive: Alumina nanoparticles (size = 13 nm) • Solvent: DMF • Voltage: 10–18 kV • Dope solution: 11–20 wt% • Dope injection: 0.1–0.6 mL/h • Needle: 18 gauge • Tip-to-collector distance: 15 cm • Collector: rotating drum (500 rpm) **Membrane** • Pore size: 0.37–0.71 μm • Thickness: 95–110 μm • Porosity: 81%–91% • Fiber diameter: 105–236 nm • Contact angle: 132–150° • LEP: 14–27 psi	• Feed temperature: 60°C • Flowrate: 8.5 L/min • Heavy metals rejection: 99.36% • Permeate flux: 16.5–20 L/m^2h	Attia et al. (2017)

(*Continued*)

TABLE 6.7 (*Continued*)
Some Examples for Application of Electrospun Membranes for AGMD Process

Application	Electrospun Membrane	Membrane Distillation	References
AGMD for brine treatment	**Electrospinning** • Polymer: PVDF (Kynar 761; Mw = 441,000 g/mol) • Solvent: DMF • Additive: LiCl and sodium dodecyl sulfate (SDS) • Dope solution: 15 wt% • Additive content: 0.004 wt% • Dope injection: 1.5 mL/h • Voltage: 18 kV • Tip-to-collector: 15 cm • Temperature: 20–25°C • Humidity: 40%–45% **Membrane** • Fiber diameter: 290–296 nm • Pore size: ~0.8 μm • Thickness: 150 μm • Porosity: 86% • Contact angle: >140° • LEP: 142–187 kPa	• Feed: RO brine • Feed temperature: 60°C • Cooling temperature: 20°C • Flowrates: 12 L/h • Permeate flux: 15.3 L/m^{2}h • Rejection: 100%	Woo et al. (2017a)

flux and solute rejection. The proposed membranes were successfully used for removing heavy metals (such as lead) using AGMD, and results were compared with pristine and commercial membranes. The results showed that pristine beadless membrane can be fabricated using 15 wt% dope solution, 0.05 wt% cationic surfactant with 6:4 DMF to acetone ratio and 14 kV high voltage. Lead rejection of 72.77%, LEP of 17 psi and water contact angle of 132° were measured for this membrane. On the other hand, the composite of 11 wt% PVDF membranes with 20 wt% of functionalized alumina (Al_2O_3) showed 150° contact angle and 27 psi as LEP which led to 99.36% of heavy metal rejection and 5.9% increase in permeate flux (Attia et al., 2017).

Recently, Makanjuola and coworkers (2018) investigated a new strategy for adding the nanoparticles to the nanofibrous structure. In this work, a superhydrophobic electrospun membrane was prepared by *in-situ* deposition of Teflon oligomer microparticles on poly(vinylidenefluoride-co-hexafluoropropylene) nanofibers. The authors explained that this new approach relies on dusting in the chamber where the threads of polymer solution are spun in the zone of fiber transition between liquid to solid. The novel technique allows bonding of the particles to the solidified fibers. The prepared membrane was characterized and then tested for DCMD desalination. It is worth noting that the proposed membrane can also be used for AGMD membrane, but this was not investigated.

A dual-layer membrane that consists of a hydrophobic layer and a hydrophilic supportive membrane can enhance MD membrane performance (Khayet et al., 2005; Bonyadi and Chung, 2007). Woo and coworkers (2017b) investigated the same strategy for electropsun membranes. The authors prepared dual-layer nanofibrous membranes for AGMD, including single- and dual-layer membranes composed of a hydrophobic polyvinylidene fluoride-co-hexafluoropropylene (PH) top layer with different supporting hydrophilic layers made of polyvinyl alcohol (PVA), nylon-6 (N6), and polyacrylonitrile (PAN) nanofibers. Heat-press post-treatment was used to improve the membranes characteristics. Study of the morphology of the fabricated membranes showed that the active layer (i.e., PH) of all electrospun membranes exhibited a rough, highly porous (80%), and hydrophobic surface (140°), while the other side was hydrophilic (<90°) with varying porosity. Heat-pressing caused the reduction of membrane thickness and smaller pore sizes (<0.27 μm). For the AGMD experiments, a co-current flow arrangement, feed temperature of 60°C and cooling temperature of 20°C were investigated. Results indicated that the single and dual-layer membranes showed a permeate flux of about 10.9–15.5 L/m^{2}h using 3.5 wt% NaCl solution as feed. This was much higher than that of the permeate flux of a commercial PVDF membrane (~5 L/m^{2}h). The authors claimed that the provision of a hydrophilic layer at the bottom layer enhanced the AGMD performance depending on the wettability and characteristics of the support layer. The PH/N6 dual-layer nanofibrous membrane prepared under the optimum conditions showed permeate flux and salt rejection of 15.5 L/m^{2}h and 99.2%, respectively, which has good potential for AGMD application.

6.4.3 VMD with Electrospun Membranes

The vacuum pressure in the permeate side means that pore-wetting risk can be more serious in VMD, mostly in the presence of low energy compounds, such as surfactants, in the feed stream. Therefore, pore size and its distribution as well as membrane porosity are more emphasized in VMD. This is arguably why very little research has been done on the fabrication of electrospun membranes for this MD configuration.

Huang and coworkers (2017) studied novel photocatalytic porous membranes consisting of ZnO supported by a PTFE substrate. The proposed membranes were fabricated by sintering electrospun PTFE/PVA/zinc acetate dehydrate composite membranes after removal of the PVA carrier. Results showed that the spinning solution exhibited good electrospinnability and can fabricate membranes with excellent flexibility, high chemical stability and a highly specific surface area. The authors claimed that the photocatalyst-ZnO particles derived from the thermal decomposition of zinc acetate dehydrate were homogeneously immobilized on the surface of the ultrafine PTFE fibers to avoid the influence of agglomeration and sedimentation. The performances of PTFE/ZnO membranes were then investigated in VMD followed by photo-degradation experiments. Results indicated that appropriate PTFE/ZnO membranes can significantly reject the salt content (>99%) while simultaneously removing dye (45%), with good self-cleaning ability. The authors also developed a cleaning procedure of the fouled membranes using 3 hours UV irradiation with a permeate flux recovery rate of >94%. Thus, it was concluded that the immobilization of ZnO on substrate of ultrafine fibrous PTFE porous membranes may provide a wide range of potential applications for the treatment of dye wastewater.

Recently, Yan and coworkers (2018) tried to enhance the hydrophobicity of electrospun membranes for VMD desalination. They used a facile surface coating technique for the fabrication of PVDF superhydrophobic nanofibrous membranes using a spraying method. The claim is as follows: the low content of carbon nanotubes within a solvent dispersion plays an important role to prevent the agglomeration of carbon nanotubes. SEM observation showed that the carbon nanotubes were interlaced on the PVDF nanofibrous membrane and formed a thin network layer. In order to achieve a good bonding force with the carbon nanotubes, the sprayed electrospun membrane was heat-pressed slightly below the PVDF melting point. The low surface energy of carbon nanotubes led to the fabrication of superhydrophobic membrane. The resultant membrane contained 20 g/m^2 carbon nanotube and exhibited highest permeate flux (~28 kg/m^2h) with a steady VMD performance for more than 26 hours. The morphology of the CNTs coated membranes had no significant change after VMD testing. The authors concluded that the carbon nanotube network coated electrospun membranes have a high potential for application in VMD desalination.

6.5 FUTURE PERSPECTIVES

Electrospun nanofibrous membranes as the next generation of MD membranes have promising features and will provide excellent opportunities for different MD configurations in the future. To enhance the morphological and topographical features as well as better solute rejection and higher permeate flux, various strategies including molecular bonding, *in-situ* polymerization and addition of molecular dopants can be combined with electrospinning technology. Moreover, surface modification techniques including nanoparticles coating, treatment with chemicals or heat, grafting and interfacial polymerization have been found to be suitable techniques to alter the surface characteristics and improve the permeate flux and solute rejection of electrospun membranes (Liao et al., 2017; Shirazi et al., 2017).

However, there are some challenges in application of electrospun membranes for MD process. Although electrospun membranes are highly porous and then highly energy efficient, the mechanical characteristics of such membranes should be significantly improved. To do this, the discussed treatment methods as well as right polymer selection should be considered. For instance, hydrophobic synthetic polymers have improved mechanical robustness. Moreover, low productivity of the *electrospinning* technique is a challenge for scaling up the fabrication and thus application of electrospun membranes. In better words, other strategies such as multiple nuzzle system, needleless electrospinning, etc. can be investigated for improving the fiber productivity. And last but not least, all efforts should lead to fabrication of membranes which are capable for long-term MD performance.

REFERENCES

Adnan, S., Hoang, M., Wang, H., Xie, Z. 2012. Commercial PTFE membranes for membrane distillation application: Effect of microstructure and support material. *Desalination* 284: 297–308.

Ahmed, F.E., Lalia, B.S., Hashaikeh, R. 2015. A review on electrospinning for membrane fabrication: Challenges and applications. *Desalination* 356: 15–30.

Alkhudhiri, A., Darwish, N., Hilal, N. 2012. Membrane distillation: A comprehensive review. *Desalination* 287: 2–18.

Alklaibi, A.M., Lior, N. 2005. Membrane-distillation desalination: Status and potential. *Desalination* 171: 111–131.

An, A.K., Guo, J., Lee, E.J., Jeong, S., Zhao, Y., Wang, Z., Leiknes, T. 2017. PDMS/PVDF hybrid electrospun membrane with superhydrophobic property and drop impact dynamics for dyeing wastewater treatment using membrane. *J. Membr. Sci.* 525: 57–67.

Anand, A., Unnikrishnan, B., Mao, J.Y., Lin, H.J., Huang, C.C. 2018. Graphene-based nanofiltration membranes for improving salt rejection, water flux and antifouling-A review. *Desalination* 429: 119–133.

Armand, S.B., Fouladitajar, A., Ashtiani, F.Z., Karimi, M. 2017. The effect of polymer concentration on electrospun PVDF membranes for desalination by direct contact membrane distillation. *Eur. Water* 58: 21–26.

Asghari, M., Harandizadeh, A., Dehghani, M., Harami, H.R. 2015. Persian Gulf desalination using air gap membrane distillation: Numerical simulation and theoretical study. *Desalination* 374: 92–100.

Ashoor, B.B., Mansour, S., Giwa, A., Dufour, V., Hasan, S.W. 2016. Principles and applications of direct contact membrane distillation (DCMD): A comprehensive review. *Desalination* 398: 222–246.

Attia, H., Alexander, S., Wright, C.J., Hilal, N. 2017. Superhydrophobic electrospun membrane for heavy metals removal by air gap membrane distillation (AGMD). *Desalination* 420: 318–329.

Bahmanyar, A., Asghari, M., Khoobi, N. 2012. Numerical simulation and theoretical study on simultaneously effects of operating parameters in direct contact membrane distillation. *Chem. Eng. Proc.: Process Intensif.* 61: 42–50.

Bhardwaj, N., Kundu, S.C. 2010. Electrospinning: A fascinating fiber fabrication technique. *Biotechnol. Adv.* 28: 325–347.

Bonyadi, S., Chung, T.S. 2007. Flux enhancement in membrane distillation by fabrication of dual layer hydrophilic-hydrophobic hollow fiber membranes. *J. Membr. Sci.* 306: 134–146.

Braghirolli, D.I., Steffens, D., Pranke, P. 2014. Electrospinning for regenerative medicine: A review of the main topics. *Drug Discov. Today* 19: 743–753.

Brogioli, D., Mantia, F.L., Yip, N.Y. 2018. Thermodynamic analysis and energy efficiency of thermal desalination processes. *Desalination* 428: 29–39.

Cai, J., Liu, X., Zhao, Y., Guo, F. 2018. Membrane desalination using surface fluorination treated electrospun polyacrylonitrile membranes with nonwoven structure and quasi-parallel fibrous structure. *Desalination* 429: 70–75.

Chang, J., Zuo, J., Zhang, L., O'Brien, G.S., Chung, T.S. 2017. Using green solvent, triethyl phosphate (TEP), to fabricate high porous PVDF hollow fiber membranes for membrane desalination. *J. Membr. Sci.* 539: 295–304.

Chew, N.G.P., Zhao, S., Loh, C.H., Permogorov, N., Wang, R. 2017. Surfactant effects on water recovery from produced water via direct-contact membrane distillation. *J. Membr. Sci.* 528: 126–134.

Deshmukh, A., Boo, C., Karanikola, V., Lin, S., Straub, A.P., Tong, T., Warsinger, D.M., Elimelech, M. 2018. Membrane distillation at the water-energy nexus: Limits, opportunities, and challenges. *Energy Environ. Sci.* doi:10.1039/C8EE00291F.

Deshmukh, A., Elimelech, M. 2017. Understanding the impact of membrane properties and transport phenomena on the energetic performance of membrane distillation desalination. *J. Membr. Sci.* 539: 458–474.

Duong, H.C., Chivas, A.R., Nelemans, B., Duke, M., Gray, S., Cath, T.Y., Nghiem, L.D. 2015. Treatment of RO brine from CGS produced water by spiral-wound air gap membrane distillation-A pilot study. *Desalination* 366: 121–129.

Duong, H.C., Chuai, D., Woo, Y.C., Shon, H.K., Nghiem, L.D., Sencadas, V. 2018. A novel electrospun, hydrophobic, and elastomeric styrene-butadiene-styrene membrane for membrane distillation applications. *J. Membr. Sci.* 549: 420–427.

Ebrahimi, A., Karimi, M., Ashtiani, F.Z. 2018. Characterization of triple electrospun layers of PVDF for direct contact membrane distillation process. *J. Polymer Res.* 25: 50–60.

Eda, G., Liu, J., Shivkumar, S. 2007. Flight path of electrospun polystyrene solutions: Effects of molecular weight and concentration. *Matter. Lett.* 61: 1451–1455.

El-Bourawi, M.S., Ding, Z., Ma, R., Khayet, M. 2006. A framework for better understanding membrane distillation separation process. *J. Membr. Sci.* 285: 4–29.

Eykens, L., De Sitter, K., Dotremont, C., Pinoy, L., Van der Bruggen, B. 2017a. Membrane synthesis for membrane distillation: A review. *Sep. Purif. Technol.* 182: 36–51.

Eykens, L., De Sitter, K., Dotremont, C., Pinoy, L., Van der Bruggen, B. 2016a. Characterization and performance evaluation of commercially available hydrophobic membranes for direct contact membrane distillation. *Desalination* 392: 63–73.

Eykens, L., De Sitter, K., Stoops, L., Dotremont, C., Pinoy, L., Van der Bruggen, B. 2017b. Development of polyethersulfone phase-inversion membranes for membrane distillation using oleophobic coatings. *J. Appl. Polymer Sci.* 134: 45516–45527.

Eykens, L., Hitsov, I., De Sitter, K., Dotremont, C., Pinoy, L., Nopens, I., Van der Bruggen, B. 2016b. Influence of membrane thickness and process conditions on direct contact membrane distillation at different salinities. *J. Membr. Sci.* 498: 353–364.

Fane, A.G. 2018. A grand challenge for membrane desalination: More water, less carbon. *Desalination* 426: 155–163.

Feng, C., Khulbe, K.C., Matsuura, T., Gopal, R., Kaur, S., Ramakrishna, S., Khayet, M. 2008. Production of drinking water from saline water by air-gap membrane distillation using polyvinylidene fluoride nanofiber membrane. *J. Membr. Sci.* 311: 1–6.

Fridrikh, S.V., Yu, J.H., Brenner, M.P., Rutledge, G.C. 2003. Controlling the fiber diameter during electrospinning. *Physic. Rev. Lett.* 90: 144502–144504

Ghaffour, N., Bundschuh, J., Mahmoudi, H., Goosen, M.F.A. 2015. Renewable energy-driven desalination technologies: A comprehensive review on challenges and potential applications of integrated systems. *Desalination* 356: 94–114.

Ghaffour, N., Missimer, T.M., Amy, G.L. 2013. Technical review and evaluation of the economics of water desalination: Current and future challenges for better water supply sustainability. *Desalination* 309: 197–207.

Gillam, W.S., McCoy, W.H. 1967. Desalination research. *Desalination* 2: 14–19.

Gryta, M. 2005. Long term performance of membrane distillation process. *J. Membr. Sci.* 265: 153–159.

Gupta, P., Elkins, C., Long, T.E., Wilkes, G.L. 2005. Electrospinning of linear homopolymers of poly(methyl methacrylate): Exploring relationships between fiber formation, viscosity, molecular weight and concentration in a good solvent. *Polymer* 46: 4799–4810.

Han, L., Tan, Y.Z., Netke, T., Fane, A.G., Chew, J.W. 2017. Understanding oily wastewater treatment via membrane distillation. *J. Membr. Sci.* 539: 284–294.

He, K., Hwang, J., Woo, M.W., Moon, I.S. 2011a. Production of drinking water from saline water by direct contact membrane distillation (DCMD). *J. Ind. Eng. Chem.* 17: 41–48.

He, K., Jung, H., Moon, I.S. 2011b. Air gap membrane distillation on the different types of membrane. *Korean J. Chem. Eng.* 28: 770–777.

He, Q., Tu, T., Yan, S., Yang, X., Duke, M., Zhang, Y., Zhao, S. 2018. Relating water vapor transfer to ammonia recovery from biogas slurry by vacuum membrane distillation. *Sep. Purif. Technol.* 191: 182–191.

Hou, D., Wang, Z., Wang, K., Wang, J., Lin, S. 2018. Composite membrane with electrospun multiscale-texture surface for robust oil-fouling resistance in membrane distillation. *J. Membr. Sci.* 546: 179–187.

Hu, X., Liu, S, Zhou, G., Huang, Y., Xie, Z., Jing, X. 2014. Electrospinning of polymeric nanofibers for drug delivery applications. *J. Controlled Release* 185: 12–21.

Huang, Q.L., Huang, Y., Xiao, C.F., You, Y.W., Zhang, C.X. 2017. Electrospun ultrafine fibrous PTFE-supported ZnO porous membrane with self-cleaning function for vacuum membrane distillation. *J. Membr. Sci.* 534: 73–82.

Hwang, H.J., He, K., Gray, S., Zhang, J., Moon, I.S. 2011. Direct contact membrane distillation (DCMD): Experimental study on the commercial PTFE membrane and modeling. *J. Membr. Sci.* 371: 90–98.

Islam, M.S., McCutcheon, J.R., Rahman, M.S. 2017. A high flux polyvinyl acetate-coated electrospun nylon 6/SiO_2 composite microfiltration membrane for the separation of oil-in-water emulsion with improved antifouling performance. *J. Membr. Sci.* 537: 297–309.

Jafari, A., Kebari, M.R.S., Rahimpour, A., Bakeri, G. 2018. Graphene quantum dots modified polyvinylidenefluride (PVDF) nanofibrous membranes with enhanced performance for air gap membrane distillation. *Chem. Eng. Proc.: Process Intensif.* 126: 222–231.

Jung, J.W., Lee, C.L., Yu, S., Kim, I.D. 2016. Electrospun nanofibers as a platform for advanced secondary batteries: A comprehensive review. *J. Mater. Chem. A* 4: 703–750.

Kaplan, R., Mamrosh, D., Salih, H.H., Dastgheib, S.A. 2017. Assessment of desalination technologies for treatment of a highly saline brine from a potential CO_2 storage site. *Desalination* 404: 87–101.

Karkhanechi, H., Salmani, S., Asghari, M. 2015. A review on gas separation applications of supported ionic liquid membranes. *ChemBioEng* 2: 290–302.

Kaur, S., Barhate, R., Sundarrajan, S., Matsuura, T., Ramakrishna, S. 2011. Hot pressing of electrospun membrane composite and its influence on separation performance on thin film composite nanofiltration membrane. *Desalination* 279: 201–209.

Kaur, S., Sundarrajan, S., Rana, D., Sridhar, R., Gopal, R., Matsuura, T., Ramakrishna, S. 2014. Review: The characterization of electrospun nanofibrous liquid filtration membranes. *J. Mater. Sci.* 49: 6143–6159.

Ke, H., Feldman, E., Guzman, P., Cole, J., Wei, Q., Chu, B., Alkhudhiri, A., Alrasheed, R., Hsiao, B.S. 2016. Electrospun polystyrene nanofibrous membranes for direct contact membrane distillation. *J. Membr. Sci.* 515: 86–97.

Khalifa, A.E. 2015. Water and air gap membrane distillation for water desalination-An experimental comparative study. *Sep. Purif. Technol.* 141: 276–284.

Khayet, M. 2011. Membranes and theoretical modeling of membrane distillation: A review. *Adv. Colloid Interface Sci.* 164: 56–88.

Khayet, M. 2013. Solar desalination by membrane distillation: Dispersion in energy consumption analysis and water production costs (a review). *Desalination* 308: 89–101.

Khayet, M., Garcia-Payo, M.C., Garcia-Fernandez, L., Contreras-Martinez, J. 2018. Dual-layered electrospun nanofibrous membranes for membrane distillation. *Desalination* 426: 174–184.

Khayet, M., Mengual, J.I., Matsuura, T. 2005. Porous hydrophobic/hydrophilic composite membranes: Application in desalination using direct contact membrane distillation. *J. Membr. Sci.* 252: 101–113.

Khulbe, K.C., Matsuura, T. 2017. Recent progresses in preparation and characterization of RO membranes. *J. Membr. Sci. Res.* 3: 174–186.

Koski, A., Yim, K., Shivkumar, S. 2004. Effect of molecular weight on fibrous PVA produced by electrospinning. *Matter. Lett.* 58: 493–497.

Lalia, B.S., Kochkodan, V., Hashaikeh, R., Hilal, N. 2013. A review on membrane fabrication: Structure, properties and performance relationship. *Desalination* 326: 77–95.

Lee, E.J., An, A.K., Hadi, P., Lee, S., Woo, Y.C., Shon, H.K. 2017. Advanced multi-nozzle electrospun functionalized titanium dioxide/polyvinylidene fluoride-co-hexafluoropropylene (TiO_2/PVDF-HFP) composite membranes for direct contact membrane distillation. *J. Membr. Sci.* 524: 712–720.

Li, X., Garcia-Payo, M.C., Khayet, M., Wang, M., Wang, X. 2017. Superhydrophobic polysulfone/polydimethylsiloxane electrospun nanofibrous membranes for water desalination by direct contact membrane distillation. *J. Membr. Sci.* 542: 308–319.

Li, X., Wang, C., Yang, Y., Wang, X., Zhu, M., Hsiao, B.S. 2014. Dual-biomimetic superhydrophobic electrospun polystyrene nanofibrous membranes for membrane distillation. *ACS Appl. Mater. Interfaces* 6: 2423–2430.

Liao, Y., Loh, C.H., Tian, M., Wang, R., Fane, A.G. 2018. Progress in electrospun polymeric nanofibrous membranes for water treatment: Fabrication, modification and applications. *Progress Polymer Sci.* 77: 69–94.

Lin, P.J., Yang, M.C., Li, Y.L., Chen, J.H. 2015. Prevention of surfactant wetting with agarose hydrogel layer for direct contact membrane distillation used in dyeing wastewater treatment. *J. Membr. Sci.* 475: 511–520.

Long, R., Lai, X., Liu, Z., Liu, W. 2018. Direct contact membrane distillation system for waste heat recovery: Modeling and multi-objective optimization. *Energy* 148: 1060–1068.

Lovineh, S.G., Asghari, M., Rajaei, B. 2013. Numerical simulation and theoretical study on simultaneous effects of operating parameters in vacuum membrane distillation. *Desalination* 314: 59–66.

Lu, F., Astruc, D. 2018. Nanomaterials for removal of toxic elements from water. *Coordination Chem. Rev.* 356: 147–164.

Luo, C.J., Nangrejo, M., Edirisinghe, M. 2010. A novel method of selecting solvents for polymer electrospinning. *Polymer* 51: 1654–1662.

Luo, C.J., Stride, E., Edirisinghe, M. 2012. Mapping the influence of solubility and dielectric constant on electrospinning polycaprolactone solutions. *Macromolecules* 45: 4669–4680.

Makanjuola, O., Janajreh, I., Hashaikeh, R. 2018. Novel technique for fabrication of electrospun membranes with high hydrophobicity retention. *Desalination* 436: 98–106.

Manee-in, J., Nithitanakul, M., Supaphol, P. 2006. Effects of solvent properties, solvent system, electrostatic field strength, and inorganic salt addition on electrospun polystyrene fibers. *Iranian Polymer J.* 15: 341–354.

Martinez-Diez, L., Vazquez-Gonzalez, M.I. 1999. Temperature and concentration polarization in membrane distillation of aqueous salt solutions. *J. Membr. Sci.* 156: 265–273.

Matsuura, T. 2001. Progress in membrane science and technology for seawater desalination—a review. *Desalination* 134: 47–54.

Milleret, V., Hefti. T., Hall, H., Vogel, V., Eberli. D. 2012. Influence of the fiber diameter and surface roughness of electrospun vascular grafts on blood activation. *Acta Biomater.* 8: 4349–4356.

Mohammadi, T., Kazemi, P. 2014. Taguchi optimization approach for phenolic wastewater treatment by vacuum membrane distillation. *Desal. Water Treat.* 52: 1341–1349.

Moore, S.E., Mirchandani, S.D., Karanikola, V., Nenoff, T.M., Arnold, R.G., Saez, E. 2018. Process modeling for economic optimization of a solar driven sweeping gas membrane distillation desalination system. *Desalination* 437: 108–120.

Moradi, G., Rajabi, L., Dabirian, F., Zinadini, S. 2017. Biofouling alleviation and flux enhancement of electrospun PAN microfiltration membranes by embedding of *para*-aminobenzoate alumoxane nanoparticles. *J. Appl. Polymer Sci.* 135: 45738–45751.

Munirasu, S., Banat, F., Durrani, A.A., Haija, M.A. 2017. Intrinsically superhydrophobic PVDF membrane by phase inversion for membrane distillation. *Desalination* 417: 77–86.

Nair, M., Kumar, D.2013. Water desalination and challenges: The middle east perspective—A review. *Desal. Water Treat.* 51: 2030–2040.

Naseri Rad, S., Shirazi, M.M.A., Kargari, A., Marzban, R. 2016. Application of membrane separation technology in downstream processing of Bacillus thuringiensis biopesticide: A review. *J. Membr. Sci. Res.* 2: 66–77.

Ozbey-Unal, B., Imer, D.Y., Keskinler, B., Koyuncu, I. 2018. Boron removal from geothermal water by air gap membrane distillation. *Desalination* 433: 141–150.

Park, S.H., Kim, J.H., Moon, S.J., Drioli, E., Lee, Y.M. 2018. Enhanced, hydrophobic, fluorine-containing, thermally rearranged (TR) nanofiber membranes for desalination via membrane distillation. *J. Membr. Sci.* 550: 545–553.

Persano, L., Camposeo, A., Tekmen, Pisignano, D. 2013. Industrial upscaling of electrospinning and applications of polymer nanofibers: A review. *Macromolecular Mater Eng.* 298: 504–520.

Phattaranawik, J., Jiraratananon, R., Fane, A.G. 2003. Heat transport and membrane distillation coefficients in direct contact membrane distillation. *J. Membr. Sci.* 212: 177–193.

Qtaishat, M., Matsuura, T., Kruczek, B., Khayet, M. 2008. Heat and mass transfer analysis in direct contact membrane distillation. *Desalination* 219: 272–292.

Qu, X., Alvarez, J.J., Li, Q. 2013. Applications of nanotechnology in water and wastewater treatment. *Water Res.* 47: 3931–3946.

Ragunath, S., Roy, S., Mitra, S. 2018. Carbon nanotube immobilized membrane with controlled nanotube incorporation via phase inversion polymerization for membrane distillation-based desalination. *Sep. Purif. Technol.* 194: 249–255.

Ray, S.S., Chen, S.S., Nguyen, N.C., Hsu, H.T., Nguyen, H.T., Chang, C.T. 2017. Poly(vinyl alcohol) incorporated with surfactant based electrospun nanofibrous layer onto polypropylene mat for improved desalination by using membrane distillation. *Desalination* 414: 18–27.

Ren, L.F., Xia, F., Chen, V., Shao, J., Chen, R., He, Y. 2017. TiO_2-FTCS modified superhydrophobic PVDF electrospun nanofibrous membrane for desalination by direct contact membrane distillation. *Desalination* 423: 1–11.

Saleem, J., Riaz, M.A., McKay, G. 2018. Oil sorbents from plastic wastes and polymers: A review. *J. Hazardous Mater.* 341: 424–437.

SalehHudin, H.S., Mohamad, E.N., Mahadi, W.N.L., Afifi, A.M. 2018. Multi-jet electrospinning methods for nanofiber processing: A review. *Mater. Manufacturing Proc.* 33: 479–498.

Sarbatly, R., Krishnaiah, D., Kamin, Z. 2016. A review of polymer nanofibers by electrospinning and their application in oil-water separation for cleaning up marine oil spill. *Mar. Pollut. Bull.* 106: 8–16.

Scholten, E., Dhamankar, H., Bromberg, L., Rutledge, G.C., Hatton, T.A. 2011. Electrospray as tool for drug micro- and nanoparticle patterning. *Langmuir* 27: 6683–6688.

Shirazi, M.J.A., Bazgir, S., Shirazi, M.M.A., Ramakrishna, S. 2013a. Coalescing filtration of oily wastewaters: Characterization and application of thermal treated electrospun polystyrene filters. *Desal. Water Treat.* 51: 5974–5986.

Shirazi, M.M.A., Bastani, D., Kargari, A., Tabatabaei, M. 2013b. Characterization of polymeric membranes for membrane distillation using atomic force microscopy. *Desal. Water Treat.* 51: 6003–6008.

Shirazi, M.M.A., Kargari, A. 2015. A review on applications of membrane distillation (MD) process for wastewater treatment. *J. Membr. Sci. Res.* 1: 101–112.

Shirazi, M.M.A., Kargari, A., Bazgir, S., Tabatabaei, M., Shirazi, M.J.A., Abdullah, M.S., Matsuura, T., Ismail, A.F. 2013c. Characterization of electrospun polystyrene membrane for treatment of biodiesel's water-washing effluent using atomic force microscopy. *Desalination* 329: 1–8.

Shirazi, M.M.A., Kargari, A., Ramakrishna, S., Doyle, J., Rajendarian, M., Babu P, R. 2017. Electrospun membranes for desalination and water/wastewater treatment: A comprehensive review. *J. Membr. Sci. Res.* 3: 209–227.

Shirazi, M.M.A., Kargari, A., Shirazi, M.J.A. 2012. Direct contact membrane distillation for seawater desalination. *Desal. Water Treat.* 49: 368–375.

Shirazi, M.M.A., Kargari, A., Tabatabaei, M. 2014a. Evaluation of commercial PTFE membranes in desalination by direct contact membrane distillation. *Chem. Eng. Proc.: Process Intensif.* 76: 16–25.

Shirazi, M.M.A., Kargari, A., Tabatabaei, M., Ismail, A.F., Matsuura, T. 2014b. Concentration of glycerol from dilute glycerol wastewater using sweeping gas membrane distillation. *Chem. Eng. Proc.: Process Intensif.* 78: 58–66.

Shojaie, R., Karimi-Sabet, J., Mousavian, S.M.A., Khadiv-Parsi, P., Moradi, R. 2017. Optimal modification of poly(vinylidene fluoride) membrane surface by using surface-modifying macromolecules for application in membrane distillation. *Desal. Water Treat.* 71: 62–78.

Silva, T.L., Morales-Torres, S., Figueiredo, J.L., Silva, A.M.T. 2015. Multi-walled carbon nanotube/PVDF blended membranes with sponge- and finger-like pores for direct contact membrane distillation. *Desalination* 357: 233–245.

Smolders, K., Franken, A.C.M. 1989. Terminology for membrane distillation. *Desalination* 72: 249–262.

Son, W.K., Youk, J.H., Lee, T.S., Park, W.H. 2004. The effect of solution properties and polyelectrolyte on electrospinning of ultrafine poly(ethylene oxide) fibers. *Polymer* 45: 2959–2966.

Subramanian, S., Seeram, R. 2013. New directions in nanofiltration applications-Are nanofibers the right materials as membranes in desalination? *Desalination* 308: 198–208.

Sun, Y., Liu, N., Shang, J., Zhang, J. 2017. Sustainable utilization of water resources in China: A system dynamics model. *J. Cleaner Prod.* 142: 613–625.

Sundaramurthy, J., Li, N., Kumar, P.S., Ramakrishna, S. 2014. Perspective of electrospun nanofibers in energy and environment. *Biofuel Res. J.* 1: 44–54.

Tan, E.P.S., Lim, C.T. 2006. Mechanical characterization of nanofibers—A review. *Composites Sci. Technol.* 66: 1102–1111.

Teo, W.E., Ramakrishna S. 2006. A review on electrospinning design and nanofibre assembles. *Nanotechnology* 17: R89–R106.

Tijing, L.D., Choi, J.S., Lee, S., Kim, S.H., Shon, H.K. 2014. Recent progress of membrane distillation using electrospun nanofibrous membrane. *J. Membr. Sci.* 453: 435–462.

Uyar, T., Besenbacher, F. 2008. Electrospinning of uniform polystyrene fibers: The effect of solvent conductivity. *Polymer* 49: 5336–5343.

Wang, P., Chung, T.S. 2015. Recent advances in membrane distillation processes: Membrane development, configuration design and application exploring. *J. Membr. Sci.* 474: 39–56.

Wang, Z., Crandall, C., Sahadevan, R., Menkhaus, T.J., Fong, H. 2017. Microfiltration performance of electrospun nanofiber membranes with varied fiber diameters and different membrane porosity and thickness. *Polymer* 114: 64–72.

Warsinger, D.M., Swaminathan, J., Guillen-Burrieza, E., Arafat, H.A., Lienhard V, J.H. 2015. Scaling and fouling in membrane distillation for desalination applications: A review. *Desalination* 356: 294–313.

Woo, Y.C., Chen, Y., Tijing, L.D., Phuntsho, S., He, T., Choi, J.S., Kim, S.H., Shon, H.K. 2017a. CF_4 plasma-modified omniphobic electrospun nanofiber membrane for produced water brine treatment by membrane distillation. *J. Membr. Sci.* 529: 234–242.

Woo, Y.C., Tijing, L.D., Park, M.J., Yao, M., Choi, J.S., Lee, S., Kim, S.H., An, K.J., Shon, H.K. 2017b. Electropsun dual-layer nonwoven membrane for desalination by air gap membrane distillation. *Desalination* 403: 187–198.

Wu, H., Shen, F., Su, Y., Chen, X., Wan, Y. 2018. Modification of polyacrylonitrile membranes via plasma treatment followed by polydimethylsiloxane coating for recovery of ethyl acetate from aqueous solution through vacuum membrane distillation. *Sep. Purif. Technol.* 197: 178–188.

Yan, K.K., Jiao, L., Lin, S., Ji, X., Lu, Y., Zhang, L. 2018. Superhydrophobic electrospun nanofiber membrane coated by carbon nanotubes network for membrane distillation. *Desalination* 437: 26–33.

Yang, F., Efome, J.E., Rana, D., Matsuura, T., Lan, C. 2018. Metal-organic frameworks supported on nanofiber for desalination by direct contact membrane distillation. *ACS Appl. Mater. Interfaces* doi:10.1021/acsami.8b01371.

Yang, Q., Li, Z., Hong, Y., Zhao, Y., Qui, S., Wang, C., Wei, Y. 2004. Influence of solvents on the formation of ultrathin uniform poly(vinylpyrrolidone) nanofibers with electrospinning. *J. Polymer Sci. Part B* 42: 3721–3726.

Yang, H.C., Zhong, W., Hou, J., Chen, V., Xu, Z.K. 2017. Janus hollow fiber membrane with a mussel-inspired coating on the lumen surface for direct contact membrane distillation. *J. Membr. Sci.* 523: 1–7.

Yao, M., Woo, Y.C., Tijing, L.D., Shim, W.G., Choi, J.S., Kim, S.H., Shon, H.K. 2016. Effect of heat-press conditions on electrospun membranes for desalination by direct contact membrane distillation. *Desalination* 378: 80–91.

Yurteri, C.U., Hartman, R.P.A., Marijnissen, J.C.M. 2010. Producing pharmaceutical particles via electrospraying with an emphasis on nano and nano-structured particles—A review. *KONA Powder Particle J.* 28: 91–115.

Zhang, J., Dow, N., Duke, M., Ostarcevic, E., Li, J.D., Gray, S. 2010. Identification of material and physical features of membrane distillation membranes for high performance desalination. *J. Membr. Sci.* 349: 295–303.

Zhang, J., Xue, Q., Pan, X., Jin, Y., Lu, W., Ding, D., Guo, Q. 2017. Graphene oxide/polyacrylonitrile fiber hierarchical-structured membrane for ultra-fas microfiltration of oil-water emulsion. *Chem. Eng. J.* 307: 643–649.

Zhang, Y., Sivakumar, M., Yang, S., Enever, K., Ramezanianpour, M. 2018. Application of solar energy in water treatment processes: A review. *Desalination* 428: 116–145.

Zhu, Z., Liu, Y., Hou, H., Shi, W., Qu, F., Gui, F., Wang, W. 2018. Dual-bioinspired design for constructing membranes with superhydrophobicity for direct contact membrane distillation. *Environ. Sci. Technol.* 52: 3027–3036.

Wu, R., [illegible] 2018. Modification [illegible] porous [illegible] superhydrophobic [illegible] vacuum and [illegible] distillation [illegible]

Yan, K., [illegible] Zhang, [illegible] 2018. [illegible] membrane coated [illegible] for membrane [illegible]

[illegible]

[illegible] 2016. [illegible] distillation [illegible]

Zhang, [illegible] 2017. [illegible] distillation. Desalination [illegible]

[illegible] 2017. Enhancing [illegible] membranes [illegible] membrane [illegible]

[illegible] 2018. [illegible] superhydrophobic [illegible] membrane distillation [illegible]

Zhong, [illegible] 2016. [illegible] distribution of [illegible]

[illegible] 2015. [illegible]

Zuo, [illegible] 2018. [illegible] composite [illegible] distillation. [illegible]

7 Nanofibrous Composite Membranes for Membrane Distillation

Yuan Liao and Rong Wang

CONTENTS

7.1 INTRODUCTION

Fresh water is a critical resource that is essential for all organisms including human beings, animals and plants. Although three quarters of the Earth's surface is covered by water, only 2.5% is fresh water (Oki and Kanae 2006). Water scarcity is among the main issues being faced by the world. In particular, more than 80 countries encounter severe water stress, and about 25% of the world population have limited access to fresh water (Karagiannis and Soldatos 2008). Fresh water is being depleted at a warning rate today.

A large number of water treatment processes have been developed to reduce or remove existing contaminants and salts in water by physical, chemical and biological approaches (Vahedi and Gorczyca 2012, Li et al. 2008, Hill et al. 1986). Among the various processes, membrane technology has been proved to be highly effective in producing clean water with high quality from surface water, brackish water, seawater and brine water (Nicolaisen 2003). The advantages of membrane processes compared with other water treatment processes include: (1) energy consumption is relatively low; (2) separation can be carried out continuously under mild conditions; (3) membrane processes can be combined with other separation processes and are easy to scale-up; (4) membrane properties are variable and can be optimized to meet special requirements. For instance, membrane distillation (MD) can use low-grade waste heat or renewable energy sources for seawater desalination with high water quality and high water recovery (Khayet 2011).

In a membrane separation process, the membrane plays a significant role in determining the performance of the entire system. A number of different techniques are available for fabricating membranes, which include sintering, stretching, track-etching and phase inversion methods (Lalia et al. 2013b). Most of commercial micro-porous membranes are manufactured by phase inversion where a polymer dope solution is transformed to a porous solid phase induced by either non-solvent or temperature difference (Wang et al. 2010). In addition to the traditional methods, in the last decades, electrospinning has become an internationally recognized method for preparing polymeric membranes composed of nanofibers with diameters down to a few nanometers (Liao et al. 2017). Compared with traditional membranes, the electrospun nanofibrous membranes (ENMs) exhibit attractive advantages such as high surface hydrophobicity, large surface-to-volume ratio and high porosity. It is also much easier to incorporate functional additives or nano-fillers into nanofibers and tailor the chemical or physical properties of the nanofibers, thereby altering the membrane surface properties and pore structure. These key features provide them with great potential in membrane separation processes (Agarwal et al. 2013).

This chapter presents a review of recent progress in fabrication and modification of ENMs for MD applications. After introducing the mechanisms and processes of different MD configurations, the key parameters of MD membranes are discussed in the following section. Next, the strategies to develop novel composite nanofibrous membranes by electrospinning and post-modification/coating are illustrated, followed by the summary of potential MD applications. Finally, brief conclusions and perspectives of novel ENMs development for MD applications are provided.

7.2 SEPARATION MECHANISM AND CONFIGURATIONS

Membrane distillation is a thermally driven membrane process, in which water vapor and volatile molecules transport through a microporous hydrophobic membrane and condense in the permeate side. The hydrophobic membrane is critical in separating the hot feed solution which contains inorganic or organic contaminants, from the permeate side. Depending on the different designs of the permeate side, MD has four basic configurations, including direct contact membrane distillation (DCMD), vacuum membrane distillation (VMD), air gap membrane distillation (AGMD) and sweeping gas membrane distillation (SGMD) (Charfi et al. 2010, Qtaishat et al. 2008, Phattaranawik et al. 2003b, Khayet et al. 2000, Lawson and Lloyd 1997). This section introduces the four basic MD configurations as well as some novel hybrid membrane systems integrated with MD.

7.2.1 Direct Contact Membrane Distillation (DCMD)

In the DCMD configuration shown in Figure 7.1, the hot feed and cold permeate solutions are in direct contact with membrane surfaces. The water vapor or volatile molecules evaporate and condense at the liquid/vapor interfaces on the membrane surfaces of both feed and permeate sides (Phattaranawik et al. 2003b). The transmembrane temperature gradient between the feed and permeate solutions produces the vapor pressure difference, which is the driving force for the transport of the water vapor or volatile molecules across the membrane. Due to the membrane hydrophobicity, the feed solution cannot penetrate through the membrane pores. Since DCMD is the simplest configuration, it has been widely investigated

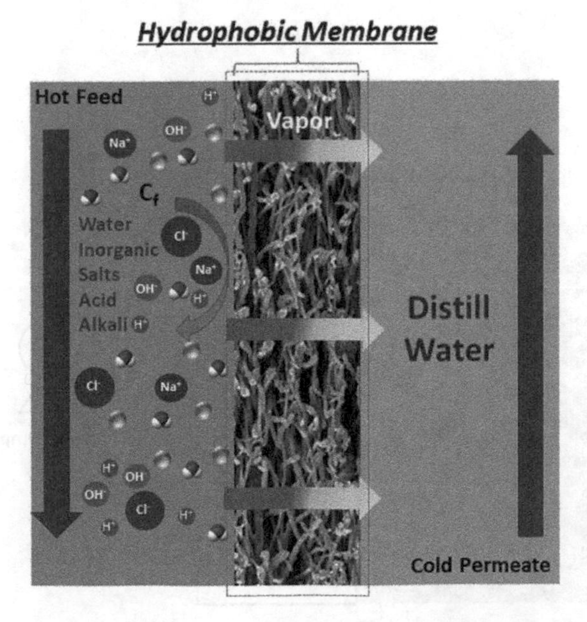

FIGURE 7.1 Configuration of direct contact membrane distillation (DCMD).

(Wang and Chung 2015). Most research evaluated ENMs in the DCMD process (Liao et al. 2017). However, as the membrane is the only barrier between the hot feed and cold permeate solutions, the heat loss in the DCMD is the highest compared to other configurations.

7.2.2 Vacuum Membrane Distillation (VMD)

In the VMD configuration, vacuum is applied at the permeate side of the membrane to remove permeate by a vacuum pump (Figure 7.2). The applied vacuum should be maintained lower than the saturation pressure of volatile molecules in the feed solutions (Lawson and Lloyd 1996). In this configuration, condensation occurs outside the membrane module. The VMD process is often applied to remove VOCs from an aqueous solution (Wang and Chung 2015). However, as an external condenser is required, the capital cost of this configuration is higher (El-Bourawi et al. 2006).

7.2.3 Air Gap Membrane Distillation (AGMD)

As shown in Figure 7.3, a thin stagnant air gap is interposed between the membrane surface at the permeate side and the cold condensation surface in the AGMD configuration (Feng et al. 2008). The evaporated molecules need to pass through the membrane pores as well as the air gap, then condense on the cold surface. Due to a significant transport resistance from the air gap, the membrane flux in the AGMD process is lower than that in the DCMD and VMD configurations (Wang and Chung 2015). The air gap resistance is affected by the air gap width (El-Bourawi et al. 2006).

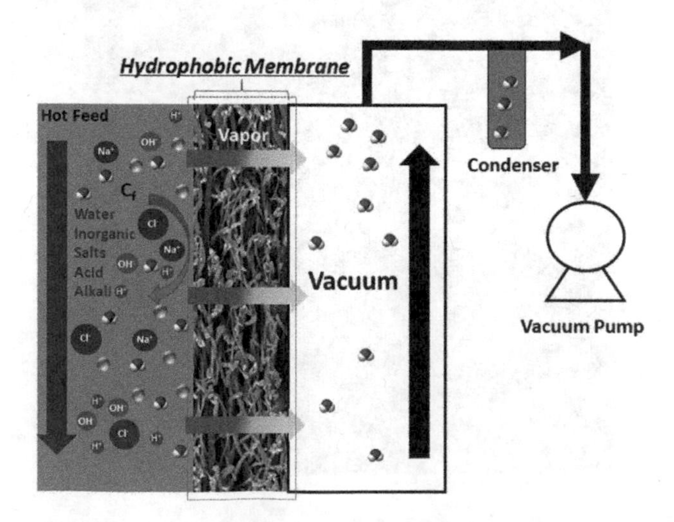

FIGURE 7.2 Configuration of vacuum membrane distillation (VMD).

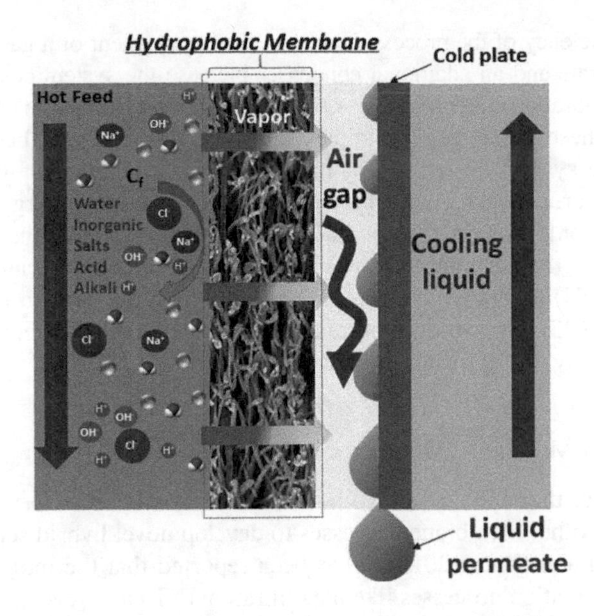

FIGURE 7.3 Configuration of air gap membrane distillation (AGMD).

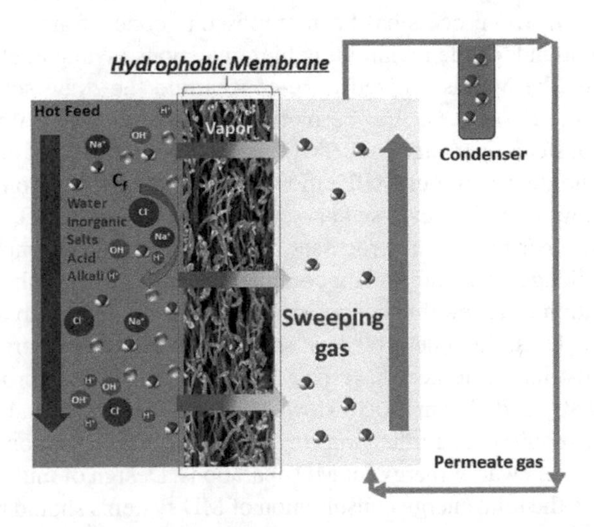

FIGURE 7.4 Configuration of sweeping gas membrane distillation (SGMD).

7.2.4 Sweeping Gas Membrane Distillation (SGMD)

The SGMD configuration shown in Figure 7.4 is developed to offer an intermediate option between the DCMD and AGMD configurations (Khayet et al. 2000). It combines the advantages of lower heat loss of the AGMD configuration and less mass transfer resistance of the DCMD configuration. The gas flow in the SGMD sweeps the evaporated molecules out of the membrane module, which effectively enhances the permeation flux and thermal efficiency of the process. However, the requirement of a gas source with a sufficient flow rate and an additional condenser increase the system cost significantly. Thus, this configuration has received less attention in comparison to the DCMD system.

Based on these basic configurations, extensive works have been conducted to improve the system design and enhance thermal efficiency of the processes. For instance, internal heat recovery has been designed to reduce energy consumption in the AGMD configuration (Yao et al. 2013). In order to reduce the mass transfer resistance due to the stagnant air layer in AGMD, different materials like sponge polyurethane (PU) and polypropylene (PP) mesh have been used to fill the air gap (Francis et al. 2013). It has been reported that the water vapor flux can be increased by 200%–800% by filling the air gap with these materials.

7.2.5 Hybrid Membrane Distillation Processes

In order to reduce the capital cost and enhance the total water recovery, MD has been integrated with other membrane processes to develop novel hybrid separation technologies (Wang and Chung 2015). It has been reported that the integration of MD with other desalination processes like nanofiltration (NF) and reverse osmosis (RO) can increase the overall production of fresh water and enhance energy efficiency of the system (Van der Bruggen 2013, Macedonio and Drioli 2008). Moreover, the forward osmosis (FO)-MD process has been proposed to concentrate proteins or pharmaceutical compounds while maintaining high rejections (Wang et al. 2011). In the FO-MD process, the MD is utilized to re-concentrate the draw solution that has been diluted in the FO process. Furthermore, as brine water with almost saturated salt concentrations can be treated in the MD process, with the aid of crystallizer, the hybrid membrane distillation-crystallization (MDC) could be a promising process to harvest clean water and recover salt crystals simultaneously (Chen et al. 2014).

Compared to other membrane processes, the high energy consumption of MD is the primary challenge impeding its full commercialization (Alkhudhiri et al. 2012). Effective utilization of renewable energy and industrial waste heat to drive MD processes is crucial. Thus, the renewable heat sources such as solar energy, geothermal energy and industrial waste heat have been explored to run MD processes (Banat et al. 2002, Sarbatly and Chiam 2013). However, more researches and developments (R&D) are still required to further improve the effective utilization of low-grade energy sources or renewable energy for MD operations. Design of multi-effective distillation to reduce the total energy consumption of MD systems should be considered.

7.3 MEMBRANE REQUIREMENTS

The critical requirements on MD membranes are outlined as follows.

7.3.1 Hydrophobicity

In the MD processes, the first requirement of an MD membrane is that it should be hydrophobic or comprised of at least one hydrophobic layer. Attributed to the unique rough surface structure constructed by overlapped nanofibers, the ENMs normally have in-air static water contact angles higher than flat sheet membranes prepared by the phase inversion (Liao et al. 2013). Recently, an increasing number of studies have been carried out to develop superhydrophobic membranes exhibiting an in-air water contact angle higher than 150° and a water roll-off angle below 10° for MD by optimizing the electrospinning solutions or post surface modifications (Liao et al. 2013a, 2014a, 2014b). The methods will be further discussed in next section.

7.3.2 Membrane Pore Size

The MD membranes usually possess pore sizes in the range of several nanometers to few micrometers (Phattaranawik et al. 2003a). To prevent the feed solution from penetrating into the membrane pores, the pore size distribution should be as narrow as possible while the maximum pore size should be sufficiently small. However, a small membrane pore size may increase membrane resistance, and thus reduces permeate flux. Thus, a balance between the membrane wetting resistance and mass transfer efficiency should be made by optimizing membrane pore size and pore size distribution.

The pore sizes of ENMs are determined by the diameter of the nanofibers that form the membrane. The membrane pores formed by thinner overlapped nanofibers are smaller than those of the ENMs composed of nanofibers with larger diameters (Liao et al. 2013b). The intrinsic properties of polymer dopes, such as polymer molecular weight, polymer concentration, solvent selections and additives, are crucial factors determining the diameter of resultant nanofibers and thus membrane pore size (Liao et al. 2017). In addition, the pore sizes of ENMs can be optimized by controlling the electrospinning parameters, including the surrounding humidity, the distance between the nozzles and the grounded collector, the applied voltage and the design of grounded collector (Liao et al. 2017).

7.3.3 Liquid Entry Pressure (LEP)

The LEP is the minimum transmembrane pressure where the liquid can penetrate into the membrane pores. The LEP can be calculated by the following Laplace equation (Tijing et al. 2014):

$$\text{LEP} = \frac{-2B\gamma_l}{r_{max}}\cos\theta \tag{7.1}$$

where B is the geometric factor, γ_l is the surface tension of the liquid, θ is the contact angle between the liquid and the membrane surface (the hydrophobicity of the membrane), r_{max} is the largest membrane pore size. To avoid membrane wetting and ensure membrane long-term stability in the MD process, the LEP should be as high as possible, especially when the feed solution contains foulants or has a high salt concentration (Gryta 2005). According to the equation, the LEP of a membrane is mainly affected by the hydrophobicity and pore size of the membrane. Thus, a material with low surface energy is preferred while the maximum pore size should be as small as possible. However, as stated above, reducing the membrane pore size can increase membrane wetting resistance but affect the membrane permeation ability negatively. Thus, an appropriate LEP should be obtained.

7.3.4 Porosity

The membrane porosity has a significant effect on the permeate flux. It is well known that a membrane with a higher porosity possesses more void spaces and better permeability (El-Bourawi et al. 2006). One of the main reasons for the ENMs being popular in MD applications is their high porosity (Liao et al. 2017).

7.3.5 Thickness

The thickness of the hydrophobic layer of MD membranes is inversely proportional to the mass transport efficiency of the membrane (El-Bourawi et al. 2006). In the case of single-layer MD membranes, the membrane thickness should have an optimized value which can balance the mass transport efficiency and thermal conductivity. For multilayer membranes, the hydrophobic layer should be as thin as possible to decrease the membrane mass transfer resistance while the hydrophilic layer should be relatively thick to reduce membrane thermal conductivity and improve membrane thermal efficiency.

ENMs with different thicknesses ranging from 100 to 1500 μm have been developed for MD by controlling electrospinning time (Essalhi and Khayet 2013). With the increase of electrospinning time and polymer concentration, an increase in membrane thickness was observed (Essalhi and Khayet 2014). An analytical model was created to estimate the optimal thickness of ENMs for the DCMD process (Wu et al. 2014). It was found that the optimal membrane thickness varied with the salt concentration of feed solution, MD operational conditions (such as feed inlet temperature) and membrane properties (such as membrane permeability). The optimal ENMs thickness could range from 2 to 739 μm for the feed concentration of NaCl varying from 0 to 24 wt% (Eykens et al. 2016).

Electrospun dual-layer hydrophobic/hydrophilic membranes have been developed for AGMD process (Woo et al. 2017). It was found that the dual-layer membrane composed of a hydrophilic nanofibrous support layer and a hydrophobic nanofibrous active layer exhibited a higher MD flux as the hydrophilic layer reduced mass transfer resistance as well as heat transfer loss.

7.3.6 Tortuosity

The tortuosity factor, which is the deviation of the pore structure from straight cylindrical pores that are normal to the surface, should be as small as possible. This factor is inversely proportional to the membrane permeability in the MD processes (Khayet 2011).

7.3.7 Thermal Conductivity

In the MD process, 50%–80% of the total heat is consumed as latent heat for permeate production while 20%–50% is lost by thermal conductivity (Khayet 2011). It has been reported that when the thermal conductivity of the membrane increased from 0.05 to 0.5 W/mK, both transmembrane flux and thermal efficiency were reduced by 26%–55%, respectively (Al-Obaidani et al. 2008). Thus, the thermal conductivity of the membrane materials should be as low as possible. Most polymers possess similar thermal conductivity coefficients from 0.1 to 0.35 W/mK (Han and Fina 2011, Khayet 2011). It is possible to diminish the membrane heat conductivity by enhancing membrane porosity, as the conductive heat transfer coefficients of air (0.026 W/mK) and water vapor (0.016 W/mK) entrapped in the pores are an order of magnitude smaller than solid polymers (Han and Fina 2011, Khayet 2011). However, a higher porosity may result in a larger membrane pore size, lower membrane wetting resistance and weaker mechanical strength. Thus, there are still research gaps in developing novel thermally isolated membranes for MD applications.

7.3.8 Mechanical Property

Although MD membranes have less requirements of mechanical stiffness than other membranes used in pressure-driven membrane processes such as RO, sufficient mechanical robustness is still required to prevent membrane deflection and rupture during operation.

Heat-pressing post-treatment has been applied to ENMs to investigate its effects on membrane mechanical properties. During the heat-pressing, the nodes of nanofibers were melted together, which created a fusion and linking point between nanofibers, and thus enhanced membrane mechanical robustness (Gopal et al. 2006, Lalia et al. 2013a, Yao et al. 2016). Previous studies demonstrated that the heat-pressing post-treatment not only improved membrane integration and mechanical properties, but also enhanced membrane long-term stability in MD operations (Liao et al. 2013). A comprehensive investigation has been conducted to study the effects of heat-pressing temperature, pressure and duration on the resultant ENMs (Yao et al. 2016). It was found that the effects of heat-pressing temperature and duration were more important than the applied pressure.

In addition, the incorporation of nanofillers such as nanocrystalline cellulose and carbon nanotubes (CNTs) in the nanofiber matrix improved the tensile strength and Young's modulus of the ENMs (Lalia et al. 2014, Tijing et al. 2016). Moreover, mechanical reinforcement can be achieved by attaching the nanofibers or nanobeads on the nonwoven support (Liao et al. 2014a). The excellent combination between the electrospun layers and the nonwoven support can effectively enhance the membrane mechanical robustness.

7.3.9 Thermal Stability

The membrane materials should exhibit good thermal stability as the MD processes may be operated at temperature close to 100°C, depending on the heat source. Graphene nanoparticles have been incorporated in nanofibers to enhance the thermal stability of the resultant composite ENMs (Woo et al. 2016).

7.3.10 Anti-Scaling/Fouling Property

The membrane surface is in direct contact with the feed solution in the MD processes and thus should possess anti-scaling/fouling properties. Anti-fouling properties of ENMs have been achieved by incorporating antimicrobial agent-loaded organosilica nanoparticles in nanofibers (Hammami et al. 2017). In addition, the membrane fouling can be mitigated by creating a superhydrophobic surface with a water contact angle higher than 160° (Hamzah and Leo 2017, Wang et al. 2017). Moreover, hydrophilic nanofibrous layers have been electrospun on a polytetrafluoroethylene (PTFE) hydrophobic substrate to achieve an underwater-superoleophobic membrane with an anti-oil-fouling property for MD processes (Hou et al. 2018a, 2018b). This electrospun in-air hydrophilic and underwater superoleophobic top surface rendered the composite membrane less attractive to oil droplets.

7.4 FABRICATION AND MODIFICATIONS OF ENMs FOR MD

7.4.1 Membrane Materials Selection

Hydrophobic polymers including polypropylene (PP), polyvinylidene fluoride (PVDF) and PTFE are the most popular materials used to fabricate MD membranes due to their low surface energy, satisfactory wetting resistance and suitability for diverse fabrication methods (Khayet 2011). PTFE is a highly crystalline polymer with a low surface energy of $9–20 \times 10^{-3}$ N/m, excellent chemical resistance and thermal stability, which represents an ideal material for MD membrane manufacturing (Francis et al. 2013). However, as PTFE is a non-polar polymer that cannot be dissolved in polar solvents, commercial PTFE membranes are usually fabricated by complicated melt-extrusion, stretching and sintering (Huang et al. 2008, Kitamura et al. 1999). Although commercial PTFE membranes with flat sheet and hollow fiber configurations have been employed in MD systems, the ENMs fabricated from PTFE have not been reported yet (Camacho et al. 2013, Wang and Chung 2015). However, PTFE micro-powders were blended in PVDF solutions as an additive to electrospin superhydrophobic composite ENMs for VMD (Dong et al. 2014). It was observed that the water contact angles and LEPs of the membranes increased significantly with the increase of PTFE micro-powder concentration. In addition, porous PTFE/ZnO composite membranes fabricated by electrospinning a PTFE/poly(vinyl alcohol) (PVA)/zinc acetate composite solution with post-calcination at 380°C showed promising performance in treating feed water contaminated by organic compounds (Huang et al. 2017).

Another highly crystalline polymer, PP, possesses a surface energy of 30.0×10^{-3} N/m (Camacho et al. 2013). Porous PP membranes are usually fabricated by melt-extrusion-stretching or thermal induced phase separation (Chandavasu et al. 2002, 2003, Yang and Perng 2001). Compared to other MD membranes, the membranes prepared from PP exhibited lower performance due to its moderate thermal stability in elevated temperature (Gryta 2007). Development of PP nanofibrous membranes for MD has not been reported.

Apart from PTFE and PP, PVDF is a semi-crystalline polymer with a low surface energy of 30.3×10^{-3} N/m, which can be dissolved in common solvents including dimethylformamide (DMF), dimethylacetamide (DMAC) and n-Methyl-2-Pyrrolidone

(NMP) at room temperature (Kang and Cao 2014). Thus, because of its low surface energy, easy process-ability, thermal and chemical stabilities, PVDF is the most popular material used to develop MD membranes in both academic research and industrial applications (Wang and Chung 2015). Most nanofibrous membranes developed for MD chose PVDF as the membrane material (Liao et al. 2013).

Other hydrophobic polymers have been utilized for MD membrane fabrication by electrospinning. Electrospun polystyrene (PS) nanofibrous membranes have been successfully fabricated for DCMD (Ke et al. 2016). Effects of membrane thickness and nanofiber diameter on membrane pore size, permeate flux and salt rejection were systematically investigated. It was confirmed that PS ENMs are promising to treat brackish and seawater in MD processes. Additionally, a superhydrophobic micro/nanofibrous PS membrane with a hierarchically rough structure was developed using electrospinning (Li et al. 2014). The PS membrane was able to maintain a stable water flux of 104 $Lm^{-2} h^{-1}$ in a 10-h continuous DCMD operation when the temperature difference was 50°C.

Polydimethylsiloxane (PDMS) is an organosilicon elastomer with a low surface energy of 20.0×10^{-3} N/m, which is promising to develop hydrophobic membranes (Li et al. 2017). However, due to its small molecular weight, PDMS itself is unsuitable for electrospinning. Thus, in order to obtain PDMS nanofibers, poly(methyl methacrylate) (PMMA) has been blended with PDMS in the electrospun solutions (Ren et al. 2017). A water contact angle of 163° was obtained on the superhydrophobic PDMS/PMMA hybrid membrane surface. The electrospun hybrid membrane showed a stable flux of 39 $Lm^{-2} h^{-1}$ and a salt rejection of 99.96% during a 24-h DCMD operation when the temperature difference was 45°C.

New materials with high hydrophobicity, low heat conductivity and high thermal stability, such as polyoxadiazoles and polytriazoles, have been synthesized to develop novel ENMs (Maab et al. 2012). Due to the satisfactory hydrophobicity and porosity of resultant ENMs, they exhibited water contact angles as high as 162° and achieved water fluxes up to 85 $Lm^{-2}h^{-1}$ in the DCMD process with rejections higher than 99.95% when the feed and permeate temperatures were 80°C–22°C, respectively. Another novel material, styrene-butadiene-styrene (SBS) with low surface energy and low thermal conductivity, was electrospun to fabricate a microporous nanofibrous membrane for MD (Duong et al. 2018). A stable permeate flux of 10 $Lm^{-2} h^{-1}$ and a salt rejection over 99.7% were achieved in a continuous DCMD operation of 120 hours when the feed and permeate temperatures were 60°C–20°C, respectively.

Hydrophilic ENMs after hydrophobic modification also have promising MD performance in terms of both permeate flux and salt rejection. The nanofibrous membranes fabricated by electrospinning hydrophilic polymer polyamide (PA) followed by surface fluorination *via* chemical vapor deposition (CVD) were able to maintain stable permeate fluxes between 2 and 11 $Lm^{-2}h^{-1}$ with NaCl rejection over 99.98% (Guo et al. 2015). A novel superhydrophobic and self-cleaning ENMs was developed by electrospinning polysulfone (PSU) nanofibers followed by PDMS post-treatment (Li et al. 2017). The PDMS adhesion and agglomeration on the ENMs surface rendered the membranes with superhydrophobic and self-cleaning properties due to their low surface energy and enhanced surface roughness. The resultant membrane showed stable flux and salt rejection over a period of 12 hours DCMD operation.

7.4.2 Post-modification on Nanofiber Surface

To enhance membrane performance in MD processes, one strategy is to conduct post-modification of nanofibers. Most reported works have focused on altering the hydrophobic surfaces of ENMs to be superhydrophobic. The methods used to achieve superhydrophobic surfaces are to create hierarchically structured surfaces on hydrophobic substrates or modify a hierarchical rough surface with a low surface energy chemical (Ma and Hill 2006). As shown in Figure 7.5, chemical modifications have been conducted on a PVDF nanofibrous membrane (Figure 7.5A) to construct a hierarchically rough surface (Figure 7.5B) with both microscale and nanoscale asperities and a nanostructured rough surface(Figure 7.5C) (Liao et al. 2013a). Both hierarchical and nanostructured surfaces provided air pockets between water droplets and the membrane surface, reduced the contact area between the liquid and membrane and consequently enhanced the in-air water contact angles of resultant membranes. Both membranes exhibited water contact angles above 150° and water roll-off angles below 10°. Compared with unmodified membrane, the integral-modified superhydrophobic membrane achieved higher and more stable MD flux.

An omniphobic membrane featuring a multilevel re-entrant surface with an ultralow surface energy would be effective for treating oily wastewater by MD

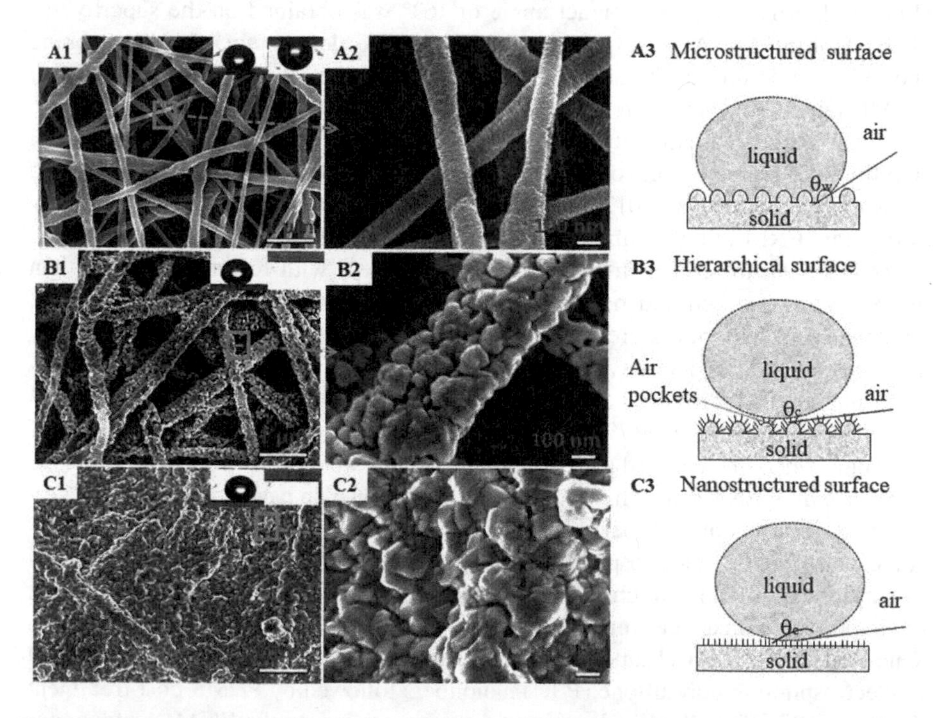

FIGURE 7.5 Surface morphologies of control (A1, A2), integral-modified (B1, B2) and surface-modified PVDF nanofibrous membranes (C1, C2); the schematic illustrations of water droplets on the surfaces of respective membranes (A3, B3 and C3). (From Liao, Y. et al., *J. Membr. Sci.*, 440, 77–87, 2013a.)

(Lee et al. 2016b). ENMs provide a suitable platform for further modification to achieve the re-entrant structure. Multilevel re-entrant structure was implemented by grafting negatively charged silica nanoparticles (NPs) on positively charged nanofibers followed by fluorination (Lee et al. 2016b). The as-fabricated membrane exhibited excellent wetting resistance against organic solvents including mineral oil, decane and ethanol. A stable desalination performance of this omniphobic membrane in the DCMD process was observed when treating sodium dodecyl sulfate (SDS)-containing saline solutions.

7.4.3 Incorporation of Functional Additives in Nanofibers

In addition to the wide selection of polymer materials, various nanomaterials can be easily incorporated into nanofibers by electrospinning blended dopes. Dope solutions prepared by dispersing modified hydrophobic SiO_2 or TiO_2 NPs in PVDF solutions were electrospun to fabricate highly porous superhydrophobic membranes (Lee et al. 2016a, Li et al. 2015, Liao et al. 2014a, Su et al. 2017). The incorporation of NPs endowed membrane surface with hierarchically rough structures and thus water contact angles above 150°. These superhydrophobic composite membranes exhibited more durable water resistance and more stable performance in long-term MD processes. In addition, superhydrophobic Al_2O_3 NPs functionalized with highly branched hydrocarbons were dispersed in PVDF solutions for ENMs fabrication (Attia et al. 2017). It was revealed that the hydrophobicity of nanofibrous membrane was improved by incorporating modified Al_2O_3 NPs in PVDF nanofibers.

Carbon-based nanofillers such as CNTs were also immobilized in nanofibers via electrospinning due to their inherent hydrophobicity, high aspect ratio, outstanding mechanical properties and excellent thermal stability, as shown in Figure 7.6 (Tian et al. 2015, Tijing et al. 2016). The resultant CNTs/PVDF ENMs exhibited robust superhydrophobicity and excellent mechanical properties (Tijing et al. 2016). In addition, they exhibited better flux than commercial PVDF membranes and maintained

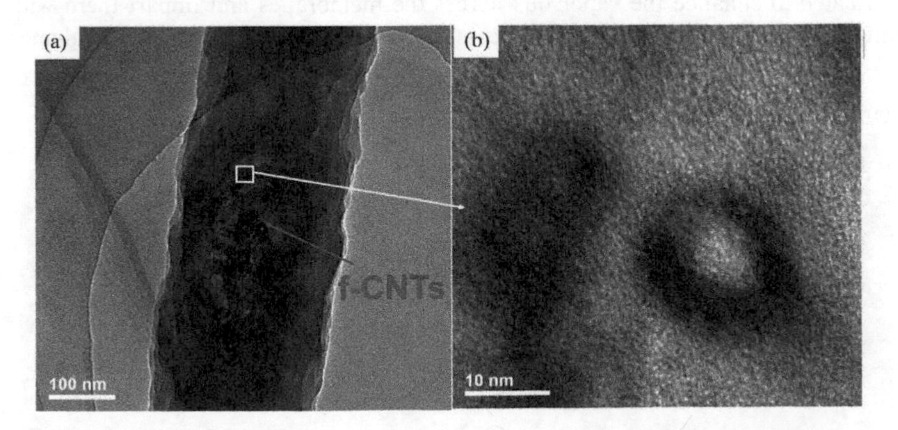

FIGURE 7.6 (a) Transmission electron microscopy (TEM) image of f-CNTs inside the polymeric nanofiber; (b) enlarged. (From Tian, M. et al., *J. Membr. Sci.*, 486, 151–160, 2015.)

a salt rejection over 99.99%. It was also reported that the well-aligned CNTs inside nanofibers not only endowed the composite membranes with superhydrophobicity, but also reduced boundary-layer resistance, enhanced repulsive energy between membrane pores and vapors and thus accelerated vapor transport in membrane pores (Kyoungjin An et al. 2017).

Two-dimensional (2D) graphene is also attractive for MD membrane fabrication due to its hydrophobic nature, capability of adsorbing water vapors and capability to endow anti-fouling properties (Allen et al. 2010). The overall membrane properties could be improved by loading graphene in PVDF nanofibers for long-term AGMD (Woo et al. 2016). The membrane possessed a superhydrophobic surface with a water contact angle above 162°, a membrane porosity over 88% and an LEP of 1.8 bar and showed more stable performance than the commercial PVDF membrane in the AGMD application. Furthermore, embedding graphene oxide (GO) into PVDF ENMs improved membrane antifouling properties due to the strong electrostatic repulsion between the membrane surface and the oily foulants (Ahmadi et al. 2017).

7.4.4 Coating on Nanofibrous Membrane Surface

Another strategy to improve ENMs performance and extend membrane life in the MD process is to develop an additional layer on the ENMs surface via electrospraying or electrospinning. As shown in Figure 7.7a, a superhydrophobic dual-layer composite membrane with a micro/nano-beaded surface was developed by electrospraying a silica-PVDF composite layer on a PVDF nanofibrous substrate (Liao et al. 2014b). As shown in Figure 7.7b, due to the microstructure consisting of modified silica-PVDF mixed islands and the nanostructure comprising of PVDF-bound silica nanoparticles on the islands' surface, the contact lines between the membrane surface and water droplets are contorted and extremely unstable, which makes the water droplets easy to move. This work also demonstrated that the silica-PVDF composite layer was able to improve the membrane stability in long-term DCMD operation.

Dual-layer hydrophilic/hydrophobic composite nanofibrous membranes have been fabricated to enhance the vapor flux across the membranes and impart them with anti-oil-fouling properties in MD processes (Ray et al. 2017). In order to enhance

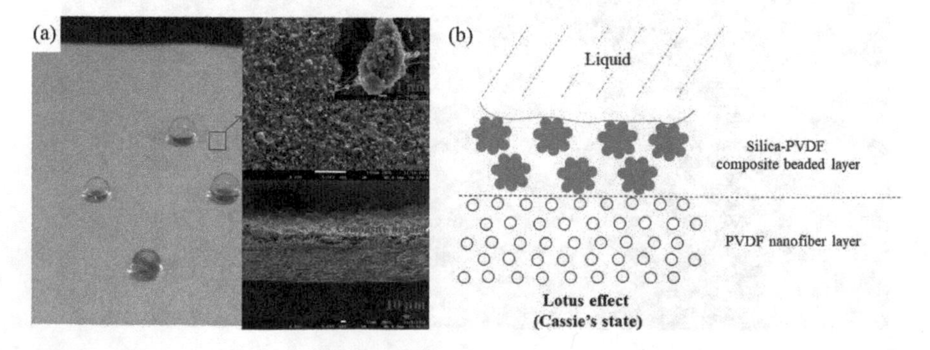

FIGURE 7.7 (a) Photo image, surface and cross-sectional morphologies of superhydrophobic silica-PVDF composite membrane and (b) corresponding schematic illustration.

membrane flux, the hydrophilic layer should face the cold permeate water to adsorb the vapor or volatile molecules from the feed, reduce membrane resistance and prevent heat loss while the hydrophobic layer should face the hot feed solution to prevent membrane pore wetting. It was found that the higher mass transfer efficiency of the dual-layer hydrophilic/hydrophobic membrane should be attributed to the shorter pathway for the vapor to move from the feed side to the permeate side (Bonyadi and Chung 2007). A hydrophilic PVA nanofibrous layer was electrospun on a PP mat to obtain PVA/PP dual-layer membrane for DCMD (Ray et al. 2017). The as-developed dual-layer membrane exhibited two times higher permeate flux compared to a single-layer PP membrane in a 15-h DCMD test. In another piece of research different hydrophilic nanofibrous layers made from either PVA, nylon-6 or polyacrylonitrile (PAN) were coated on a hydrophobic PVDF nanofibrous layer by electrospinning (Woo et al. 2017). According to this work, it was confirmed that an additional thin hydrophilic layer at the permeate side was able to enhance the membrane performance in AGMD systems. The PVDF/nylon-6 dual-layer composite nanofibrous membrane showed the most promising properties due to a linked and intertwined PVDF/nylon-6 interface.

In contrast, an in-air hydrophilic and underwater superoleophobic nanofibrous layer was developed to face contaminated feed solutions to mitigate oil fouling (Hou et al. 2018b). The strong integration between water and the hydrophilic nanofibers via hydrogen bonding could form a barrier preventing the attachment of hydrophobic contamination on the membrane surfaces. A dual-layer composite membrane composed of a PTFE hydrophobic substrate facing permeate water and a thin underwater oleophobic nanofibrous layer was fabricated by electrospinning cellulose acetate (CA)/silica NPs (Hou et al. 2018b). The hydrophilic composite layer drastically decreased membrane surface attraction to oil and consequently enhanced the membrane oil-fouling resistance. While a PTFE membrane was fouled in a DCMD process using a 1000 mg/L crude oil-in-water as feed, the dual-layer membrane was able to achieve a stable performance over 30 hours.

Furthermore, an additional PVDF layer was coated on a nanofibrous poly (vinylidene fluoride-co-hexafluoropropylene) (PVDF-HFP) mat by non-solvent induced phase separation (NIPS) to reduce membrane surface pore size (Shaulsky et al. 2017). The coating layer integrated within the electrospun mat decreased the membrane pore size without sacrificing the surface hydrophobicity. The dual-layer membrane exhibited stable desalination performance in DCMD, which was attributed to the smaller pore size and enhanced surface wetting resistance. In another study, PVDF nanofibers were coated on microporous PVDF flat-sheet membranes prepared by the phase inversion method to increase the membrane hydrophobicity (Efome et al. 2016, Prince et al. 2013). The nanofiber-based top layer increased water contact angles and the LEPs of the resultant composite membranes.

7.5 APPLICATIONS

MD processes can be used for seawater desalination, removal of volatile compounds from wastewater, water reuse, concentration of food productions and water removal from blood and protein solutions for biomedical applications (García-Payo et al. 2000,

Gunko et al. 2006, Hausmann et al. 2014). Among these applications, MD for desalination has been studied extensively as theoretically, 100% rejection of inorganic ions can be achieved in the MD processes. Due to the excellent chemical stability of hydrophobic microporous membranes, MD processes were also applied to treat wastewater under harsh conditions such as acid solutions (Tomaszewska 1993, Tomaszewska et al. 1995). In addition, MD offers a significant advantage over the pressure-driven membrane processes such as RO, as it can extract distilled water from brines with extremely high osmotic pressure (Chen et al. 2014).

7.6 CONCLUSIONS AND PERSPECTIVES

The unique characteristics of ENMs such as high hydrophobicity, high porosity of up to 90%, high feasibility of constructing multi-layer membranes and easy incorporation of functionalized NPs into nanofibers make them highly attractive for MD applications. Significant efforts have been made to enhance membrane hydrophobicity by constructing superhydrophobic surfaces, improve membrane stability by reducing membrane pore size to prevent pore wetting, increase membrane flux by developing dual-layer hydrophilic/hydrophobic composite membranes and endow membranes with excellent anti-fouling properties by coating an additional hydrophilic layer on the membrane surface.

However, the authors argued that novel membranes with combined anti-wetting and anti-fouling properties should be developed. Oils and other organic liquids usually possess a higher surface attraction due to their lower surface tension compared to water (Tuteja et al. 2007). It is highly desirable to develop ENMs with a superamphiphobic (both superhydrophobic and superoleophobic) surface, which can exhibit excellent anti-wetting and anti-fouling performance when treating oil-contaminated feed water (Deng et al. 2012). Nevertheless, it is noted that such surfaces are extremely rare in nature, and there are limited studies on this topic (Liu et al. 2013). It is challenging but meaningful to devote efforts in design and development of superamphiphobic ENMs for MD in future research.

Similarly to other membrane processes, fouling and scaling in MD processes are of particular importance, as they reduce membrane flux significantly, cause contamination in permeate water, and increase the costs of membrane cleaning and replacement (Warsinger et al. 2015). Although some research have found that the superhydrophobic surfaces could mitigate membrane scaling/fouling in MD processes, more and comprehensive research is still required to study the anti-fouling/scaling properties of different surfaces including superhydrophilic, superhydrophobic and superamphiphobic surfaces toward combating complex foulants/scalants in real water (Razmjou et al. 2012). Furthermore, there is a general lack of information about the effects of polymer type on membrane fouling/scaling in MD processes. Different composite nanofibrous membranes can be developed to comprehensively investigate the effects of polymer materials on membrane anti-fouling/scaling performance in MD processes.

While most research in developing anti-fouling/scaling membranes is focused on tailoring the morphology and chemistry of membrane surfaces, the impacts of membrane surface charge are often ignored (Wang et al. 2016). Electrostatic interaction is an important determinant of fouling/scaling phenomena as suggested by the extended Derjaguin Landau Verwey Pverbeek (x-DVLO) theory (Van Oss 1993, Van Oss et al. 1990). Systematic studies revealing the impacts of nanofibrous membrane surface charges on fouling/scaling in MD processes should be conducted.

ENMs have shown great potential for MD application. The scale-up of ENMs production and membrane module designs are equally important to realize the practical application of the ENMs for MD, though these two topics are not covered in this Chapter. Joint efforts from membrane scientists and engineers are needed to facilitate the ENM application in MD processes.

REFERENCES

Agarwal, Seema, Andreas Greiner, and Joachim H. Wendorff. 2013. "Functional materials by electrospinning of polymers." *Progress in Polymer Science* 38 (6):963–991. doi:10.1016/j.progpolymsci.2013.02.001.

Ahmadi, Arsalan, O. Qanati, Mir Saeed Seyed Dorraji, Mohammad Hossein Rasoulifard, and Vahid Vatanpour. 2017. "Investigation of antifouling performance a novel nanofibrous S-PVDF/PVDF and S-PVDF/PVDF/GO membranes against negatively charged oily foulants." *Journal of Membrane Science* 536:86–97. doi:10.1016/j.memsci.2017.04.056.

Al-Obaidani, Sulaiman, Efrem Curcio, Francesca Macedonio, Gianluca Di Profio, Hilal Al-Hinai, and Enrico Drioli. 2008. "Potential of membrane distillation in seawater desalination: Thermal efficiency, sensitivity study and cost estimation." *Journal of Membrane Science* 323 (1):85–98. doi:10.1016/j.memsci.2008.06.006.

Alkhudhiri, Abdullah, Naif Darwish, and Nidal Hilal. 2012. "Membrane distillation: A comprehensive review." *Desalination* 287:2–18. doi:10.1016/j.desal.2011.08.027.

Allen, Matthew J., Vincent C. Tung, and Richard B. Kaner. 2010. "Honeycomb carbon: A review of graphene." *Chemical Reviews* 110 (1):132–145. doi:10.1021/cr900070d.

Attia, Hadi, Shirin Alexander, Chris J. Wright, and Nidal Hilal. 2017. "Superhydrophobic electrospun membrane for heavy metals removal by air gap membrane distillation (AGMD)." *Desalination* 420:318–329. doi:10.1016/j.desal.2017.07.022.

Banat, Fawzi, Rami Yousef Jumah, and Maysoon A. Garaibeh. 2002. "Exploitation of solar energy collected by solar stills for desalination by membrane distillation." *Renewable Energy* 25 (2):293–305. doi:10.1016/S0960-1481(01)00058-1.

Bonyadi, Sina, and Tai Shung Chung. 2007. "Flux enhancement in membrane distillation by fabrication of dual layer hydrophilic–hydrophobic hollow fiber membranes." *Journal of Membrane Science* 306 (1):134–146. doi:10.1016/j.memsci.2007.08.034.

Camacho, Lucy, Ludovic Dumée, Jianhua Zhang, Jun-de Li, Mikel Duke, Juan Gomez, and Stephen Gray. 2013. "Advances in membrane distillation for water desalination and purification applications." *Water* 5 (1):94.

Chandavasu, Chaya, Marinas Xanthos, Kamalesh Sirkar, and Costas G. Gogos. 2002. "Polypropylene blends with potential as materials for microporous membranes formed by melt processing." *Polymer* 43 (3):781–795. doi:10.1016/S0032-3861(01)00654-1.

Chandavasu, C., M. Xanthos, K. K. Sirkar, and C. G. Gogos. 2003."Fabrication of microporous polymeric membranes by melt processing of immiscible blends." *Journal of Membrane Science* 211 (1):167–175. doi:10.1016/S0376-7388(02)00032-7.

Charfi, Kais, Mohamed Khayet, and Mohamed Jomâa Safi. 2010. "Numerical simulation and experimental studies on heat and mass transfer using sweeping gas membrane distillation." *Desalination* 259 (1–3):84–96. doi:10.1016/j.desal.2010.04.028.

Chen, Guizi, Yinghong Lu, William B. Krantz, Rong Wang, and Anthony G. Fane. 2014. "Optimization of operating conditions for a continuous membrane distillation crystallization process with zero salty water discharge." *Journal of Membrane Science* 450:1–11. doi:10.1016/j.memsci.2013.08.034.

Deng, Xu, Lena Mammen, Hans-Jürgen Butt, and Doris Vollmer. 2012. "Candle soot as a template for a transparent robust superamphiphobic coating." *Science* 335 (6064):67–70. doi:10.1126/science.1207115.

Dong, Zhe-Qin, Xiao-hua Ma, Zhen-Liang Xu, Wen-Ting You, and Fang-bing Li. 2014. "Superhydrophobic PVDF–PTFE electrospun nanofibrous membranes for desalination by vacuum membrane distillation." *Desalination* 347:175–183. doi:10.1016/j.desal.2014.05.015.

Duong, Hung Cong, Dan Chuai, Yun Chul Woo, Ho Kyong Shon, Long Duc Nghiem, and Vitor Sencadas. 2018. "A novel electrospun, hydrophobic, and elastomeric styrene-butadiene-styrene membrane for membrane distillation applications." *Journal of Membrane Science* 549:420–427. doi:10.1016/j.memsci.2017.12.024.

Efome, Johnson E., Dipak Rana, Takeshi Matsuura, and Christopher Q. Lan. 2016. "Enhanced performance of PVDF nanocomposite membrane by nanofiber coating: A membrane for sustainable desalination through MD." *Water Research* 89:39–49. doi:10.1016/j.watres.2015.11.040.

El-Bourawi, Mohamed, Zhongwei Ding, Runyu Ma, and Mohamed Khayet. 2006. "A framework for better understanding membrane distillation separation process." *Journal of Membrane Science* 285 (1):4–29. doi:10.1016/j.memsci.2006.08.002.

Essalhi, Mohamed, and Mohamed Khayet. 2013. "Self-sustained webs of polyvinylidene fluoride electrospun nanofibers at different electrospinning times: 1. Desalination by direct contact membrane distillation." *Journal of Membrane Science* 433:167–179. doi:10.1016/ j.memsci.2013.01.023.

Essalhi, Mohamed, and Mohamed Khayet. 2014. "Self-sustained webs of polyvinylidene fluoride electrospun nano-fibers: Effects of polymer concentration and desalination by direct contact membrane distillation." *Journal of Membrane Science* 454:133–143. doi:10.1016/j.memsci.2013.11.056.

Eykens, Lies, I. Hitsov, Kristien De Sitter, Chris Dotremont, Luc Pinoy, Ingmar Nopens, and Bart van der Bruggen. 2016. "Influence of membrane thickness and process conditions on direct contact membrane distillation at different salinities." *Journal of Membrane Science* 498:353–364. doi:10.1016/j.memsci.2015.07.037.

Feng, C. Y., Kailash Khulbe, Takeshi Matsuura, Renuga Gopal, Satinderpal Kaur, Seeram Ramakrishna, and Mohamed Khayet. 2008. "Production of drinking water from saline water by air-gap membrane distillation using polyvinylidene fluoride nanofiber membrane." *Journal of Membrane Science* 311 (1):1–6. doi:10.1016/j.memsci.2007.12.026.

Francis, Lijo, Noreddine Ghaffour, Ahmad A. Alsaadi, and Gary L. Amy. 2013. "Material gap membrane distillation: A new design for water vapor flux enhancement." *Journal of Membrane Science* 448:240–247. doi:10.1016/j.memsci.2013.08.013.

García-Payo, M. C., M. Amparo Izquierdo-Gil, and Cristóbal Fernández-Pineda. 2000. "Air gap membrane distillation of aqueous alcohol solutions." *Journal of Membrane Science* 169 (1):61–80. doi:10.1016/S0376-7388(99)00326-9.

Gopal, Renuga, Satinderpal Kaur, Zuwei Ma, Casey Chan, Seeram Ramakrishna, and Takeshi Matsuura. 2006. "Electrospun nanofibrous filtration membrane." *Journal of Membrane Science* 281 (1):581–586. doi:10.1016/j.memsci.2006.04.026.

Gryta, Marek. 2005. "Long-term performance of membrane distillation process." *Journal of Membrane Science* 265 (1):153–159. doi:10.1016/j.memsci.2005.04.049.

Gryta, Marek. 2007. "Influence of polypropylene membrane surface porosity on the performance of membrane distillation process." *Journal of Membrane Science* 287 (1):67–78. doi:10.1016/j.memsci.2006.10.011.

Gunko, Sergey, Svetlana Verbych, Mykhaylo Bryk, and Nidal Hilal. 2006. "Concentration of apple juice using direct contact membrane distillation." *Desalination* 190 (1):117–124. doi:10.1016/j.desal.2005.09.001.

Guo, Fei, Amelia Servi, Andong Liu, Karen K. Gleason, and Gregory C. Rutledge. 2015. "Desalination by membrane distillation using electrospun polyamide fiber membranes with surface fluorination by chemical vapor deposition." *ACS Applied Materials & Interfaces* 7 (15):8225–8232. doi:10.1021/acsami.5b01197.

Hammami, Mohammed Amen, Jonas G. Croissant, Lijo Francis, Shahad K. Alsaiari, Dalaver H. Anjum, Noreddine Ghaffour, and Niveen M. Khashab. 2017. "Engineering hydrophobic organosilica nanoparticle-doped nanofibers for enhanced and fouling resistant membrane distillation." *ACS Applied Materials & Interfaces* 9 (2):1737–1745. doi:10.1021/acsami.6b11167.

Hamzah, Norhaziyana, and Choe Peng Leo. 2017. "Membrane distillation of saline with phenolic compound using superhydrophobic PVDF membrane incorporated with TiO2 nanoparticles: Separation, fouling and self-cleaning evaluation." *Desalination* 418:79–88. doi:10.1016/j.desal.2017.05.029.

Han, Zhidong, and Alberto Fina. 2011. "Thermal conductivity of carbon nanotubes and their polymer nanocomposites: A review." *Progress in Polymer Science* 36 (7):914–944. doi:10.1016/j.progpolymsci.2010.11.004.

Hausmann, Angela, Peter Sanciolo, Todor Vasiljevic, Ulrich Kulozik, and Mikel Duke. 2014. "Performance assessment of membrane distillation for skim milk and whey processing." *Journal of Dairy Science* 97 (1):56–71. doi:10.3168/jds.2013-7044.

Hill, N. P., Alun E. McIntyre, Roger A. Perry, and John N. Lester. 1986. "Behaviour of chlorophenoxy herbicides during the activated sludge treatment of municipal waste water." *Water Research* 20 (1):45–52. doi:10.1016/0043-1354(86)90212-5.

Hou, Deyin, Chunli Ding, Kuiling Li, Dichu Lin, Dewu Wang, and Jun Wang. 2018a. "A novel dual-layer composite membrane with underwater-superoleophobic/hydrophobic asymmetric wettability for robust oil-fouling resistance in membrane distillation desalination." *Desalination* 428:240–249. doi:10.1016/j.desal.2017.11.039.

Hou, Deyin, Zhangxin Wang, Kunpeng Wang, Jun Wang, and Shihong Lin. 2018b. "Composite membrane with electrospun multiscale-textured surface for robust oil-fouling resistance in membrane distillation." *Journal of Membrane Science* 546:179–187. doi:10.1016/j.memsci.2017.10.017.

Huang, Lei-Ti, Ping-Shun Hsu, Chun-Yin Kuo, Shia-Chung Chen, and Juin-Yih Lai. 2008. "Pore size control of PTFE membranes by stretch operation with asymmetric heating system." *Desalination* 233 (1):64–72. doi:10.1016/j.desal.2007.09.028.

Huang, Qing-Lin, Yan Huang, Chang-Fa Xiao, Yan-Wei You, and Chao-Xin Zhang. 2017. "Electrospun ultrafine fibrous PTFE-supported ZnO porous membrane with self-cleaning function for vacuum membrane distillation." *Journal of Membrane Science* 534:73–82. doi:10.1016/j.memsci.2017.04.015.

Kang, Guo-dong, and Yi-ming Cao. 2014. "Application and modification of poly(vinylidene fluoride) (PVDF) membranes—A review." *Journal of Membrane Science* 463:145–165. doi:10.1016/j.memsci.2014.03.055.

Karagiannis, Ioannis C., and Petros G. Soldatos. 2008. "Water desalination cost literature: Review and assessment." *Desalination* 223 (1–3):448–456. doi:10.1016/j.desal.2007.02.071.

Ke, Huizhen, Emma Feldman, Plinio Guzman, Jesse Cole, Qufu Wei, Benjamin Chu, Abdullah Alkhudhiri, Radwan Alrasheed, and Benjamin S. Hsiao. 2016. "Electrospun polystyrene nanofibrous membranes for direct contact membrane distillation." *Journal of Membrane Science* 515:86–97. doi:10.1016/j.memsci.2016.05.052.

Khayet, Mohamed. 2011. "Membranes and theoretical modeling of membrane distillation: A review." *Advances in Colloid and Interface Science* 164 (1):56–88. doi:10.1016/j.cis.2010.09.005.

Khayet, Mohamed, Paz Godino, and Juan I. Mengual. 2000. "Theory and experiments on sweeping gas membrane distillation." *Journal of Membrane Science* 165 (2):261–272. doi:10.1016/S0376-7388(99)00236-7.

Kitamura, Taketo, Ken-Ichi Kurumada, Masataka Tanigaki, Masahiro Ohshima, and Shin-Ichi Kanazawa. 1999. "Formation mechanism of porous structure in polytetrafluoroethylene (PTFE) porous membrane through mechanical operations." *Polymer Engineering & Science* 39 (11):2256–2263. doi:10.1002/pen.11613.

Kyoungjin An, Alicia, Eui-Jong Lee, Jiaxin Guo, Sanghyun Jeong, Jung-Gil Lee, and Noreddine Ghaffour. 2017. "Enhanced vapor transport in membrane distillation via functionalized carbon nanotubes anchored into electrospun nanofibres." *Scientific Reports* 7:41562. doi:10.1038/srep41562. https://www.nature.com/articles/srep41562#supplementary-information.

Lalia, Boor Singh, Elena Guillen-Burrieza, Hassan A. Arafat, and Raed Hashaikeh. 2013a. "Fabrication and characterization of polyvinylidenefluoride-co-hexafluoropropylene (PVDF-HFP) electrospun membranes for direct contact membrane distillation." *Journal of Membrane Science* 428:104–115. doi:10.1016/j.memsci.2012.10.061.

Lalia, Boor Singh, Elena Guillen, Hassan A. Arafat, and Raed Hashaikeh. 2014. "Nanocrystalline cellulose reinforced PVDF-HFP membranes for membrane distillation application." *Desalination* 332 (1):134–141. doi:10.1016/j.desal.2013.10.030.

Lalia, Boor Singh, Victor Kochkodan, Raed Hashaikeh, and Nidal Hilal. 2013b. "A review on membrane fabrication: Structure, properties and performance relationship." *Desalination* 326 (0):77–95. doi:10.1016/j.desal.2013.06.016.

Lawson, Kevin W., and Douglas R. Lloyd. 1996. "Membrane distillation. I. Module design and performance evaluation using vacuum membrane distillation." *Journal of Membrane Science* 120 (1):111–121. doi:10.1016/0376-7388(96)00140-8.

Lawson, Kevin W., and Douglas R. Lloyd. 1997. "Membrane distillation." *Journal of Membrane Science* 124 (1):1–25. doi:10.1016/S0376-7388(96)00236-0.

Lee, Eui-Jong, Alicia Kyoungjin An, Tao He, Yun Chul Woo, and Ho Kyong Shon. 2016a. "Electrospun nanofiber membranes incorporating fluorosilane-coated TiO2 nanocomposite for direct contact membrane distillation." *Journal of Membrane Science* 520:145–154. doi:10.1016/j.memsci.2016.07.019.

Lee, Jongho, Chanhee Boo, Won-Hee Ryu, André D. Taylor, and Menachem Elimelech. 2016b. "Development of omniphobic desalination membranes using a charged electrospun nanofiber scaffold." *ACS Applied Materials & Interfaces* 8 (17):11154–11161. doi:10.1021/acsami.6b02419.

Li, Qilin, Shaily Mahendra, Delina Y. Lyon, Lena Brunet, Michael V. Liga, Dong Li, and Pedro Jose J. Alvarez. 2008. "Antimicrobial nanomaterials for water disinfection and microbial control: Potential applications and implications." *Water Research* 42 (18):4591–4602. doi:10.1016/j.watres.2008.08.015.

Li, X., M. C. García-Payo, M. Khayet, M. Wang, and X. Wang. 2017. "Superhydrophobic polysulfone/polydimethylsiloxane electrospun nanofibrous membranes for water desalination by direct contact membrane distillation." *Journal of Membrane Science* 542:308–319. doi:10.1016/j.memsci.2017.08.011.

Li, Xiong, Ce Wang, Yin Yang, Xuefen Wang, Meifang Zhu, and Benjamin S. Hsiao. 2014. "Dual-biomimetic superhydrophobic electrospun polystyrene nanofibrous membranes for membrane distillation." *ACS Applied Materials & Interfaces* 6 (4):2423–2430. doi:10.1021/am4048128.

Li, Xiong, Xufeng Yu, Cheng Cheng, Li Deng, Min Wang, and Xuefen Wang. 2015. "Electrospun superhydrophobic organic/inorganic composite nanofibrous membranes for membrane distillation." *ACS Applied Materials & Interfaces* 7 (39):21919–21930. doi:10.1021/acsami.5b06509.

Liao, Yuan, Chun-Heng Loh, Miao Tian, Rong Wang, and Anthony G. Fane. 2017. "Progress in electrospun polymeric nanofibrous membranes for water treatment: Fabrication, modification and applications." *Progress in Polymer Science*. doi:10.1016/j.progpolymsci.2017.10.003.

Liao, Yuan, Chun-Heng Loh, Rong Wang, and Anthony G. Fane. 2014a. "Electrospun superhydrophobic membranes with unique structures for membrane distillation." *ACS Applied Materials & Interfaces* 6 (18):16035–16048. doi:10.1021/am503968n.

Liao, Yuan, Rong Wang, and Anthony G. Fane. 2013a. "Engineering superhydrophobic surface on poly(vinylidene fluoride) nanofiber membranes for direct contact membrane distillation." *Journal of Membrane Science* 440:77–87. doi:10.1016/j.memsci.2013.04.006.

Liao, Yuan, Rong Wang, and Anthony G. Fane. 2014b. "Fabrication of bioinspired composite nanofiber membranes with robust superhydrophobicity for direct contact membrane distillation." *Environmental Science & Technology* 48 (11):6335–6341. doi:10.1021/es405795s.

Liao, Yuan, Rong Wang, Miao Tian, Changquan Qiu, and Anthony G. Fane. 2013b. "Fabrication of polyvinylidene fluoride (PVDF) nanofiber membranes by electrospinning for direct contact membrane distillation." *Journal of Membrane Science* 425–426:30–39. doi:10.1016/j.memsci.2012.09.023.

Liu, Kesong, Ye Tian, and Lei Jiang. 2013. "Bio-inspired superoleophobic and smart materials: Design, fabrication, and application." *Progress in Materials Science* 58 (4):503–564. doi:10.1016/j.pmatsci.2012.11.001.

Ma, Minglin, and Randal M. Hill. 2006. "Superhydrophobic surfaces." *Current Opinion in Colloid & Interface Science* 11 (4):193–202. doi:10.1016/j.cocis.2006.06.002.

Maab, Husnul, Lijo Francis, Ahmad Al-saadi, Cyril Aubry, Noreddine Ghaffour, Gary Amy, and Suzana P. Nunes. 2012. "Synthesis and fabrication of nanostructured hydrophobic polyazole membranes for low-energy water recovery." *Journal of Membrane Science* 423–424:11–19. doi:10.1016/j.memsci.2012.07.009.

Macedonio, Francesca, and E. Drioli. 2008. "Pressure-driven membrane operations and membrane distillation technology integration for water purification." *Desalination* 223 (1):396–409. doi:10.1016/j.desal.2007.01.200.

Nicolaisen, Bjarne. 2003. "Developments in membrane technology for water treatment." *Desalination* 153 (1–3):355–360. doi:10.1016/S0011-9164(02)01127-X.

Oki, Taikan, and Shinjiro Kanae. 2006. "Global hydrological cycles and world water resources." *Science* 313 (5790):1068–1072. doi:10.1126/science.1128845.

Phattaranawik, Jirachote, Ratana Jiraratananon, and A. G. Fane. 2003a. "Effect of pore size distribution and air flux on mass transport in direct contact membrane distillation." *Journal of Membrane Science* 215 (1):75–85. doi:10.1016/S0376-7388(02)00603-8.

Phattaranawik, Jirachote, Ratana Jiraratananon, and A. G. Fane. 2003b. "Heat transport and membrane distillation coefficients in direct contact membrane distillation." *Journal of Membrane Science* 212 (1):177–193. doi:10.1016/S0376-7388(02)00498-2.

Prince, J. A., Vanangamudi Anbharasi, T. S. Shanmugasundaram, and Gurdev Singh. 2013. "Preparation and characterization of novel triple layer hydrophilic–hydrophobic composite membrane for desalination using air gap membrane distillation." *Separation and Purification Technology* 118:598–603. doi:10.1016/j.seppur.2013.08.006.

Qtaishat, M., T. Matsuura, B. Kruczek, and M. Khayet. 2008. "Heat and mass transfer analysis in direct contact membrane distillation." *Desalination* 219 (1–3):272–292. doi:10.1016/j.desal.2007.05.019.

Ray, Saikat Sinha, Shiao-Shing Chen, Nguyen Cong Nguyen, Hung-Te Hsu, Hau Thi Nguyen, and Chang-Tang Chang. 2017. "Poly(vinyl alcohol) incorporated with surfactant based electrospun nanofibrous layer onto polypropylene mat for improved desalination by using membrane distillation." *Desalination* 414:18–27. doi:10.1016/j.desal.2017.03.032.

Razmjou, Amir, Ellen Arifin, Guangxi Dong, Jaleh Mansouri, and Vicki Chen. 2012. "Superhydrophobic modification of TiO2 nanocomposite PVDF membranes for applications in membrane distillation." *Journal of Membrane Science* 415–416:850–863. doi:10.1016/j.memsci.2012.06.004.

Ren, Long-Fei, Fan Xia, Jiahui Shao, Xiaofan Zhang, and Jun Li. 2017. "Experimental investigation of the effect of electrospinning parameters on properties of superhydrophobic PDMS/PMMA membrane and its application in membrane distillation." *Desalination* 404:155–166. doi:10.1016/j.desal.2016.11.023.

Sarbatly, Rosalam, and Chel-Ken Chiam. 2013. "Evaluation of geothermal energy in desalination by vacuum membrane distillation." *Applied Energy* 112:737–746. doi:10.1016/j.apenergy.2012.12.028.

Shaulsky, Evyatar, Siamak Nejati, Chanhee Boo, François Perreault, Chinedum O. Osuji, and Menachem Elimelech. 2017. "Post-fabrication modification of electrospun nanofiber mats with polymer coating for membrane distillation applications." *Journal of Membrane Science* 530:158–165. doi:10.1016/j.memsci.2017.02.025.

Su, Chunlei, Junjun Chang, Kexin Tang, Fang Gao, Yuping Li, and Hongbin Cao. 2017. "Novel three-dimensional superhydrophobic and strength-enhanced electrospun membranes for long-term membrane distillation." *Separation and Purification Technology* 178:279–287. doi:10.1016/j.seppur.2017.01.050.

Tian, Miao, Rong Wang, Kunli Goh, Yuan Liao, and Anthony G. Fane. 2015. "Synthesis and characterization of high-performance novel thin film nanocomposite PRO membranes with tiered nanofiber support reinforced by functionalized carbon nanotubes." *Journal of Membrane Science* 486:151–160. doi:10.1016/j.memsci.2015.03.054.

Tijing, Leonard D., June-Seok Choi, Sangho Lee, Seung-Hyun Kim, and Ho Kyong Shon. 2014. "Recent progress of membrane distillation using electrospun nanofibrous membrane." *Journal of Membrane Science* 453:435–462. doi:10.1016/j.memsci.2013.11.022.

Tijing, Leonard D., Yun Chul Woo, Wang-Geun Shim, Tao He, June-Seok Choi, Seung-Hyun Kim, and Ho Kyong Shon. 2016. "Superhydrophobic nanofiber membrane containing carbon nanotubes for high-performance direct contact membrane distillation." *Journal of Membrane Science* 502:158–170. doi:10.1016/j.memsci.2015.12.014.

Tomaszewska, M., M. Gryta, and A. W. Morawski. 1995. "Study on the concentration of acids by membrane distillation." *Journal of Membrane Science* 102:113–122. doi:10.1016/0376-7388(94)00281-3.

Tomaszewska, Maria. 1993. "Concentration of the extraction fluid from sulfuric acid treatment of phosphogypsum by membrane distillation." *Journal of Membrane Science* 78 (3):277–282. doi:10.1016/0376-7388(93)80007-K.

Tuteja, Anish, Wonjae Choi, Minglin Ma, Joseph M. Mabry, Sarah A. Mazzella, Gregory C. Rutledge, Gareth H. McKinley, and Robert E. Cohen. 2007. "Designing superoleophobic surfaces." *Science* 318 (5856):1618–1622. doi:10.1126/science.1148326.

Vahedi, Arman, and Beata Gorczyca. 2012. "Predicting the settling velocity of flocs formed in water treatment using multiple fractal dimensions." *Water Research* 46 (13):4188–4194. doi:10.1016/j.watres.2012.04.031.

Van der Bruggen, Bart. 2013. "Integrated membrane separation processes for recycling of valuable wastewater streams: Nanofiltration, membrane distillation, and membrane crystallizers revisited." *Industrial & Engineering Chemistry Research* 52 (31):10335–10341. doi:10.1021/ie302880a.

Van Oss, C. J. 1993. "Acid—base interfacial interactions in aqueous media." *Colloids and Surfaces A: Physicochemical and Engineering Aspects* 78:1–49. doi:10.1016/0927-7757(93)80308-2.

Van Oss, C. J., R. F. Giese, and Patricia M. Costanzo. 1990. "DLVO and non-DLVO interactions in hectorite." *Clays Clay Miner* 38 (2):151–159.

Wang, Kai Yu, May May Teoh, Adrian Nugroho, and Tai-Shung Chung. 2011. "Integrated forward osmosis–membrane distillation (FO–MD) hybrid system for the concentration of protein solutions." *Chemical Engineering Science* 66 (11):2421–2430. doi:10.1016/j.ces.2011.03.001.

Wang, Peng, and Tai-Shung Chung. 2015. "Recent advances in membrane distillation processes: Membrane development, configuration design and application exploring." *Journal of Membrane Science* 474:39–56. doi:10.1016/j.memsci.2014.09.016.

Wang, Rong, Lei Shi, Chuyang Y. Tang, Shuren Chou, Changquan Qiu, and Anthony G. Fane. 2010. "Characterization of novel forward osmosis hollow fiber membranes." *Journal of Membrane Science* 355 (1):158–167. doi:10.1016/j.memsci.2010.03.017.

Wang, Zhangxin, Jian Jin, Deyin Hou, and Shihong Lin. 2016. "Tailoring surface charge and wetting property for robust oil-fouling mitigation in membrane distillation." *Journal of Membrane Science* 516:113–122. doi:10.1016/j.memsci.2016.06.011.

Wang, Ziyi, Yuanyuan Tang, and Baoan Li. 2017. "Excellent wetting resistance and anti-fouling performance of PVDF membrane modified with superhydrophobic papillae-like surfaces." *Journal of Membrane Science* 540:401–410. doi:10.1016/j.memsci.2017.06.073.

Warsinger, David M., Jaichander Swaminathan, Elena Guillen-Burrieza, Hassan A. Arafat, and John H. Lienhard V. 2015. "Scaling and fouling in membrane distillation for desalination applications: A review." *Desalination* 356:294–313. doi:10.1016/j.desal.2014.06.031.

Woo, Yun Chul, Leonard D. Tijing, Myoung Jun Park, Minwei Yao, June-Seok Choi, Sangho Lee, Seung-Hyun Kim, Kyoung-Jin An, and Ho Kyong Shon. 2017. "Electrospun dual-layer nonwoven membrane for desalination by air gap membrane distillation." *Desalination* 403:187–198. doi:10.1016/j.desal.2015.09.009.

Woo, Yun Chul, Leonard D. Tijing, Wang-Geun Shim, June-Seok Choi, Seung-Hyun Kim, Tao He, Enrico Drioli, and Ho Kyong Shon. 2016. "Water desalination using graphene-enhanced electrospun nanofiber membrane via air gap membrane distillation." *Journal of Membrane Science* 520:99–110. doi:10.1016/j.memsci.2016.07.049.

Wu, Ho Yan, Rong Wang, and Robert W. Field. 2014. "Direct contact membrane distillation: An experimental and analytical investigation of the effect of membrane thickness upon transmembrane flux." *Journal of Membrane Science* 470:257–265. doi:10.1016/j.memsci.2014.06.002.

Yang, M. C., and J. S. Perng. 2001. "Microporous polypropylene tubular membranes via thermally induced phase separation using a novel solvent—Camphene." *Journal of Membrane Science* 187 (1):13–22. doi:10.1016/S0376-7388(00)00587-1.

Yao, Kun, Yingjie Qin, Yingjin Yuan, Liqiang Liu, Fei He, and Yin Wu. 2013. "A continuous-effect membrane distillation process based on hollow fiber AGMD module with internal latent-heat recovery." *AIChE Journal* 59 (4):1278–1297. doi:10.1002/aic.13892.

Yao, Minwei, Yun Chul Woo, Leonard D. Tijing, Wang-Geun Shim, June-Seok Choi, Seung-Hyun Kim, and Ho Kyong Shon. 2016. "Effect of heat-press conditions on electrospun membranes for desalination by direct contact membrane distillation." *Desalination* 378:80–91. doi:10.1016/j.desal.2015.09.025.

LIST OF ABBREVIATIONS AND SYMBOLS

2D	two-dimensional
AGMD	air gap membrane distillation
CA	cellulose acetate
CNTs	carbon nanotubes
CVD	chemical vapor deposition
DCMD	direct contact membrane distillation
DMAC	dimethylacetamide
DMF	dimethylformamide
ENM	electrospun nanofibrous membrane
FO	forward osmosis
GO	graphene oxide
LEP	liquid entry pressure
MD	membrane distillation
MDC	membrane distillation-crystallization
NF	nanofiltration
NIPS	non-solvent induced phase separation
NMP	n-methyl-2-pyrrolidone
NPs	nanoparticles
PA	polyamide
PDMS	polydimethylsiloxane
PMMA	poly(methyl methacrylate)
PP	polypropylene
PS	polystyrene
PSU	polysulfone
PTFE	polytetrafluoroethylene
PU	polyurethane
PVA	poly(vinyl alcohol)
PVDF	polyvinylidene fluoride
PVDF-HFP	poly(vinylidene fluoride-co-hexafluoropropylene)
RO	reverse osmosis
SBS	styrene-butadiene-styrene
SDS	sodium dodecyl sulfate
SGMD	sweeping gas membrane distillation
TEP	triethyl phosphate
VMD	vacuum membrane distillation
γL	surface tension of the liquid
B	geometric factor
θ	contact angle between the liquid and the membrane surface
rmax	largest membrane pore size

8 Electrospun Nanofiber Membranes as Efficient Adsorbent for Removal of Heavy Metals

Faten Ermala Che Othman and Norhaniza Yusof

CONTENTS

8.1 INTRODUCTION

In recent years, the world has faced serious threats of land, air, and water pollutions due to various human activities. Particularly, water pollution due to drought and misuse, as well as the production of large volume of wastewater, has caused serious environmental effects. Due to that, more stringent environmental legislation has been issued. The wastewater was contaminated by different types of pollutants; however, inorganic pollutants such as heavy metals (cadmium, lead, arsenic, chromium, mercury, etc.) are the most dangerous due to their toxicity and biodegradability (Gupta and Ali, 2013). The intense exposure of heavy metals above their tolerance levels could lead to adverse effects to the aquatic flora and fauna and may constitute serious public health disorders (Sorme and Lagerkvist, 2002). Therefore, it is very crucial to treat metal-contaminated wastewater prior to its discharge to the environment. Due to the shortage of water resources and demands for potable water are rising every year, there is a clear need for the development of effective, safe and economical technologies and materials to meet these challenges. Up to now, heavy metal removal from inorganic effluent in the wastewater can be achieved by numerous treatment approaches including chemical precipitation, ion exchange, membrane filtration, adsorption, photocatalysis, and many more (Fu and Wang, 2011).

Among the available approaches, adsorption has been found as the best alternative method in treating the heavy metals in wastewater with better removal performances as compared to the other available treatments. Adsorption is a process that using adsorbent as adsorption materials to adsorb heavy metals. This method is known to be one of the best technologies for water decontamination due to its effectiveness, economical, and eco-friendly treatment method (Babel and Kurniawan, 2003). Therefore, the search for low-cost adsorbent that has metal-binding capacities, large surface area, and high adsorption capacities has intensified and drawn major attention over the world. According to Kurniawan et al. (2006), the adsorbents could be organic or biological origin, mineral, agricultural wastes, industrial by-products, zeolites, biomass, and polymeric materials.

Although there are many approaches can be utilized for inorganic pollutant treatments, the ideal treatment should be not only suitable, appropriate, and applicable to the local conditions, also able to achieve the maximum contaminant level (MCL) standard established. Recently, low-cost agricultural waste by-products such as coconut husk, rice husk, oil palm shell, sawdust, and neem bark have been utilized for elimination of heavy metals in the contaminated wastewater (Ajmal et al., 2003). However, these low-cost agricultural-based materials were unable to show significant adsorption capabilities, which is the main principle in adsorption method. Even though cost is an important parameter for comparing the adsorbent materials, the cost information is seldom reported, and it is believed the production expenses of individual sorbents vary depending on the degree of required process and local availability.

Part to that, adsorbents that possessed micro- or nano-structure have received major attention due to their large surface area as well as high adsorption capabilities (Kurniawan and Babel, 2003; Barakat, 2011). Even the improved sorption capacity of this new material may compensate the cost of additional processing. Nanofiber membrane (NFM) is one of the nano-adsorbents that has been utilized in recent approaches of heavy metal treatments. The development of the NFMs can be conducted by using electrospinning method by electrospinning polymers into nanofibers and the uniqueness of these NFMs are discussed in details in later sub-topic. Of all the commercially available polymeric materials that utilized in fabrication of NFMs, polyacrylonitrile (PAN) is one of the most widely investigated due to its excellent intrinsic physicochemical properties (discussed in later sub-topic). Thus, in this chapter focus will be given to the overview of the fabrication and importance of NFMs as effective adsorbent for heavy metal removal in wastewater.

8.2 WASTEWATER POLLUTANTS

There are various types of contaminants that contribute to the quality of water resources including thermal pollution, infectious agents, oxygen demanding wastes, radioactive pollutants, sediments and suspended solids, also organic and inorganic pollutants (Abdel-Raouf and Abdul-Raheim, 2017). Among these pollutants, the inorganic effluents are extremely harmful due to their high toxicity and biodegradability. Key examples of inorganic pollutants are heavy metals which have become the most concerning, due to their toxicities to human health and also the environment.

8.2.1 Heavy Metals in Wastewater

Industrial wastewater streams containing heavy metals are effluent from different industries. For instance, the metal surface treatment processes, electroplating, and printed circuit board (PCB) manufacturing are the main contributors that produce a large quantity of wastewaters, residues and sludges containing heavy metals. They can be categorized as hazardous wastes that need extensive waste treatment. Figure 8.1 shows the different heavy metals that are commonly found in industrial wastewaters including cadmium (Cd), zinc (Zn), lead (Pb), chromium (Cr), nickel (Ni), copper (Cu), platinum (Pt), mercury (Ag), and titanium (Ti). Heavy metal is a term that refers to any naturally occurring metallic chemical element that has a relatively high atomic weight and density which exceeds 5 g per cubic centimeter (Srivastava and Majumder, 2008). These types of metals cannot be degraded or destroyed as they are natural components of the Earth's crust. Some metals such as iron (Fe), manganese (Mn), magnesium (Mg), copper (Cu), zinc (Zn), cobalt (Co), and molybdenum (Mo) may be dense or toxic, yet are required micronutrients for various biochemical and physiological functions in human or other organisms. These essential heavy metals may be needed to support key enzymes, act as cofactors, or act in oxidation-reduction reactions. Inadequate supply of these micronutrients results in a variety of deficiency diseases or syndromes.

Certain heavy metals, however, are of grave concern because they are poisonous or toxic to humans and the environment even at low exposure levels as they can cause cellular damage and diseases as shown in Figure 8.1 and described in Table 8.1.

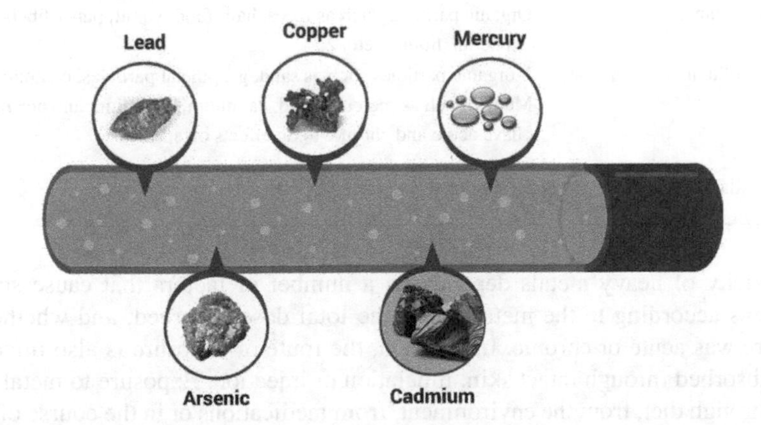

FIGURE 8.1 Different types of heavy metals in wastewater. (From Metals, https://www.fda.gov/food/foodborneillnesscontaminants/metals/, 2016; Reprinted with permission from Shapin, S., The Scientific Rev*olution*, University of Chicago Press, Chicago, IL, pp. 15–64, 2006.)

TABLE 8.1
Various Types of Wastewater Pollutants and Their Effects to on the Environment and/or Humans

Types of Pollutants	Sources and Effects
Thermal pollution	Power stations and industrial manufacturers are the biggest sources of thermal pollution. These plants draw water from nearby source to keep machines cool and then release back to the source with higher temperature. The alteration of the natural environment leads to decrease in the quality of life for these aquatic based life forms and can ultimately destroy habitats.
Infectious agents	Bacteria, viruses, and disease-causing pathogens can pollute beaches and contaminate shellfish populations, leading to restrictions on human recreation, drinking water consumption, and shellfish consumption.
Oxygen demanding wastes	Decaying organic matter and debris can use up the dissolved oxygen (oxygen concentration goes to 0) in a lake so fish and others aquatic biota cannot survive.
Radioactive pollutants	Radioactive waste has been created by military weapons production and testing, mining, electrical power generation, medical diagnosis, manufacturing and treatment, biological and chemical research, and other industrial uses. Radiation exposure can cause cancer, birth defects, and other abnormalities, depending on the time of exposure, amount of radiation, and the decay mechanism.
Sediments and suspended solids	These solids include anything drifting or floating in the water, from sediment, silt, and sand to plankton and algae. As algae, plants, and animals decay, the decomposition process allows small organic particles to break away and enter the water column as suspended solids. Excessive suspended sediment can impair water quality for aquatic and human life.
Organic pollutants	Organic particles such as feces, hair, food, vomit, paper fibers, plant material, humus, etc.
Inorganic pollutants	Inorganic particles such as sand, grit, metal particles, ceramic, etc. Metals, such as mercury, lead, cadmium, chromium and arsenic, can have acute and chronic toxic effects on species.

Source: Adams, R., *Econ. Dev. Soc. Change*, 47, 155–173, 2010.

The toxicity of heavy metals depends on a number of factors that cause specific symptoms according to the metal types, the total dose adsorbed, and whether the exposure was acute or chronic. In addition, the route of exposure is also important either absorbed through intact skin, inhalation or injection. Exposure to metals may occur through diet, from the environment, from medications or in the course of work or play. In recent years, there has been an increasing ecological and global public health concern associated with environmental contamination by these metals.

The tendency of heavy metals to bioaccumulate is the main reason why it is very harmful to human bodies. Bioaccumulation means an increase in the concentration of a chemical in a biological organism over time, compared to the chemical's concentration in the environment (Malik et al., 2010). Compounds accumulate in living things any time they are taken up and stored faster than they are broken down or excreted (Vinodhini and Narayanan, 2008). The wide distribution and concentration of heavy metals in the environment are commonly as a result of human activities, including sources from mining and industrial wastes, domestic, agricultural, medical, and technological applications. Moreover, atmospheric deposition, soil erosion of metal ions, metal corrosion, sediment re-suspension, and metal evaporation from water resources to soil and ground water are some of the processes that lead to the environmental contamination. In addition, natural phenomenon such as volcanic eruption and weathering is also the main contributor to heavy metal pollution. Industrial activities such as coal burning in power plants, nuclear power stations, metal processing in refineries, petroleum combustion, wood preservation, high tension lines, textiles, plastics, microelectronics, and paper processing plants also lead to heavy metal pollution (Figure 8.2).

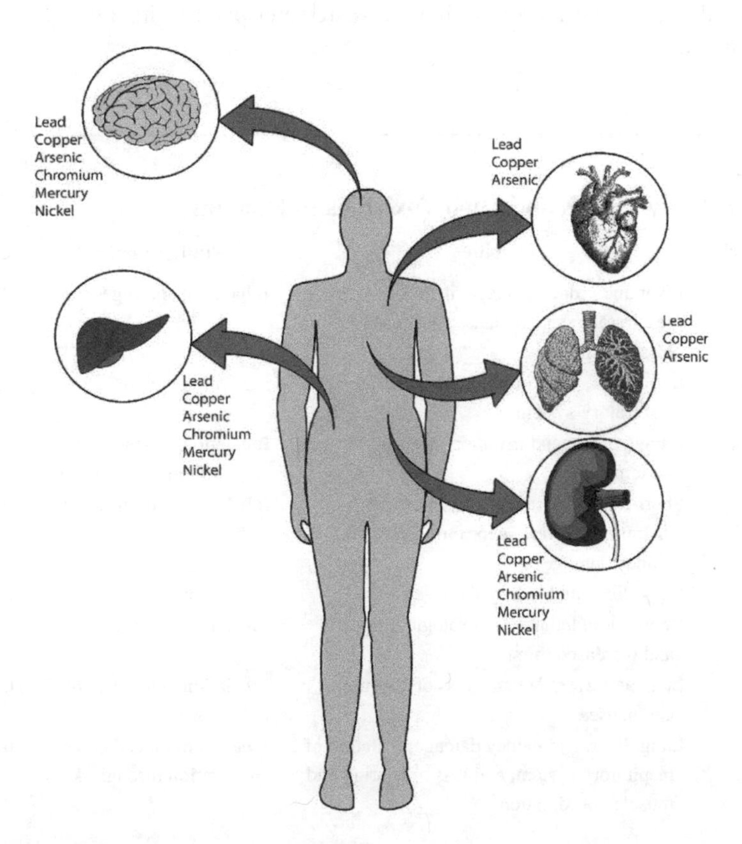

FIGURE 8.2 Effects of extensively exposure heavy metals above tolerance levels to human health. (Reprinted with permission from Shapin, S., *The Scientific Revolution*, University of Chicago Press, Chicago, IL, pp. 15–64, 2006.)

Table 8.2 tabulates the various types of entry, toxicities, and tolerance level of different heavy metals on human health. From the table, it can be seen that long exposure to small amounts of heavy metals can lead to severe health problems for living organisms especially humans. For instance, a little amount of mercury (≈0.00003 mg/L) exposure that accumulates in the human body can lead to high risk of damage to the lungs, liver, and kidneys as well as irritation of the respiratory system, loss of hearing, and muscle coordination. Mercury is known as global pollutant with complicated and extraordinary physical and chemical properties that is highly toxic and can cause neurotoxicological disorders (Abdel-Raouf and Abdul-Raheim, 2017). Mercury can enter human bodies through several pathways by inhalation, ingestion or adsorption through skin. However, the main pathway is ingestion which is through the food chain. Because of the solubility of mercury compounds, mercury is easily found in aquatic environments released from various natural sources and phenomenon, thus there is a high chance for mercury to bioaccumulate and concentrate in living organisms, especially fish. A type of mercury, monomethylmercury can lead to brain and central nervous system damage and also congenital malfunction and abortion. Therefore, it is critical to treat metal-contaminated wastewater prior to its discharge into the environment.

TABLE 8.2
Types of Heavy Metals and Their Toxicities to Humans

Heavy Metal	Toxicities	Route of Entry	MCL(mg/L)
Arsenic	Liver and kidney damage, irritation of respiratory system, nausea and vomiting, loss of appetite	Inhalation and ingestion	0.05
Cadmium	Lung, liver, and kidney damage; irritation of respiratory system	Inhalation and digestion	0.01
Chromium	Lung damage and irritation or respiratory system	Inhalation, ingestion, and adsorption through skin	0.05
Copper	Short-term exposure: gastrointestinal distress; long-term exposure: liver and kidney damage	Inhalation and ingestion	0.25
Nickel	Lung, liver, and kidney damage	Inhalation	0.20
Zinc	Depression, lethargy, neurological signs, and increased thirst	Inhalation and ingestion	0.80
Lead	Lung and liver, damage, loss of appetite, and nausea	Inhalation and ingestion	0.006
Mercury	Lung, liver, and kidney damage; irritation of respiratory system; and loss of hearing and muscle coordination	Inhalation, ingestion, and adsorption through skin	0.00003

Source: Abas, S.N.A. et al., *World Appl. Sci. J.*, 28, 1518–1530, 2013; Adams, R., *Econ. Dev. Soc. Change*, 47, 155–173, 2010.

8.2.2 Available Methods for Heavy Metal Removal

Heavy metals have become environmental priority pollutants that are faced with more and more stringent regulations. Indeed, heavy metals are becoming one of the most serious environmental problems. Hence, the removal of these toxic heavy metals from the sources of pollution has become the main priority in order to protect people and the environment. Recently, numerous approaches have been studied for the development of cheaper and effective technologies in order to improve the quality of the treated pollutant. The approaches function either physicochemically or biologically in removing heavy metals from the environment, including wastewater, soil, and air. Ion exchange, adsorption, membrane filtration, reverse osmosis, ultrafiltration, photocatalysis, chemical precipitation, electrodialysis, and electrochemical deposition are among the physicochemical processes for removal of heavy metals (Fu and Wang, 2011). Biological treatment processes are an alternative for removing heavy metals by using microorganisms to settle solids in the solution, especially in wastewater. For instance, trickling filters, activated sludge, and stabilization ponds are widely used for treating industrial wastewater. Moreover, new bioadsorbents such as forest waste, agricultural waste, industrial waste, and algae are used for maximum removal of heavy metals from wastewater (Tarley and Arruza-Zezzi, 2004; Soto et al., 2016). However, among the methods listed, the physicochemical treatment methods are the most suitable for toxic inorganic compounds produced from various industries. The advantages and disadvantages of physicochemical treatments have been tabulated in Table 8.3.

According to Kurniawan et al. (2006), the ion exchange method is a treatment that employs a reversible interchange of ions between solid and liquid phases as its basic principle. It starts with ion-exchange reactions where the heavy metal ions are physically adsorbed and form a complex between the counter-ion and a functional group. The last stage of the process is when the hydration occurs at the surface of the solution or pores of the adsorbent. Different factors including pH, temperature, initial concentration of the adsorbent and sorbate, time of contact, and anions can affect the ion exchange operation (Kang et al., 2004; Gode and Pehlivan, 2006). Adsorption is another physicochemical method for removing heavy metals that has been successful in reducing the metal concentration in wastewater. Adsorption is a mass transfer between a liquid phase and a solid phase on sorbent materials known as adsorbents (Kurniawan and Babel, 2003; Babel and Kurniawan, 2004). For sorption of a pollutant onto an adsorbent, three stages are involved: (i) the penetration of the pollutant from the bulk solution to the adsorbent surface; (ii) adsorption of the pollutant on the adsorbent surface; and (iii) penetration in the adsorbent structure (Barakat, 2011). These adsorbents provide large surface area and high adsorption capacity. A key point is that the process can run in reversible mode and the adsorbents will be regenerated by desorption.

Membrane filtration has also been utilized due to the simple separation method (Murthy and Chaudhari, 2009; Padaki et al., 2015; Razavi et al., 2016). The separation takes place by controlling the passage of particles, letting the solvent pass through and catching the solute, using a semipermeable membrane with different pore sizes. There are different types and pore sizes of membranes for removal of heavy metals including ultrafiltration, reverse osmosis, and electrodialysis. These technologies show

TABLE 8.3
The Main Advantages and Disadvantages of the Various Physicochemical Methods for Treatment of Heavy Metal in Wastewater

Treatment Method	Advantages	Disadvantages	References
Ion exchange	• High treatment capacity • High removal efficiency • Fast kinetics	• High maintenance cost	Hubicki and Kolodynska (2012); Al-Enezi et al. (2004)
Adsorption	• Low cost • Easy operating conditions • Wide pH range • High metal binding capacities	• Low selectivity • Production of waste products	Maximous et al. (2010); Kurniawan and Babel (2003); Barakat (2011)
Membrane filtration	• Small space requirement • Low pressure • High separation selectivity	• High operational cost due to membrane fouling	Fu and Wang (2011); Maximous et al. (2010)
Electrodialysis	• High separation selectivity	• High operational cost due to membrane fouling and energy consumption	Choi and Jeoung (2002); Gering and Scamehorn (2006)
Reverse osmosis	• Very high removal efficiency	• High consumption due to pumping pressures and restoration of membranes.	Singh (2015); Al-Jlil and Alharbi (2010)
Ultrafiltration	• High removal efficiency • Surfactant can be recovered and reused, make it more economically and feasible. • High binding selectivity • Highly concentrated metal concentrates for reuse	• Removal efficiency depends on the characteristics and concentrations of metals and surfactants, pH of solution, and ionic strength • Removal efficiency depends on metal and polymer type, ratio of metal to polymer, pH, and existence of other metal ions	Lam et al. (2018); Maximous et al. (2010); Bernat et al. (2007)

(Continued)

TABLE 8.3 (*Continued*)
The Main Advantages and Disadvantages of the Various Physicochemical Methods for Treatment of Heavy Metal in Wastewater

Treatment Method	Advantages	Disadvantages	References
Photocatalysis	• Removal of metals and organic pollutant simultaneously, less harmful by-products	• Long duration time, limited applications	Kabra et al. (2008); Chowdhury et al. (2014); Molinari et al. (2015)
Electrochemical deposition	• Practical and efficient • No further reagents are necessary • No sludge will be produced during the process • High selectivity • Low costs • Can be applied to non-aqueous solutions or solutions containing chelating agents	• Conventional aqueous solutions face some problems including low thermal stability, liberation of hydrogen gas molecules, and narrow electrochemical window • Some technical issues in commercialization of non-aqueous solutions like cell heat balance • Low current efficiency and corrosion of cell components	Tonini and Ruotolo (2016); Chen and Lim (2005); Jayakrishnan (2012)
Chemical precipitation	• Low capital cost • Simple operation	• Sludge generation • Extra operational cost for sludge disposal	Fu and Wang (2011); Wang et al. (2005)

Source: Barakat, M., *Arab. J. Chem.*, 4, 361–377, 2011; Adams, R., *Econ. Dev. Soc. Change*, 47, 155–173, 2010.

impressive ability due to membrane efficiency, ease of operation, and space saving (Van der Bruggen and Vandecasteele, 2003; Sutherland, 2008). Typically, ultrafiltration (UF) is based on molecular sieving using porous membranes (Cassano and Basile, 2011; Shenvi et al., 2015; Lam et al., 2018) within a pressure-driven process to discard any dissolved macromolecules, colloidal materials, and sometimes suspended particles that are larger than the pore size (Singh, 2015; Montana et al., 2013). Another separating technology in the treatment of heavy metals is reverse osmosis (RO), which uses a semipermeable membrane working under hydraulic operating pressure (Malaeb and Ayoub, 2011; Singh, 2015) by. This type of membrane does not possess any discrete pores to allow the particles to pass through, so the transition has to take place by diffusion (Greenlee et al., 2009). A novel hybrid membrane process with ion-exchange membranes (IEMs) has been introduced in separating heavy metals from wastewater by using electrodialysis. IEMs are unable to transport cations and anions simultaneously. Thus, currently, there are two types of membrane available in this process which are cation-exchange membrane (CEM) and anion-exchange membranes (AEM) (Fu and Wang, 2011). The basic principle of ion transportation in this process is based on electrical potential or concentration gradients (Moon and Yun, 2014).

An advanced oxidation process (AOP) in purification of water is photocatalysis. Fundamentally, this method employs light to acts as an activator. The non-toxic semiconductors that harness light with appropriate wavelength have been used instead of chemical compounds (Robertson, 1996; Molinari et al., 2015). As the wastewater containing heavy metals is introduced into a photocatalytic system, the heavy metals are transferred to the surface from the aqueous phase. The heavy metals are then adsorbed by the semiconductor surface and the photocatalytic reactions takes place in the adsorbed phase. Then, the products are decomposed and removed from the interface region (Kabra et al., 2008). Another method, electrochemical deposition (ED) transforms dissolved metal ions into solid particles by deposition on ionic conductor to prevent corrosion (Chen and Lim, 2005; Jayakrishnan, 2012). ED is a one-step clean method based on oxidation and reduction of heavy metal ions in a cell. In this method, heavy metals are reduced and electroplate onto the cathode (Chang, 2009; Jayakrishnan, 2012). In addition to the above methods, chemical precipitation is also another method used for wastewater treatment. This process utilizes chemical precipitant agents to react with heavy metal ions and convert them into insoluble solid particles (Wang et al., 2005; Fu and Wang, 2011). Then, the solid particles are separated from the solution by filtration or sedimentation (Matis et al., 2004; Zamboulis et al., 2004).

Among those existing techniques for removing heavy metals from the environment, especially wastewater, major research focus has been directed to the adsorption technique. Adsorbents used in the technique can be derived from various sources either from natural material or polymeric precursors (Ferreira et al., 1999). Some of the most used adsorbents in this process are activated carbon (AC), zeolites, carbon nanotubes (CNTs), and sawdust. However, there are some limitations of the existing adsorbents, and to overcome these nanofiber membranes (NFMs) have been introduced. This new adsorbent material exhibits a large surface area-to-volume ratio, flexible, and fine pore structure due to the nanoscale fiber diameters. The details of the advantages and disadvantages of available adsorbents are tabulated in Table 8.4, while discussion on NFMs and the associated adsorption mechanisms are discussed in the next section.

TABLE 8.4
The Advantages and Disadvantages of Different Types of Adsorbents for Heavy Metal Removal

Adsorbents	Advantages	Disadvantages	References
Activated carbon	• Possesses large micropore and mesopore volumes as well as high surface area. • Very high heavy metal removal efficiency	• Relatively high capital cost • The effectiveness adsorption depends on the diameter of pores	Karnib et al. (2014); Al-Malack and Basaleh (2016)
Natural zeolites	• Low-cost adsorbent • Cation exchange capacity and ion selectivity • Naturally abundant	• Low purity • Poor separation performance	Abdel Salam et al. (2011); Ackley et al. (2002)
Carbon nanotubes	• Chemical, mechanical, and thermal stable • High surface area • Quite high removal efficiency	• Expensive • Discharged to the water environment and poses a risk to humans	Cai et al. (2017); Yadav and Srivastava (2017); Sadegh et al. (2016)
Nanofiber membranes	• Super-high surface area • High porosity • Fast and high rate of adsorption • High removal capacity • Metal selective	• Expensive • Time consuming	Kamplanonwat and Supaphol (2014)
Bioadsorbent	• Low cost • Availability • Profitability • Ease of operation • High efficiency	• Limited to low metal concentration	Saeed et al. (2005); Javanbakht et al. (2014)

Source: Adams, R., *Econ. Dev. Soc. Change*, 47, 155–173, 2010.

8.2.3 Fabrication of Nanofiber Membranes via Electrospinning

There are numerous approaches for fabrication of nanofibers membranes (NFMs) available including self-assembly, template synthesis, solution blow spinning, electrospinning, and draw spinning methods. Among these available methods for production of NFMs, electrospinning is widely used in fiber-making industries due to its simplicity for generating fibers in nanostructures (Li et al., 2006). The major difference between the fibers collected from electrospinning in comparison to the other spinning methods is the uniform fibers diameter, mainly in the nano-range scale with higher mechanical strength. In addition, this method also possesses many advantages such as simple production operation and also the capability to produce bulky fibers (Subbiah et al., 2005). Electrospinning is based on electrostatic interaction driving forces, which make it different from the other methods that are based on stretching and pulling. The most fascinating point about electrospinning is that this technique is capable of producing fibers down to tens of nanometers. Furthermore, NFMs that are produced from electrospinning possess several advantages including extremely high surface area, tunable porosity, flexibility to fit in to a wide range of shapes and sizes, and the capability to control the NFs composition to attain the desired results from its properties and functionality (Zhang et al., 2002). Thus, due to its exceptional features as well as its simplicity and versatility, it has received enormous research interest over the past decades. Figure 8.3 is a simplified diagram of the fabrication NFMs and their advantages as adsorbents.

Electrospinning is a process that uses electrostatic forces that makes it a unique method to produce superfine NFMs with controllable fiber fineness, orientation, surface morphology, and cross-sectional configuration. Besides that, there are other several techniques for nanofibers development that have been reported including phase separation, template synthesis, self-assembly, and other spinning methods (dry, melt, and wet spinning) (Wang et al., 2011). There are three main components for an electrospinning setup as shown in the generic design below (Figure 8.3) which are a syringe pump, grounded collector, and a high voltage power supply (Demir et al., 2004) (Figure 8.4).

In simple electrospinning processes, a metallic needle and syringe are used as a spinneret. The polymer solution is loaded in the syringe that is connected to the metallic needle while aluminum foil is used as a grounded collector (Li et al., 2006) Typically, the polymer solution is prepared before electrospinning by dissolving

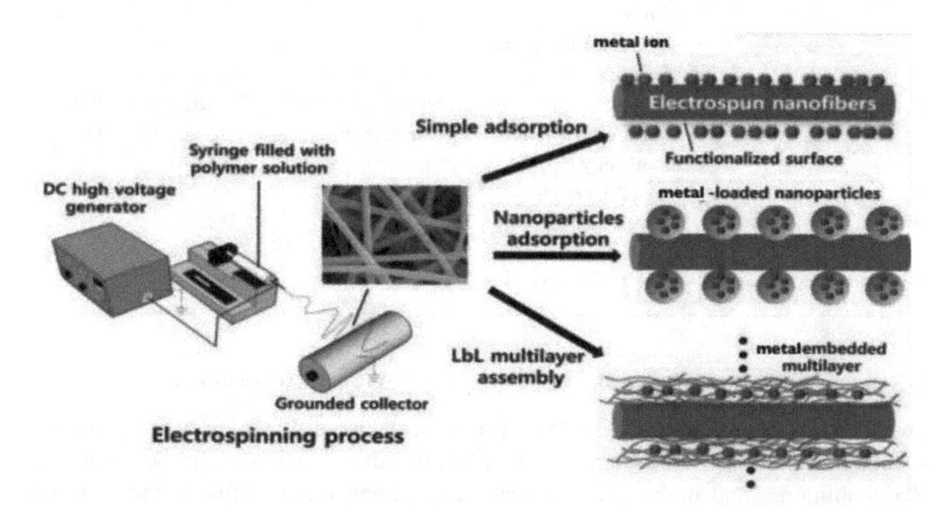

FIGURE 8.3 Overview process of fabrication of nanofiber membranes via electrospinning method. (Reprinted with permission from Shapin, S., *The Scientific Revolution*, University of Chicago Press, Chicago, IL, pp. 15–64, 2006.)

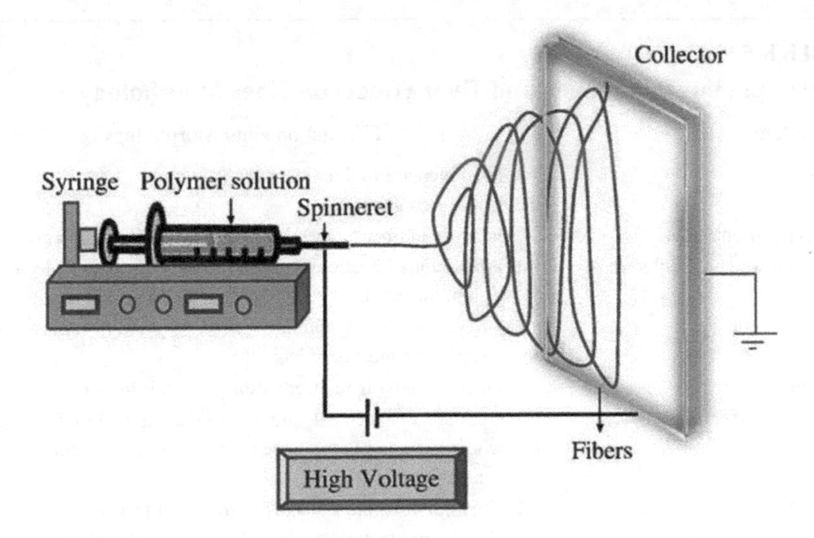

FIGURE 8.4 Schematic diagram of electrospinning setup. (From Li, D. et al., *J. Am. Ceram. Soc.*, 89, 1861–1869, 2006; Reprinted with permission from Shapin, S., *The Scientific Revolution*, University of Chicago Press, Chicago, IL, pp. 15–64, 2006.)

the polymer in an appropriate solvent (Bhardwaj and Kundu, 2010). The process is conducted within a chamber with a proper ventilation system as some of the polymers and solvents may emit harmful or even unpleasant odors (Huang et al., 2003). However, despite the advantages offered by electrospinning, there are various parameters that need to take into l consideration including the viscosity, molecular weight, surface tension, applied voltage, flow rate, and distance between the tip to collector as each of these parameters significantly affect the produced fiber morphology. Several important electrospinning parameters with their effects on fiber morphology have been discussed in Table 8.5. By optimization of these parameters, nanofibers with desired morphology and diameter can be obtained.

The increased attention focused on electrospinning technology for NFMs is due to their high porosity, large specific surface area per unit mass, high permeability, flexibility, wide diversity of surface functional groups, and high adsorption capacities. There is no doubt in stating that electrospun polymer nanofiber-based materials have exhibited great advantages over other conventional and commercial adsorbents for heavy metal remediation. Even so, there still remains a lot of challenges. First, the difficulty in mass production of high-quality nanofibers directly restricts their practical applications. Second, the selection of suitable polymers and modifiers to introduce new functionality to meet specific needs is another critical problem too be solved. And most important of all, the control of the mechanical properties of the electrospun membranes needs dramatic improvement if their utility is to be increased.

TABLE 8.5
Electrospinning Parameters and Their Effects on Fiber Morphology

Parameters	Effect(s) on Fiber Morphology
Viscosity	1. High increase in fiber diameter 2. Low-beads generation
Polymer concentration	1. Increase in fiber diameter with increase of concentration
Molecular weight of polymer	1. Production of beads and droplet decreased as the molecular weight increased
Surface tension	1. No conclusive link with fiber morphology, high surface tension results in instability of jets.
Applied voltage	1. Fiber diameter is reduced as the voltage is increased.
Distance between tip and collector	1. Optimum distance is required for production of uniform fibers, where too small or too large distance will lead to beads generation.
Flow rate	1. Too high flow rate leads to generation of beads.
Humidity	1. High humidity results in circular pores on the fibers.
Temperature	1. Fiber diameter reduces as temperature increases.

Source: Bhardwaj, N. and Kundu, S.C., *Biotechnol. Adv.*, 28, 325–347, 2010; Adams, R., Econ. Dev. Soc. Change, 47, 155–173, 2010.

8.3 ADSORPTION MECHANISM OF HEAVY METALS BY NANOFIBER MEMBRANES

One of the best and most-recognized technologies that can effectively remove heavy metals from wastewater in economic and eco-friendly way is adsorption. This method involves a mass transfer process where a metal ionic transferred from the liquid phase to the surface of a solid (sorbent) and bound together by physical or chemical interactions (Fu and Wang, 2011) as shown in Figure 8.5. All adsorption mechanisms are

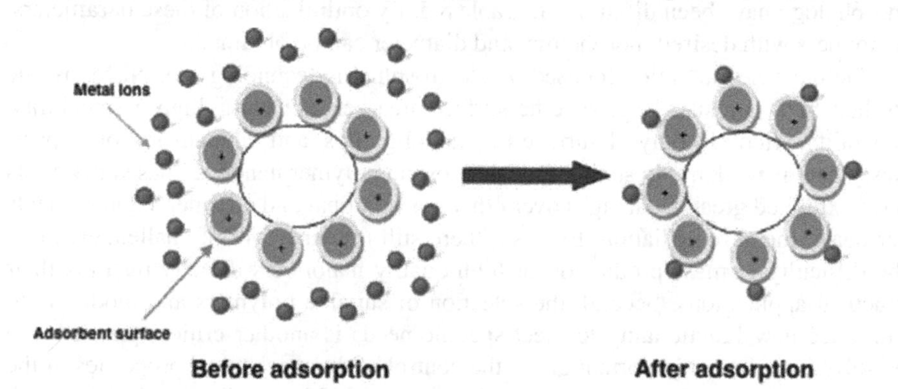

FIGURE 8.5 The basic principle of adsorption mechanism. (Reprinted with permission from Shapin, S., *The Scientific Revolution*, University of Chicago Press, Chicago, IL, pp. 15–64, 2006.)

dependent on the solid-liquid equilibrium and on mass transfer rates. The removal of heavy metal ions from various metal effluents, using nanofiber membranes (NFMs), is based on the interactions between the heavy metal ions and functional sites on the NFMs surface, including electrostatic interactions and physical affinity or complexation and chemical chelating (Rivas and Maureira, 2008). It is noteworthy to mention that the functional sites that exist or are attached to the surface of NFMs are the major factors determining the capability of heavy metal removal.

NFMs can be referred to as a perfect microfiltration medium due to their micro and nanoporous structure. The combination of strong interactions and effective microfiltration makes NFMs an optimal membrane for removal of heavy metals. Commonly, in removal of toxic metals, there are two types of coupling processes which are static adsorption and dynamic adsorption. Static adsorption is based on the rate of adsorption of the functional surface and capacity of the sorption while dynamic adsorption is related to permeation flux, pressure drop, and a membrane's operating life. Up to now, most of heavy metal removal is applied using the static adsorption mode (Huang et al., 2014).

Table 8.6 summarizes the adsorption capabilities of several different types of adsorbents for different types of heavy metals in wastewater. The adsorbents identified in the table were derived from agricultural wastes, natural and also polymeric materials. The heavy metals such as lead, cadmium, zinc, copper, chromium, and nickel were chosen for study and their adsorption capabilities were compared by adsorption to different types of adsorbents. It can be seen that NFMs consistently showed high adsorption capabilities for all heavy metals as compared to other adsorbents which are 94.3, 121.9, 70.9, and 56 mg/g for Pb^{2+}, Cd^{2+}, Cu^{2+} and Cr^{6+}, respectively.

The recently developed and enhanced NFMs have been introduced and widely studied in order to overcome the drawbacks of other conventional and available adsorbents. These NFMs have great potential in various adsorption applications especially in removal of heavy metal contaminants from waters. As stated previously in Table 8.6, these NFMs possess significantly high surface area, abundant

TABLE 8.6
Adsorption Capabilities of Different Adsorbents for Different Heavy Metals

	Adsorption Capacity (mg/g)					
Adsorbent	Pb^{2+}	Cd^{2+}	Zn^{2+}	Cu^{2+}	Cr^{6+}	Ni^{2+}
Rice husk	—	2.0	—	—	0.79	—
Zeolite	1.6	2.4	0.5	1.64	—	0.4
Clay	—	—	63.2	83.3	—	—
Cross-linked chitosan	—	150	—	164	—	—
Activated carbon	90	61	—	—	—	—
Carbon nanotubes	104.1	75.8	58	50.4	—	49.3
Nanofiber membranes	94.3	121.9	—	70.9	56	—

Source: Adams, R., *Econ. Dev. Soc. Change*, 47, 155–173, 2010.

macro and micropore volumes, fast rates of adsorption as well as high selectivity for heavy metals. All these make them great adsorbent materials for removal of heavy metals. The outstanding characteristics of the NFMs are contributed to by several factors including the type of the polymers used and also the method of fabrication. In the past few decades, the main focus has been on the fabrication of NFMs through electrospinning.

8.3.1 Surface Modification on Nanofiber Membranes

Surface area-to-volume-ratio of the NFMs is a major factor determining the adsorption capacity of the adsorbent material. The efficiency of the adsorbent with an aqueous phase can be enhanced by several modifications either physically or chemically. Physical modification involves the enhancement on the surface area of NFMs by pulverization or phase separation while chemical modification involves the incorporation of desired functional groups on the surface of NFMs. In the study of Martin et al. (2018), the main research on enhancing the property of NFMs was focused on modifying the specific chemical functionalities by introduction of chelating agents, complexing agents or cross-linkers in the polymer-based NFMs.

There are various polymer-based precursors that have been utilized in fabrication of NFMs such as polyethersulfone (PES), polyvinyl alcohol (PVA), polyacrylonitrile (PAN), polyethyleneimine (PEI), and polyvinylchloride (PVC). For instance, PES was blended with PEI to form micro-nano structured nanofibrous affinity membranes fabricated by electrospinning followed by solvent etching with a cross-linking solution (Min et al., 2011). In this solution GA (glutaraldehyde) was mixed with water and acetone, thus PEI was not only solvent etched by the solvent water but also simultaneously cross-linked by GA. The influence of the components of the cross-linking bath on the morphology of the resulting PES/PEI nanofibers was investigated (Min et al., 2011). A series of static adsorption experiments indicated that the unique micro-nano structure had much higher surface area and efficiency for Cu(II) removal from aqueous solution. In another study, PEI was used as a chelating agent for removal of heavy metal cations from wastewater and the results obtained for the NFM showed high affinity for various metal ions (Molinari et al., 2004). The heavy metal ions studied in adsorption experiments were Pb(II), Cu(II), and Cd(II) and the PEI affinity membrane showed strong adsorbability with the maximum adsorption capacities of 40.34, 70.92, and 121.95 mg/g, respectively.

Polyacrylic acid (PAA) has also been widely used as a metal ion complexing agent as it contains carboxyl groups (Hayashi and Komatsu, 1992; Huang et al., 2014). For example, Xiao et al. (2010) investigated the stability and morphology of electrospun ultrafine PAA/PVA (polyvinyl alcohol) nanofibers and how these were influenced by the processing parameters. They found that a uniformly distributed fiber diameter of PAA/PVA can be achieved with a heating treatment of 145°C and high concentration of 25%. The nanofibrous mats were able to quickly eliminate Cu(II) ions from an aqueous environment with 91% removal within 3 h and excellent selectivity under the co-presence of Ca(II) ions. In a study conducted by Sang et al. (2008), a fibrous membrane was prepared via electrospinning of polyvinyl chloride (PVC).

This type of membrane is preferred to be used as separator and adsorbent of heavy metals from groundwater due to the excess chlorine on the fiber surface, which acts as negatively charged adsorption sites. The maximum adsorption capacities of Pb(II), Cd(II), and Cu(II) were evaluated in static adsorption tests with results of 5.03, 5.35, and 5.65 mg/g, respectively.

Kampalanonwat and Supaphol (2010) modified PAN nanofibers, to create NFMs for heavy metal removal, containing amidino diethylene diamine (DETA), which were termed as aminated PAN (APAN). They found that the amounts of the metal ions adsorbed onto the PAN nanofibers can be affected by the initial concentration and initial pH of the metal ions solutions. The maximal adsorption capacities of the Fe(II), Pb(II), Cu(II), and Ag(I) on the APAN nanofibers were 11.5, 60.6, 150.6, and 155.5 mg/g, respectively. In addition, the modification of PAN-based nanofibers surfaces can also be obtained by applying polyethyleneaminetetraacetic (EDTA) and using ethylenediamine (EDA) as the cross-linker during the electrospinning process. In a study by Chaúque et al. (2016), the surface morphology changes of chemically modified PAN-based nanofibers were monitored. As shown in Figure 8.6, the average diameter of sole PAN, EDA-PAN, and EDTA-EDA-PAN nanofibers were 289 ± 31, 29 ± 55, and 288 ± 68 nm, respectively. The results obtained demonstrate that chemical modification on the PAN surface was found to produce increase in the average diameter of nanofibers. Subsequently, the modified EDTA-EDA-PAN nanofibers were applied in the treatment of wastewater for Cd(II) and Cr(VI) removal and showed effective sorption affinity with maximum adsorption capacities of 32.68–66.24 mg/g, respectively at 25°C (Azimi et al., 2017).

Among the polymer precursors used, PAN shows the greatest performance in maximal adsorption capacities for heavy metals. However, few studies have examined PAN NFMs for heavy metal removal. The fact that PAN materials have promising physicochemical properties such as good mechanical strength, good chemical resistance, and high thermal stability means that they are highly suitable for further development of NFMs in future studies.

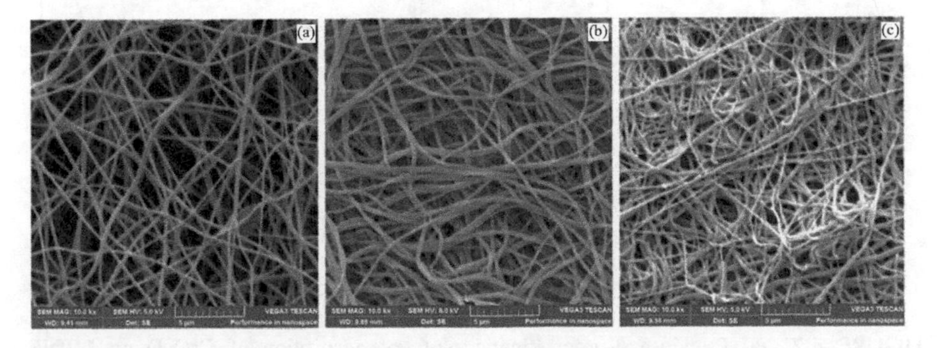

FIGURE 8.6 SEM micrograph of the (a) electrospun PAN, (b) EDA-PAN, and (c) EDTA-EDA-PAN nanofibers at 10,000x magnification. (Reprinted with permission from Shapin, S., *The Scientific Revolution*, University of Chicago Press, Chicago, IL, pp. 15–64, 2006.)

8.3.2 Kinetics Modeling

In order to obtain further insight into the adsorption behaviors and mechanism of the adsorption of heavy metals such as Cr(VI), Cd(II), Ni(II), Cu(II), and Pb(II) ions from aqueous solutions onto the NFMs surface, experimental data can be regressed against the pseudo-first-order and pseudo-second-order model represented by the following linear equations (Ho and McKay, 1999; Long et al., 2011):

$$\log(q_e - q_t) = \log q_e - \left(\frac{k_1}{2.303}\right)t \tag{8.1}$$

$$\frac{t}{q_t} = \frac{1}{k_2 q_e^2} + \frac{1}{q_e}(t) \tag{8.2}$$

where q_e and q_t are the adsorption capacity at equilibrium and time (t), respectively. k_1 and k_2 are the rate constants of the pseudo-first-order (min^{-1}) and pseudo-second-order (g mg^{-1} min^{-1}) kinetics, respectively. In the first-order kinetic model, $\log(q_e - q_t)$ is plotted against time, t, while the linear curves of the second-order kinetic model obtained when t/q_t is plotted against time for the adsorption of heavy metals onto NFMs.

Theoretically, the adsorption rates can be estimated by using these rate equations and the characteristic of possible reaction mechanisms can also be predicted. With the agreement between the theoretical and experimental uptake values and linear correlation coefficient (R^2) values, the best fit model can be selected (Moussavi and Khosravi, 2010). A recent example is the research by Brungesh et al. (2017), who successfully removed toxic Cr(VI) ions from aqueous solution by using MnO_2 coated polyaniline nanofiber (MPNF) adsorbents by a batch adsorption technique (Abdulsalam et al., 2015). They plotted linearized pseudo-first-order and second-order kinetic models as shown in Figure 8.7 and the correlation coefficients (R^2) values

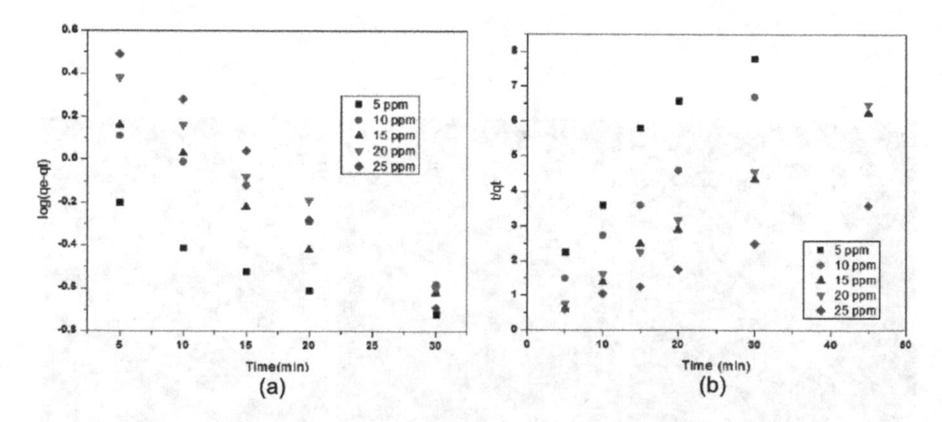

FIGURE 8.7 (a) Pseudo-first-order kinetic plots of adsorption of Cr(VI) on MPNF; (b) pseudo-second-order kinetic plots of adsorption of Cr(VI) on MPNF. (Reprinted with permission from Shapin, S., *The Scientific Revolution*, University of Chicago Press, Chicago, IL, pp. 15–64, 2006.)

are tabulated in Table 8.7. From the table, it can be seen that the regression coefficient for the pseudo-second-order model ($R^2 > 0.99$) was higher than pseudo-first-order, indicating the adsorption kinetic data obeyed the pseudo-second-order kinetics.

Another study conducted by Najim and Salim (2017) also revealed the adsorption of the Cr(VI) by polyaniline nanofibers is best fitted with the pseudo-second-order model based on the R^2 values obtained as shown in Table 8.8. This conclusion came from the good correlation coefficients attained from the second-order plot with a good match between the theoretical and experimental values.

In 2013, the effect of contact time on the adsorption of different heavy metal ions such as Ni(II), Cu(II), Cd(II), and Pb(II) onto NFM surfaces was also studied by Aliabadi et al. (2013). The parameters of both kinetic models are tabulated in Table 8.7. It can be seen that both equations, pseudo-first-order ($R^2 > 0.99$) and pseudo-second-order ($R^2 > 0.99$), fitted the experimental data very well. From the kinetic modeling studies, it can also be concluded that greater adsorption capacity of the modified NFMs should be attributed to the large surface area of the NFMs as compared to other adsorbents. Thus, the modification of NFMs structure to produce NFMs with greater surface area for better adsorption capacity has become a major challenge.

TABLE 8.7
Kinetic Parameters for Metal Ions Adsorption onto Nanofiber Adsorbents

	Experiment	Pseudo-First-Order Model			Pseudo-Second-Order Model		
Metal Ions	q_e (mg/g)	q_e (mg/g)	k_1 (min^{-1})	R^2	q_e (mg/g)	k_2 (g mg^{-1} min^{-1})	R^2
Ni(II)	175.1	183.4	0.02638	0.998	242.9	0.000097	0.996
Cd(II)	143.8	150.3	0.02783	0.997	198.6	0.000125	0.992
Pb(II)	135.4	143.4	0.02403	0.998	195.1	0.000104	0.994
Cu(II)	163.7	172.3	0.02593	0.998	229.2	0.000100	0.996
Cr(VI)	130.5	55.2	0.11300	0.9931	134.3	0.0232	0.997

Source: Brungesh, K.V. et al., *J. Environ. Anal. Toxicol.*, 7, 1–8, 2017; Aliabadi, M. et al., *Chem. Eng. J.*, 220, 237–243, 2013; Adams, R., *Econ. Dev. Soc. Change*, 47, 155–173, 2013.

TABLE 8.8
Kinetic Parameters of Adsorption Cr(VI) onto Polyaniline Nanofibers

	Experiment	Pseudo-First-Order Model			Pseudo-Second-Order Model		
Metal ions	q_e (mg/g)	q_e (mg/g)	k_1 (min^{-1})	R^2	q_e (mg/g)	k_2(g/mgmin)	R^2
Cr(VI)	7.47	1.98	0.035	0.9653	7.61	0.047	0.9995

Source: Adams, R., *Econ. Dev. Soc. Change*, 47, 155–173, 2010.

8.3.3 Adsorption Isotherms

The distribution of adsorbate molecules between solution and adsorbent surfaces when the adsorption process reaches an equilibrium state is used to define adsorption isotherms. This provides greater understanding of the sorption equilibrium and the relationship with the surface properties and the affinity of adsorbent. Two adsorption isotherm models that can be used to fit the adsorption equilibrium are Langmuir and Freundlich models, and correlation coefficients (R^2) are used to evaluate the suitability of the two models.

8.3.3.1 Langmuir Isotherm

The most widely known equation in evaluating adsorption isotherms is the Langmuir equation. Generally, the Langmuir equation is based on the physical hypothesis that maximum adsorption capacity consists of monolayer adsorption between adsorbent surface and adsorbate, containing an identical and finite number of active sites and is energetically equivalent. The linear expression of the Langmuir equation is as follows:

$$\frac{C_e}{q_e} = \frac{1}{K_L \text{x} q_{max}} + \frac{C_e}{q_{max}} \tag{8.3}$$

where C_e (mg/L) is the equilibrium metal ions concentration in solution, q_e (mg/g) is the metal uptake capacity on the adsorbent at equilibrium, q_{max} is the maximum amount of metal ions adsorbed in a complete monolayer (mg/g), K_L is the Langmuir constant related to the affinity of the binding site (L/mg). Thus, if a graph is plotted between C_e/q_e versus C_e, a straight line with slope q_e and intercept K_L is obtained.

8.3.3.2 Freundlich Isotherm

In order to describe the sorption of metal ions on an heterogeneous adsorbent surface and to fit the adsorption equilibrium data, a linear form of the Freundlich expression is as follows:

$$\ln q_e = \ln K_F + \frac{1}{n} \ln \text{C}_e \tag{8.4}$$

where K_F represents the Freundlich constant related to the adsorption capacity and n is the intensity of the sorbent.

To date, there are many studies that have determined the adsorption isotherm model related to the sorption of heavy metal ions onto NFM surfaces. Different types of NFMs and their modifications, as well as types of metal ions, will show different isotherm results. Aliabadi et al. (2013) found that the adsorption of different metal ions using PEO/chitosan NFMs obeyed the Langmuir isotherm model as the R^2 value obtained was higher than that for the Freundlich model as tabulated in Table 8.9. This result suggests that the adsorption of metal ions was a homogeneous and monolayer adsorption process.

TABLE 8.9
Adsorption Isotherm Model Parameters for Sorption of Different Metal Ions on Different Types of Adsorbents

		Langmuir			Freundlich		
Adsorbent	**Metal ion**	q_{max}	K_L (L/mg)	R^2	K_F (mg/g)	n	R^2
PEO/chitosan NFMs	Ni(II)	357.1	0.12	0.983	122.9	5.879	0.902
	Cu(II)	310.2	0.09	0.986	101.3	5.675	0.904
	Cd(II)	248.1	0.12	0.963	101.2	7.117	0.821
	Pb(II)	237.2	0.07	0.991	81.52	6.023	0.885
MPNF	Cr(VI)	158.2	0.35	0.920	39.3	1.620	0.9703

Source: Brungesh, K.V. et al., *J. Environ. Anal. Toxicol.*, 7, 1–8, 2017; Adams, R., *Econ. Dev. Soc. Change*, 47, 155–173, 2010.

In contrast, another study conducted by Brungesh et al. (2017) on adsorption of Cr(VI) ions on an MPNF adsorbent demonstrated that this is best fitted by a Freundlich isotherm model (R^2 = 0.9703) as compared to a Langmuir model (R^2 = 0.9204). This experimental data indicates that the adsorption sites are heterogeneous with multilayer adsorption occurring at equilibrium. The possible reasons for this are the large surface area and high porosity of NFMs. Furthermore, the smaller size of metal ions compared to that of the pores of adsorbent means that it was easier for the metal ions to diffuse into the pores and adsorb onto the internal surface of the adsorbent (Chen et al., 2008). The higher adsorption capacity of the adsorbent used in these studies may be coming from the pore size properties of the NFMs.

8.4 FUTURE PERSPECTIVES AND CHALLENGES

Over the past decades, the rebirth of electrospun nanofiber membranes (NFMs) through polymer electrospinning has shown their versatility and applicability, across many fields. However, the total number of publications or how far NFMs have gone at the large industrial scale, does not compete with other smart materials and nanoparticles. Moreover, the misconception that electrospun NFMs are not promising materials, as previously believed, is not valid anymore. The future outlook for using electrospun NFMs, especially with those capable of dealing with water remediation challenges, is extremely promising. Nonetheless, different from water filtration membranes, the requirement for higher specifications for water adsorption membrane is greater. This infers that the control of NFMs pore size and its distribution, uniformity, support layer, and barrier layer thickness must be attained by electrospinning. Although the strength of their mechanical properties lag behind other materials and adsorbents, NFMs provide an unprecedented flexibility and modularity in design.

At the present day, almost all developing adsorbent materials are facing similar developmental challenges including those of the manufacturing process in terms of novel design, quality control, and reproducibility. Intrinsically, these challenges can be overcome by NFMs if given sustained research effort and enough time. For heavy metal ion adsorption from wastewater, the high surface to volume ratio is the major advantage of the electrospun NFMs, which greatly enhance their adsorption capacity. More than that, improvements in NFMs durability and strength as well as their incorporation with in composite adsorbent materials needs to be considered in order to compete with current adsorbent technology. In addition, NFM's improved capabilities compared to existing technologies in such areas as water purification and separation will provide expansive opportunities for emerging technologies. The authors believe that the electrospinning method is full of potential for functional materials and continued research interest is expected to impact positively in most areas of life in the near future.

8.5 CONCLUDING REMARKS

The advances in nanoscale science and engineering have become new alternatives in the development of environmentally acceptable and more economical water treatment technology. Nanofiber membranes (NFMs) are widely utilized in various applications especially in wastewater purification, due to their enhanced physicochemical properties that make them excellent candidates for removing heavy metals. Recent research has indicated that NFMs as adsorbents are useful tools for pollutants removal, due to their unique structure and surface characteristics. In this chapter, we mainly focused on recent developments related to the removal of metal ions from wastewater using NFMs. A review of the literature on the different methods of surface modification of NFMs revealed that different methods have been used to enhance the adsorption performance of NFMs. To develop environmentally friendly and inexpensive adsorbents is the key for improved removal of heavy metals. With the continued developments of nanotechnology, the application of new efficient adsorption materials is essential and will continue indefinitely.

REFERENCES

Abas, S.N.A., Ismail, M.H.S., Kamal, M.L., Izhar, S. 2013. Adsorption process of heavy metals by low cost adsorbent: A review. *World Applied Science Journal* 28: 1518–1530.

Abdel-Raouf, M.S., Abdul-Raheim, A.R.M. 2017. Removal of heavy metals from industrial waste water by biomass-based materials: A review. *Journal of Pollution Effects and Control* 5(1): 1–13.

Abdel Salam, O.E., Reiad, N.A., ElShafei, M.M. 2011. A study of the removal characteristics of heavy metals from wastewater by low-cost adsorbents. *Journal of Advanced Research* 2(4): 297–303.

Abdulsalam, A.S., Omar, A.H., Ahmed, M.A. 2015. Catalytic activity of polyaniline-MnO_2 composites towards the oxidative decolorization of organic dyes. *Applied Catalysis B: Environmental* 80: 6–115.

Ackley, M.W., Rege, S.U., Saxena, H. 2002. Application of natural zeolites in the purification and separation of gases. *Microporous and Mesoporous Materials* 61: 25–42.

Adams, R. 2010. Investment and rural assets in Pakistan. *Economic Development and Social Change* 47(1): 155–173.

Ajmal, M., Rao, R.A., Anwar, S., Ahmad, J., Ahmad R. 2003. Adsorption studies on rice husk: Removal and recovery of Cd(II) from wastewater. *Bioresources Technology* 86: 147–149.

Al-Enezi, G., Hamoda, M.F., Fawzi, N. 2004. Ion exchange extraction of heavy metals from wastewater sludges. *Journal of Environmental Science and Health. Part A: Toxic/hazardous Substances and Environmental Engineering* 39(2): 455–464.

Al-Jlil, S.A., Alharbi, O.A. 2010. Comparative study on the use of reverse osmosis and adsorption process for heavy metals removal from wastewater in Saudi Arabia. *Research Journal of Environmental Sciences* 4(4): 400–406.

Al-Malack, M.H., Basaleh, A.A. 2016. Adsorption of heavy metals using activated carbon produced from municipal organic solid waste. *Desalination and Water Treatment* 57(51): 24519–24531.

Aliabadi, M., Irani, M., Ismaeili, J., Piri, H., Parnian, M.J. 2013. Electrospun nanofiber membrane of PEO/chitosan for the adsorption of nickel, cadmium, lead and copper ions from aqueous solution. *Chemical Engineering Journal* 220: 237–243.

Azimi, A., Azari, A., Rezakazemi, M., Ansarpour, M. 2017. Removal of heavy metals from industrial wastewaters: A review. *ChemBioEng Reviews* 4(1): 37–59.

Babel, S., Kurniawan, T.A. 2003. Low-cost adsorbents for heavy metals uptake from contaminated water: A review. *Journal of Hazardous Materials* 97: 219–243.

Babel, S., Kurniawan, T.A. 2004. Cr(VI) removal from synthetic wastewater using coconut shell charcoal and commercial activated carbon modified with oxidizing agents and/or chitosan. *Chemosphere* 54: 951–967.

Barakat, M. 2011. New trends in removing heavy metals from industrial wastewater. *Arabian Journal of Chemistry* 4(4): 361–377.

Bernat, X., Sanchez, I., Stuber, F., Bengoa, C., Fortuny, A., Fabregat, A., Font, J. 2007. Removal of heavy metals by ultrafiltration. *Proceedings of European Congress of Chemical Engineering (ECCE-6)*, Copenhagen, Denmark.

Bhardwaj, N., Kundu, S.C. 2010. Electrospinning: A fascinating fiber fabrication technique. *Biotechnology Advances* 28: 325–347.

Brungesh, K.V., Nagabhushana, B.M., Harish, M.N.K., Hari Krishna, R. 2017. An efficient removal of toxic Cr(VI) from aqueous solution by MnO_2 coated polyaniline nanofibers: Kinetic and thermodynamic study. *Journal of Environmental and Analytical Toxicology* 7(2): 1–8.

Cai, Z., Song, X., Zhang. Q., Zhai, T. 2017. Electrospun polyindole nanofibers as a nano-adsorbent for heavy metal ion adsorption for wastewater treatment. *Fibers and Polymers* 18(3): 502–513.

Cassano, A., Basile, A. 2011. *Advanced Membrane Science and Technology for Sustainable Energy and Environmental Applications*: Basile A, Nunes SP, editor. Woodhead Publishing, Cambridge, 647–679.

Chang, J.H., Ellis, A.V., Yan, C.T., Tung, C.H. 2009. The electrochemical phenomena and kinetics of EDTA-copper wastewater reclamation by electrodeposition and ultrasound. *Separation and Purification Technology* 68(2): 216–221.

Chowdhury, P., Elkamel, A., Ray, A. 2014. *Heavy Metals in Water: Presence, Removal and Safety*; Chapter 2: Photocatalytic Processes for the Removal of Toxic Metal Ions. Royal Society of Chemistry.

Chaúque, E.F.C., Dlamini, L.N., Adelodun, A.A., Greyling, C.J., Ngila, J.C. 2016. Modification of electrospun polyacrylonitrile nanofibers with EDTA for the removal of Cd and Cr ions from water effluents. *Applied Surface Science* 369: 19–28.

Chen, A.H., Liu, S.C., Chen, C.Y. 2008. Comparative adsorption of Cu(II), Zn(II), and Pb(II) ions in aqueous solution on the crosslinked chitosan with epichlorohydrin. *Journal of Hazardous Materials* 154: 184–1991.

Chen, J.P., Lim, L.L. 2005. Recovery of precious metals by an electrochemical deposition method. *Chemosphere* 60(10): 1384–1392.

Choi, K., Jeoung, T.Y. 2002. Removal of zinc ions in wastewater by electrodialysis. *Korean Journal of Chemical Engineering* 19(1): 107–113.

Demir, M.M., Gulgun, M.A., Menceloglu, Y.Z., Erman, B., Abramchuk, S.S., Makhaeva, E.E., Khokhlov, A.R., Matveev, V.G., Sulman, M.G. 2004. Palladium nanoparticles by electrospinning from poly (acrylonitrile-co-acrylic acid)-$PdCl_2$ solutions. Relations between preparation conditions, particle size, and catalytic activity. *Macromolecules* 37: 1787–1792.

Ferreira, S.L.C., de Brito, C.F., Dantas, A.F., de Araujo, N.M.L., Costa, A.S. 1999. Nickel determination in saline matrices by ICP-AES after sorption on amberlite XED-2 loaded with PAN. *Talanta* 48(5): 1173–1177.

Fu, F., Wang, Q. 2011. Removal of heavy metal ions from wastewaters: A review. *Journal of Environmental Management* 92: 407–418.

Gering, K.L., Scamehorn, J.F. 2006. Use of electrodialysis to remove heavy metals from water. *Journal of Separation Science and Technology* 23(14–15): 2231–2267.

Gode, F., Pehlivan, E. 2006. Removal of chromium (III) from aqueous solutions using Lewatit S 100: The effect of pH, time, metal concentration and temperature. *Journal of Hazardous Materials* 136(2): 330–337.

Greenlee, L.F., Lawler, D.F., Freeman, B.D., Marrot, B., Moulin, P. 2009. Reverse osmosis desalination: Water sources, technology, and today's challenges. *Water Resources* 43(9): 2317–2348.

Gupta, V.K., Ali, I. 2013. Chapter 2—Water treatment for inorganic pollutants by adsorption technology. In: Ali VKG, editor. *Environmental Water*, Elsevier, Amsterdam, the Netherlands, 29–91.

Hayashi, H., Komatsu, T. 1992. Complex formation of polyacrylic acid with Mg, Mn, and Pb ions in dioxane-water mixed solvents. *Bulletin of the Chemical Society of Japan* 65(6): 1500–1505.

Ho, Y.S., Mckay, G. 1999. The adsorption of lead(II) ions on peat. *Water Resources* 33: 578–584.

Huang, Y., Miao, Y., Liu, T. 2014. Electrospun fibrous membranes for efficient heavy metal removal. *Journal of Applied Polymer Science* 131(19): 1–12.

Huang, Z.M., Zhang, Y.Z., Kotaki, M., Ramakrishna, S. 2003. A review on polymer nanofibers by electrospinning and their applications in nanocomposites. *Composites Science Technology* 63: 2223–2253.

Hubicki, Z., Kolodynska, D. 2012. Selective removal of heavy metal ions from waters and waste waters using ion exchange methods. In: Kilislioglu A, editor. *Ion Exchange Technologies*, IntechOpen Limited, London, UK.

Javanbakht, V., Alavi, S.A., Zilouei, H. 2014. Mechanisms of heavy metal removal using microorganisms as bioadsorbent. *Water Science and Technology* 69(9): 1775–1787.

Jayakrishnan, D.S. 2012. *Corrosion Protection and Control Using Nanomaterials*. Saji VS, Cook R, editor. Woodhead Publishing, Cambridge, UK, Chapter 5, 86–125.

Kabra, K., Chaudhary, R., Sawhney, R.L. 2008. Solar photocatalytic removal of metal ions from industrial wastewater. *Environmental Progress and Sustainable Energy* 27(4): 487–495.

Kampalanonwat, P., Supaphol, P. 2010. Preparation and adsorption behavior of aminated electrospun polyacrylonitrile nanofiber mats for heavy metal ion removal. *ACS Applied Materials and Interfaces* 2(12): 3619–3627.

Kamplanonwat, P., Supaphol, P. 2014. The study of competitive adsorption of heavy metal ions from aqueous solution by aminated polyacrylonitrile nanofiber mats. *Energy Procedia* 56: 142–151.

Kang, S.Y., Lee, J.U., Moon, S.H., Kim, K.W. 2004. Competitive adsorption characteristics of Co^{2+}, Ni^{2+} and Cr^{33+} by IRN-77 cation exchange resin in synthesized wastewater. *Chemosphere* 56(2): 141–147.

Karnib, M., Kabbani, A., Holail, H., Olama, Z. 2014. Heavy metals removal using activated carbon, silica and activated carbon composite. *Energy Proceedia* 50: 113–120.

Kurniawan, T., Babel, S. 2003. A research study on Cr(VI) removal from contaminated wastewater using low-cost adsorbents and commercial activated carbon. *Second International Conference of Energy and Technology towards a Clean Environment*, Phuket, 2: 1110–1117.

Kurniawan, T.A., Chan, G.Y.S., Lo, W.H., Babel, S. 2006. Physico-chemical treatment techniques for wastewater laden with heavy metals. *Chemical Engineering Journal* 118: 83–98.

Lam, B., Deon, S., Morin-Crini, N., Crini, G., Fievet, P. 2018. Polymer-enhanced ultrafiltration for heavy metal removal: Influence of chitosan and carboxymethyl cellulose on filtration performances. *Journal of Cleaner Production* 171: 927–933.

Li, D., McCann, J.T., Xia, Y. 2006. Electrospinning: A simple and versatile technique for producing ceramic nanofibers and nanotubes. *Journal of the American Ceramic Society* 89: 1861–1869.

Long, F., Gong, J.L., Zeng, G.M., Chen, L., Wang, J.H., Deng, Niu, Q.Y., Zhang, H.Y., Zhang, X.R. 2011. Removal of phosphate from aqueous solution by magnetic Fe-Zr binary oxide. *Chemical Engineering Journal* 171: 448–455.

Malaeb, L., Ayoub, G.M. 2011. Reverse osmosis technology for water treatment: State of the art review. *Desalination* 267(1): 1–8.

Malik, N., Biswas, A.K., Qureshi, T.A., Borana, K., Virha, R. 2010. Bioaccumulation of heavy metals in fish tissues of a freshwater lake of Bhopal. *Environmental Monitoring and Assessment* 160: 267–276.

Martin, D.M., Faccini, M., Garcia, M.A., Amantia, D. 2018. Highly efficient removal of heavy metal ions from polluted water using ion-selective polyacrylonitrile nanofibers. *Journal of Environmental Chemical Engineering* 6(1): 236–245.

Matis, K.A., Zouboulis, A.I., Gallios, G.P., Erwe, T., Blocher, C. 2004. Application of flotation for the separation of metal-loaded zeolites. *Chemosphere* 55(1): 65–72.

Maximous, N.N., Nakhla, G.F., Wan, W.K. 2010. Removal of heavy metals from wastewater by adsorption and membrane processes: A comparative study. *International Journal of Environmental and Ecological Engineering* 4(4): 125–130.

Metals (September 21, 2016). Retrieved February 7, 2018, from https://www.fda.gov/food/foodborneillnesscontaminants/metals/.

Min, M., Wang, X., Hsiao, B.S. 2011. Fabrication of micro-nano structure nanofibers by solvent etching. *Journal of Nanoscience and Nanotechnology* 11(8): 6919–6925.

Molinari, R., Argurio, P., Lavorato, C. 2015. *Membrane Reactors for Energy Applications and Basic Chemical Production*, Basile A, Hai LDPL, Piemonte V, editors. Woodhead Publishing, Cambridge, UK, 605–639.

Molinari, R., Gallo, S., Argurio, P. 2004. Metal ions removal from wastewater or washing water from contaminated soil by ultrafiltration-complexation. *Water Resources* 38(3): 593–600.

Montana, M., Camacho, A., Serrano, I., Devesa, R., Matia, L., Valles, I. 2013. Removal of radionuclides in drinking water by membrane treatment using ultrafiltration, reverse osmosis and electrodialysis reversal. *Journal of Environmental Radioactivity* 125: 86–92.

Moon, S.H., Yun, S.H. 2014. Progress integration of electrodialysis for a cleaner environment. *Current Opinion in Chemical Engineering* 4: 25–31.

Moussavi, G., Khosravi, R. 2010. Removal of cyanide from wastewater by adsorption onto pistachio hull wastes: Parametric experiments, kinetics and equilibrium analysis. *Journal of Hazardous Materials* 183: 724–730.

Murthy, Z.V.P., Chaudhari, L.B. 2009. Separation of binary heavy metals from aqueous solutions by nanofiltration and characterization of the membrane using Spiegler-Kedem model. *Chemical Engineering Journal* 150(1): 181–187.

Najim, T.S., Salim, A.J. 2017. Polyaniline nanofibers and nanocomposites: Preparation, characterization, and application for Cr(VI) and phosphate ions removal from aqueous solution. *Arabian Journal of Chemistry* 10(2): 3459–3467.

Padaki, M., Murali, R.S., Abdullah, M.S., Misdan, N., Moslehyani, A., Kassim, M.A., Hilal, N., Ismail, A.F. 2015. Membrane technology enhancement in oil-water separation. A review. *Desalination* 357: 197–207.

Razavi, S.M.R., Rezakazemi, M., Albadarin, A.B., Shirazian, S. 2016. Simulation of CO_2 adsorption by solution of ammonium ionic liquid in hollow-fiber contactors. *Chemical Engineering Process* 108: 27–34.

Rivas, B.L., Maureira, A. 2008. Water-soluble polyelectrolytes containing sulfonic acid groups with metal ion binding ability by using the liquid phase polymer based retention technique. *Macromolecular Symposia* 270: 143.

Robertson, P.K.J. 1996. Semiconductor photocatalysis: An environmentally acceptable alternative production technique and effluent treatment process. *Journal of Clean Production* 4(3–4): 203–212.

Sadegh, H., Ghoshekandi, R.S., Masjedi, A., Mahmoodi, Z., Kazemi, M. 2016. A review on carbon nanotubes adsorbents for the removal of pollutants from aqueous solutions. *International Journal of Nano Dimension* 7(2): 109–120.

Saeed, A., Akhter, M.W., Iqbal, M. 2005. Removal and recovery of heavy metals from aqueous solution using papaya wood as a new adsorbent. *Separation and Purification Technology* 45(1): 25–31.

Sang, Y.M., Li, F.S., Gu, Q.B., Liang, C.Z., Chen, J.Q. 2008. Heavy metal—Contaminated groundwater treatment by a novel nanofiber membrane. *Desalination* 223: 349.

Shapin, S. 2006. *The Scientific Revolution.* University of Chicago Press, Chicago, IL, 15–64.

Shenvi, S.S., Isloor, A.M., Ismail, A.F. 2015. A review on RO membrane technology: Development and challenges. *Desalination* 368: 10–26.

Singh R. 2015. *Membrane Technology and Engineering for Water Purification*, 2nd ed., Butterworth-Heinemann, Oxford, UK, Chapter 3, 179–281.

Sorme, L., Lagerkvist, R. 2002. Sources of heavy metals in urban wastewater in Stockholm. *Science Total Environment* 298: 131–145.

Soto, D., Urdaneta, J., Pernia, K., Leon, O., Munoz-Bonilia, A. 2016. Removal of heavy metal ions in water by starch esters. *Starch* 68: 37–46.

Srivastava, N.K., Majumder, C.B. 2008. Novel biofiltration methods for the treatment of heavy metals from industrial wastewater. *Journal of Hazardous Materials* 151: 1–8.

Subbiah, T., Bhat, G.S., Tock, R.W., Parameswaran, S., Ramkumar, S.S. 2005. Electrospinning of nanofibers. *Journal of Applied Polymer Science* 96(2): 557–569.

Sutherland, K. 2008. What is nanofiltration? *Filtration and Separation* 45(8): 32–35.

Tarley, C.R.T., Arruda Zezzi, M.A.Z. 2004. Biosorption of heavy metals using ricemilling by-products. Characterisation and application for removal of metals from aqueous solutions. *Chemosphere* 54: 987–995.

Tonini, G.A., Ruotolo, L.A.M. 2016. Heavy metal removal from simulated wastewater using electrochemical technology: optimization of copper electrodeposition in a membraneless fluidized bed electrode. *Clean Technologies and Environmental Policy* 19(2): 403–415.

Van der Bruggen, B., Vandecasteele, C. 2003. Removal of pollutants from surface water and groundwater by nanofiltration: Overview of possible applications in the drinking water industry. *Environmental Pollution* 122(3): 435–445.

Vinodhini, R., Narayanan, M. 2008. Bioaccumulation of heavy metals in organs of fresh water wish cyprinus carpio (common carp). *International Journal of Environmental Science and Technology* 5(2): 179–182.

Wang, L.K., Hung, Y.T., Shammas, N.K. 2005. *Physicochemical Treatment Processes*, Springer, Heidelberg, Germany, 141–197.

Wang, X.F., Min, M.H., Liu, Z.Y., Yang, Y., Zhou, Z., Zhu, M.F., Chen, Y.M., Hsiao, B.S. 2011. Poly(ethyleneimine) nanofibrous affinity membrane fabricated via one step wet–electrospinning from poly(vinyl alcohol)-doped poly(ethyleneimine) solution system and its application. *Journal of Membrane Science* 379: 191.

Xiao, S., Shen, M., Ma, H., Guo, R., Zhu, M., Wang, S., Shi, X. 2010. Fabrication of water-stable electrospun polyacrylic acid-based nanofibrous mats for removal of copper (II) ions in aqueous solution. *Journal of Applied Polymer Science* 116(4): 2409–2417.

Yadav, D.K., Srivastava, S. 2017. Carbon nanotubes as adsorbent to remove heavy metal ion (Mn^{+7}) in wastewater treatment. *Materials Today: Proceedings* 4(2): 4089–4094.

Zhang, Z.Q., Liu, Y.W., Huang, Y.D., Liu, L., Bao, J.W. 2002. The effect if carbon-fibe surface properties on the electron-beam curing of epoxy-resin composites. *Composites Science and Technology* 62(3): 331–337.

Zamboulis, D., Pataroudi, S.I., Zouboulis, A.I., Matis, K.A. 2004. The application of sorptive flotation for the removal of metal ions. *Desalination* 162: 159–168.

9 Surface Functionalized Electrospun Nanofibers for Removal of Toxic Pollutants in Water

Brabu Balusamy, Anitha Senthamizhan, and Tamer Uyar

CONTENTS

9.1 INTRODUCTION

A major public health threat that has developed over the last decades is environmental pollution due to industrialization and societal development, calling for technological support and new policy guidelines. According to the World Health Organization (WHO), 12.6 million deaths every year are due to a polluted environment and water pollution [1]. This demands immediate attention toward eliminating water pollution since water is a vital resource for human existence as well as industrial processes. Thus, WHO has devised a standard for the quality of water and prioritized water sanitation and hygiene. A recent report by WHO has revealed that there are approximately 2.1 billion people who lack safe drinking water at home [2]. Water pollution can arise due to various biological agents, heavy metals, organic dyes, pesticides, herbicides, and organic solvents, which can cause a hazardous impact on human and environmental health. To date, various approaches have been adopted for purifying water including filtration, reverse osmosis, precipitation, flocculation, electrodialysis, bio-sorption, and adsorption [3–9]. However, these techniques are expensive, require complex setup, and consume high amounts of energy. This enforces the need for new eco-friendly and affordable methods to deal with water pollution.

Over the last two decades, nanotechnology has emerged as a significant field in all branches of science and technology, owing to the unique characteristics of nanomaterials in terms of improved physical and chemical properties. Nanotechnology has also be applied in devising an effective water treatment technique. To date, various nanomaterials have been extensively studied for treating polluted water by adopting numerous approaches [10–19]. One of them includes electrospun nanofibers, a unique type of nanomaterial, that has high surface-to-volume ratios and excellent mechanical properties, porosity, easy production, and surface functionalization potentials. In recent years, much effort has been made using electrospun nanofibers, in applications from the healthcare to environmental sectors [20–25]. In particular, electrospun nanofibers have been studied for effective removal of toxic water pollutants including heavy metal ions, organic pollutants, and microorganisms [26]. This chapter demonstrates the application of surface functionalized electrospun nanofibers for removing aquatic pollutants using different approaches, with special attention to the removal of heavy metals and dye molecules.

9.2 SURFACE FUNCTIONALIZED ELECTROSPUN NANOFIBERS FOR POLLUTANTS REMOVAL

9.2.1 Removal of Organic Dye Pollutants in Water

The world's first commercial dye was discovered by Henry Perkin in 1856. Since then, thousands of dye molecules have been produced and used for multiple purposes. Dyes are organic compounds, comprised of two major components, namely chromophores and auxochromes. The chromophores are responsible for determining and producing color, whereas auxochromes support the solubility in water and controls the color intensity [27,28]. Organic dyes can be classified into various types including nitro dyes, azo dyes, xanthene dyes, direct dyes, and mordant dyes, based on their chemical structure and methods of application [29,30]. These dyes are considered to be major organic contaminants of water, majorly released from dye manufacturing, textile, paper, leather, plastic, cosmetics, ceramic, food processing, and ink industries [31–34]. Although these brightly colored compounds have multiple functions, they also add a negative impact to the environment and human life. This has resulted in the development of numerous technologies for removing these dye pollutants from water including physical, chemical, biological, radiation, and electrical processes. A detailed description of the methods used for the removal of dyes from waste water is presented in Figure 9.1. But, each of these methods comes with their own disadvantages as well [35,36].

Increasing development has been observed in recent years on using nanotechnology for the removal of dyes in contaminated water. We have seen the use of carbon, metal, metal oxide, and nanofiber-based materials in effectively eliminating dye pollutants from water [37–41]. Interestingly, electrospun nanofibers have gained considerable attention as a source material for removing these dye pollutants [42–48]. Un-functionalized or structurally modified, these electrospun nanofibers have proved to provide potential impact in removing dye pollutants, through adsorption ascribed from their excellent surface characteristics. Although there has been evident potential shown by various other electrospun nanocomposite membranes, we are going to discuss in detail the development and application of surface functionalized electrospun nanofibers for removing dye pollutants from water.

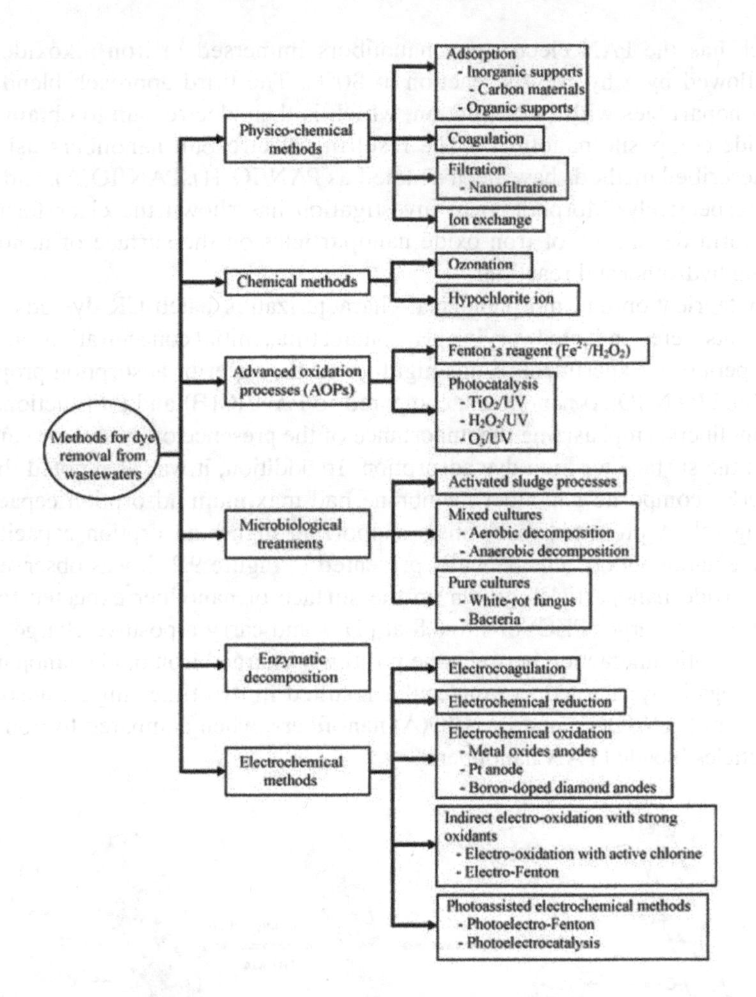

FIGURE 9.1 Main methods used for the removal of organic dyes from wastewaters. (Reprinted from *Appl. Catal., B*, Martínez-Huitle, C. A. and Brillas, E. Decontamination of wastewaters containing synthetic organic dyes by electrochemical methods: A general review, 87, 105–145, Copyright (2009), with permission from Elsevier.)

Starting with their structure, the optimum electrospun nanofibers for pollutant removal are composed of metal nanostructures on the surface, which have a remarkably higher interactive ability toward pollutants. This has been evidently pointed out by Patel et al. (2016) that iron oxide nanoparticles grown on a polyacrylonitrile (PAN) electrospun nanofiber surface show an improved dye removal capacity [49]. Three varied approaches have been applied to prepare PAN/iron oxide composite nanofibers and an in-depth comparison of their removal efficiencies for the adsorption of Congo Red (CR) dye from aqueous solution has been performed. The first approach has the electrospun PAN/iron(III) acetylacetonate composite nanofibers prepared by electrospinning, followed by hydrothermal method for *in situ* growth of iron oxide nanoparticles on the surface of PAN nanofibers. The second approach has the PAN electrospun nanofibers immersed in iron alkoxide solution, followed by a hydrolysis reaction at 80°C. The third approach blends iron oxide nanoparticles with PAN solution, which is then electrospun to obtain PAN/iron oxide composite nanofibers. The resulting electrospun nanofibers using the above described methods have been denoted as PAN/IO(H), PAN/IO(A), and PAN/IO(B), respectively. Morphological investigation has shown the clear formation and uniform decoration of iron oxide nanoparticles on the surface of nanofibers following hydrothermal reactions.

After fabrication and morphological characterization batch CR dye adsorption experiments were conducted varying pH, contact time, initial concentration variation, and temperature. Experiment results highlighted the superior adsorption properties of both and PAN/IO(A) nanofibers, compared to PAN/IO(B) and un-functionalized PAN nanofibers, emphasizing the importance of the presence of iron oxide nanoparticles on the surface for high dye adsorption. In addition, it was also noted that the PAN/IO(H) composite nanofiber membrane had maximum adsorption capacity of 52.08 mg g^{-1}. A proposed mechanism supporting higher adsorption capacities of composite nanofibers is schematically presented in Figure 9.2. It was observed that the iron oxide nanoparticles grown on the surface of nanofiber expected to have Point of Zero Charge (PZC) of 7.8–8.8 at pH 7 and carry a positive charge. Thus, the electrostatic interaction between the positively charged iron oxide nanoparticles and the negatively charged dye molecules resulted in two times higher adsorption capacities of PAN/IO(H) and PAN/IO(A) nanofibers, when compared to iron oxide nanoparticles blended PAN nanofibers.

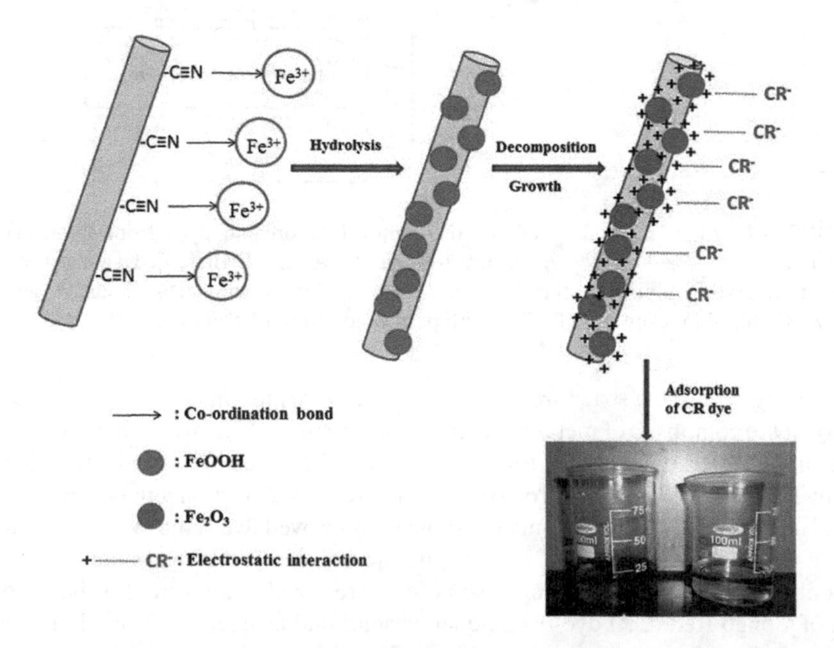

FIGURE 9.2 Schematic representation of iron oxide nanoparticles grown on the surface of PAN nanofibers and mechanism of CR dye removal. (Patel, S. and Hota, G. *RSC Adv.*, 6, 15402–15414, 2016. Reproduced by permission of The Royal Society of Chemistry.)

Similarly, zerovalent iron nanoparticles (ZVI NPs) immobilized onto electrospun poly(acrylic acid) (PAA)/poly(vinyl alcohol) (PVA) nanofibrous materials, through *in situ* reduction of Fe(III) ions, demonstrated an increased dye removal performance compared to ZVI NPs [50]. The decolorization efficiency of the VI NPs/polymer nanofibrous mats were evaluated with acid fuchsine as a model contaminant, which is a common organic dye in the textile industry containing a benzene ring. The ZVI NP-containing polymer nanofibrous mats were found to decolorize ~95% of the acid fuchsine with a concentration of 60 mg/L, but this was not observed in the dye solution when treated alone with ZVI NPs at comparable concentration. There was a sudden decrease in the red color of the dye solution after 5 min, with further decolorization after 40 minutes, with confirmation by UV–vis spectroscopic measurements. Figure 9.3 shows a comparison of the degradation efficiencies of ZVI NP-containing polymer nanofibrous mats and ZVI NPs. The enhanced decolorization efficiency can be attributed to the uniform distribution of ZVI NPs in the fibrous membrane that attracts reactive degradation [50].

Yet another approach was taken to fabricate hierarchical SiO_2@γ-AlOOH (Boehmite) core/sheath fibers by combining electrospinning and hydrothermal reaction [51]. The γ-AlOOH (Boehmite) nanoplatelets were uniformly anchored on the surface of SiO_2 fibers to improve the adsorption efficiency of the SiO_2 fiber membrane

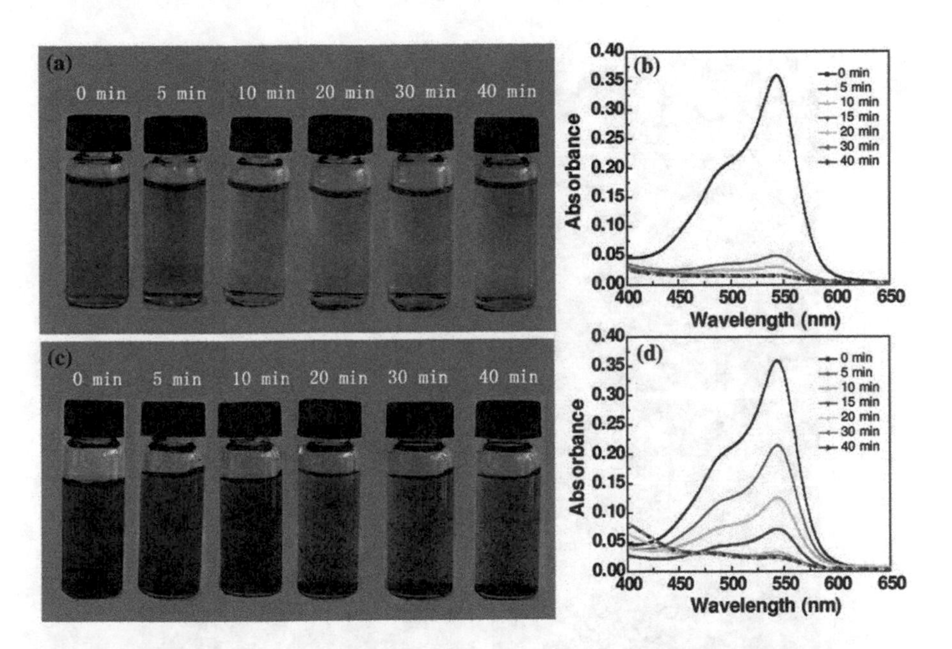

FIGURE 9.3 Photographs of the acid fuchsine solution treated with ZVI NP-containing polymer nanofibrous mats (a) and ZVI NPs synthesized using literature method (c) at different time intervals. UV–vis spectra of a solution of acid fuchsine (60 mg/L, 20 mL) in the presence of ZVI NP-containing polymer nanofibrous mats (b) and ZVI NPs synthesized using literature method (d) at a time interval of 0, 5, 10, 20, 30, and 40 min. (Reprinted with permission from Xiao, S. et al., *J. Phys. Chem. C*, 113, 18062–18068, 2009. Copyright American Chemical Society.)

for organic dyes and microorganisms. In the first stage, silica gel was prepared using hydrolysis and polycondensation, and then PVA solution was added, mixed, electrospun and calcined at 800°C in air for 2 h. The obtained SiO_2 fiber was then immersed in a bath of H_2SO_4/H_2O_2 (3:1, v/v) for about 1 h and thoroughly rinsed. This was followed by treating the membrane hydrothermally using sodium aluminate and urea, washed and dried to obtain SiO_2@γ-AlOOH (Boehmite) core/sheath fiber. In-depth observation of the SiO_2@γ-AlOOH (Boehmite) core/sheath fibers (See Figure 9.4) showed that there was perpendicular growth of the Boehmite nanoplatelets on the surface of the SiO_2 fibers. Interestingly, the removal of Congo red dye happened within 90 min of the adsorption experiment [51].

Algal, bacterial, and fungal cells have been used for the bioremediation of various dye pollutants in water, for many years. Among them, microalgae are considered to be an ideal candidate for removing dye pollutants owing to their expense, availability, and higher affinity toward pollutants. This has attracted greater attention, over recent years, to attach microbial agents on electrospun nanofibers for improving their capability to remove dye pollutants from water. This has also been previously reported by San Keskin et al. (2015), where *Chlamydomona reinhardtii* microalgal cells have been immobilized in a polysulfone nanofibrous web (PSU-NFW), which enhanced the efficiency of removing two reactive dyes,

FIGURE 9.4 Low- and high-magnification SEM images of (a and b) SiO_2 fibers and (c and d) SiO_2@γ-AlOOH (Boehmite) core/sheath fibers. (Reprinted with permission from Miao, Y.E., *ACS Appl. Mater. Interfaces*, 4, 5353–5359, 2012. Copyright American Chemical Society.)

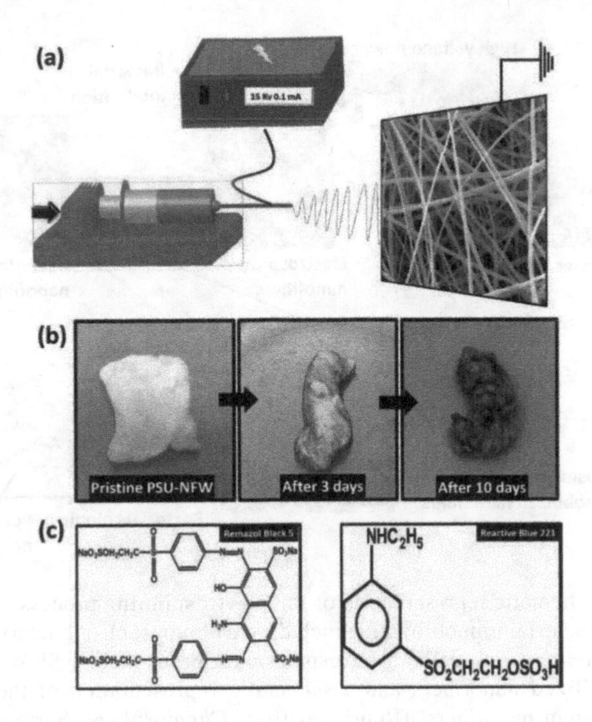

FIGURE 9.5 (a) Schematic representation of the electrospinning process for PSU-NFW. (b) Photos of attachment of free microalgae cells on PSU-NFW for 3 and 10 days attachment processes. (c) Chemical structures of Remazol Black 5 (RB5) and Reactive Blue 221 (RB221). (Reprinted with permission from San Keskin, N.O. et al., *Ind. Eng. Chem. Res.*, 54, 5802–5809, 2015. Copyright American Chemical Society.)

Remazol Black 5 (RB5) and Reactive Blue 221 (RB221) [52]. A schematic representation of the electrospinning process, the attachment of microalgal cells on the 3rd and 10th day, and the chemical structures of the reactive dyes used in the study are presented in Figure 9.5. Using time and dye as functional parameters, the decolorization experiment was conducted. The results indicated that microalgae/PSU-NFW has a decolorization rate of 72.97% ± 0.3% for RB5 and 30.2% ± 0.23% for RB221, in addition to their excellent reusable characteristics. In contrast, pristine PSU-NFW was observed to remove very little reactive dyes compared to microalgae/PSU-NFW. An important factor revealed from the study highlighted the use of microalgae/PSU-NFW as a promising membrane material for industrial wastewater treatment [52].

Another study proposed by Sarioglu et al. (2017) showed bacterial cell encapsulated electrospun nanofibers display capabilities to remove methylene blue (MB) dye [53]. The *Pseudomonas aeruginosa* bacterial cells were mixed with polyvinyl alcohol (PVA) and polyethylene oxide (PEO) electrospinning solution and electrospun to get nanofibrous mats. Encapsulation and viability of bacterial cells in nanofibrous webs were analyzed by SEM and fluorescence microscopy. Both bacteria/PVA and bacteria/PEO webs displayed great potential in the remediation of MB

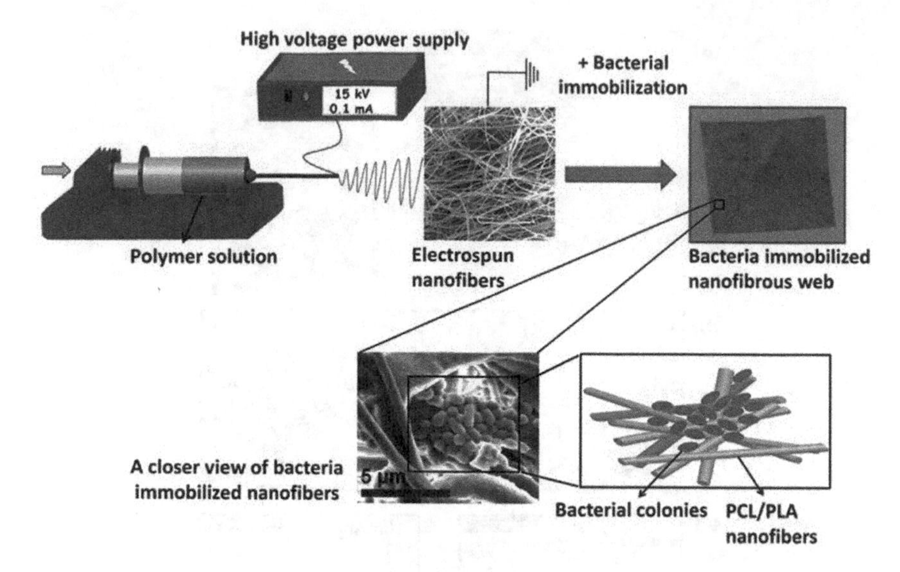

FIGURE 9.6 Schematic representation of the electrospinning process and representative images for bacterial immobilization including a photograph of bacteria immobilized (*Clavibacter michiganensis* cells) electrospun nanofibrous web, a SEM micrograph of bacteria immobilized nanofibers and a schematic representation of the immobilized cells on electrospun nanofibers. (Reprinted from *Chemosphere*, Sarioglu, O.F. et al., Bacteria immobilized electrospun polycaprolactone and polylactic acid fibrous webs for remediation of textile dyes in water, 184, 393–399, Copyright (2017), with permission from Elsevier.)

dyes, and the webs were stable for 3 months at 4°C [53]. Similarly, textile dyes remediating the bacterial isolate *Clavibacter michiganensis* were immobilized on electrospun polycaprolactone (PCL) and polylactic acid (PLA) nanofibrous polymeric webs and used for removal of reactive textile dye Setazol Blue BRF-X, as shown in Figure 9.6 [54]. Both bacteria/PCL and bacteria/PLA nanofibrous webs have shown high removal capabilities of Setazol Blue BRF-X dye within 48 h in addition to their reusability [54].

The electrospun nanofibers have shown to be promising for the surface modification of suitable functional groups to remove specific pollutants. Chaúque et al. (2017) performed surface modification of electrospun poly-acrylonitrile (PAN) nanofibers with ethylenediaminetetraacetic acid (EDTA) and ethylenediamine (EDA) cross-linker [55]. The resulting functionalized EDTA-EDA-PAN nanofibers were systematically characterized, which revealed the impregnation of EDA and EDTA chelating agents on the surface of PAN significantly changing the distribution of nanofibers. On the other hand, the pristine PAN nanofibers were not altered substantially. Adsorption studies show that the modified PAN fibers possess enhanced sorption capacity toward methyl orange (MO) and reactive red (RR), supported by the maximum adsorption capacities of 99.15 and 110.0 mg g^{-1}, respectively [55]. In another technique, an active functional electrospun adsorbent membrane was fabricated by combining a hydrophilic biopolymer filler, such as chitin nanowhiskers (ChNW) that contains two functional groups, and a hydrophobic polymer matrix such as polyvinylidene fluoride (PVDF) using an electrospinning technique [56]. SEM imaging of the PVDF/ChNW electrospun nanofibers confirmed the alteration of the surface roughness and water contact-angle measurements revealed enhanced hydrophilicity. Further, indigo carmine (IC) dye adsorption experiments demonstrated improved removal capabilities (88.9%) and adsorption capacities (72.6 mg g^{-1}) compared to neat PVDF. With findings from FTRI studies depicted in Figure 9.7, the improvement in the removal of the dyes can be evidently seen. The two functional groups of the chitin moiety serve as active sites of physical adsorption that could attribute to the hydrogen bonding and electrostatic attraction. The hydrophilic nature of the membrane aids in easy adsorption of dissolved dye molecules from water. The miscibility of the dye molecules and polymers depends on the molecular configuration and presence of the ionic groups in the dye molecule [56].

In another study, Yan et al. (2015) prepared free standing poly(vinyl alcohol)/poly(acrylic acid) membranes with polydopamine coating (PVA/PAA@PDA) using a combination of electrospinning and self-polymerization of dopamine [57]. In brief, the PVA and PAA solutions at 10 and 12 wt% were prepared and 1:1 ratio of the solutions were used for the electrospinning process. The electrospun membrane was further subjected to heat at 120°C for 3 h to induce cross-linking between PVA and PAA. The cross-linking was induced by an esterification reaction between the carboxylic acid groups of PAA and hydroxyl groups of PVA. The resultant water insoluble membrane was immersed in an aqueous solution of dopamine under mild stirring at 50°C with varied reaction times, followed by repeated washing to remove non-adhered PDA, and then subjected to drying. The membranes were characterized through FESEM, where the PDA layer was formed on the surface of the fiber with thin to thick coating, upon increasing the

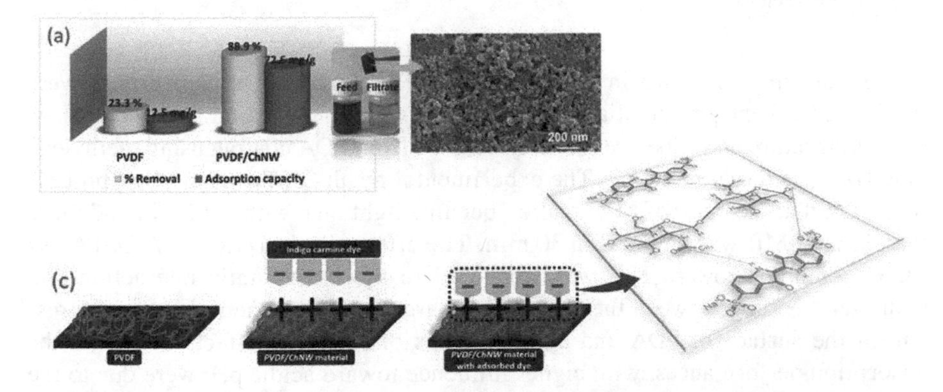

FIGURE 9.7 (a) Percentage of removal and adsorption capacity (qe) of IC on PVDF and PVDF/ChNW electrospun; (b) SEM image of indigo carmine on PVDF/ChNW membrane; (c) graphical representation of adsorption mechanism. (Reprinted from *Carbohydr. Polym.*, Gopi, S. et al., Chitin nanowhisker (ChNW)-functionalized electrospun PVDF membrane for enhanced removal of Indigo carmine. 165, 115–122, Copyright (2017), with permission from Elsevier.)

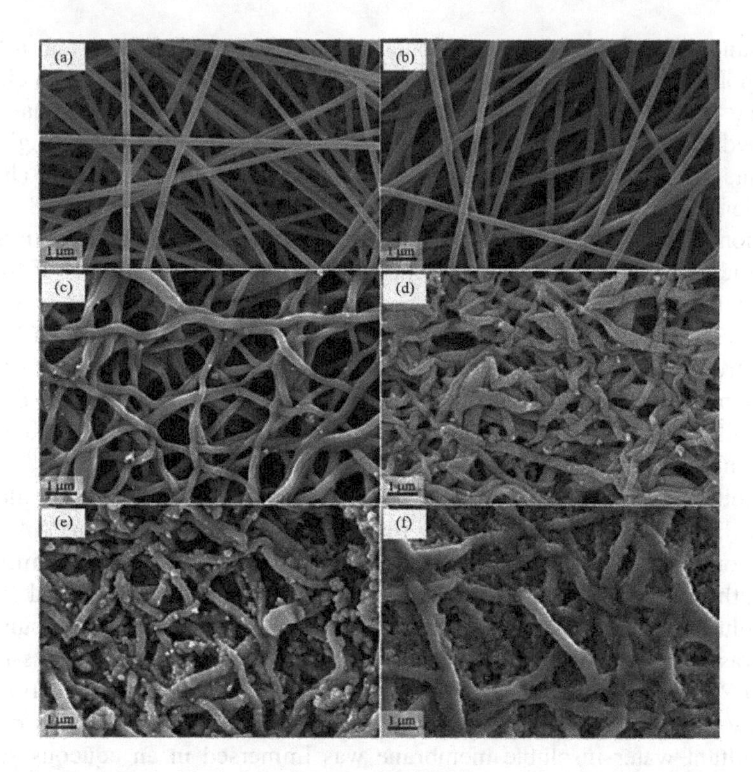

FIGURE 9.8 FESEM images of original PVA/PAA membrane (a), PVA/PAA membrane after heat treatment (b), PVA/PAA@PDA-5 (c), PVA/PAA@PDA-15 (d), PVA/PAA@PDA-30 (e), and PVA/PAA@PDA-45 (f) membranes. (Reprinted from *J. Hazard. Mater.*, Yan, J. et al., Polydopamine-coated electrospun poly(vinyl alcohol)/poly(acrylic acid) membranes as efficient dye adsorbent with good recyclability, 283, 730–739, Copyright (2015), with permission from Elsevier.)

immersion time as shown in Figure 9.8. The PVA/PAA@PDA membranes were used as adsorbent for potential applications in water purification because of their porous structure and adhesive property derived from PDA coating using methylene blue (MB) as target pollutant. The experimental results of the adsorption process indicated that the color of MB solution became light blue within 10 min and more than 93% of MB was removed in 30 min. The effective adsorption of PVA/PAA@PDA membrane toward MB was attributed to the electrostatic interaction and hydrogen bonding between the negatively charged oxygen functional groups present on the surface of PDA and amine groups of MB. The effects of pH on the adsorption performances, with higher influence toward acidic pH, were due to the positive charge present in the adsorbate, because the amino groups present in MB were protonated to form amino cations. On the other hand, the negatively charged adsorbent species lead to more adsorption upon decreasing pH [57].

A hierarchical bioinspired nanocomposite material of poly(vinyl alcohol)/poly(acrylic acid)/carboxylate graphene oxide nanosheet@polydopamine (PVA/PAA/GO-COOH@PDA) was successfully prepared by using electrospinning,

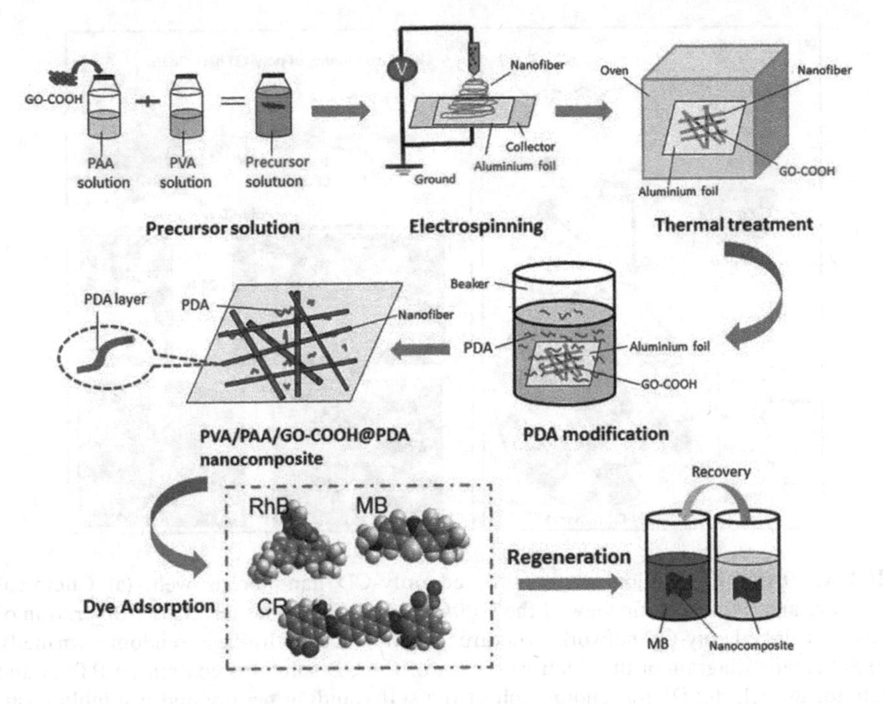

FIGURE 9.9 Schematic illustration of the fabrication and dye adsorption of PVA/PAA/GO-COOH@PDA nanocomposites by electrospinning and thermal treatment. (Reprinted with permission from Xing, R. et al. *ACS Sustainable Chem. Eng.*, 5, 4948–4956, 2017. Copyright American Chemical Society.)

thermal treatment, and polydopamine modification [58]. The schematic representation of the preparation and modification is illustrated in Figure 9.9. The resultant composite membranes are composed of polymeric nanofibers with carboxylate graphene oxide nanosheets anchored on the fibers, by heat-induced cross-linking reaction. First, GO-COOH was prepared, freeze-dried, and mixed with PAA solution. An equal volume of PVA and PAA solution containing GO-COOH was mixed, electrospun, and further heated at 120°C for 3 h for heat-induced cross-linking between carboxyl acid groups in PAA/GO-COOH components and hydroxyl groups in PVA molecules. The adsorption properties of the prepared PVA/PAA/GO- COOH@ PDA nanocomposite membranes were investigated with solutions of three model dyes (Methylene blue, Rhodamine B, and Congo red). The results indicated that the adsorption capacities of PVA/PAA/GO-COOH@PDA nanocomposites are higher for MB when compared to the other two dyes. This is because of the strong π–π stacking and electrostatic interaction between nanocomposites and MB molecules, in addition to PDA offering abundant amino and hydroxyl groups resulting in improved adsorption performance [58].

Celebioglu et al. (2017) produced a highly efficient molecular filter membrane based on cyclodextrins (CD) through electrospinning without addition of any polymeric carrier [59]. Hydroxypropyl-β-cyclodextrin (HPβCD) was dissolved in water at room temperature, 1,2,3,4-butanetetracarboxylic acid (BTCA) was added to the

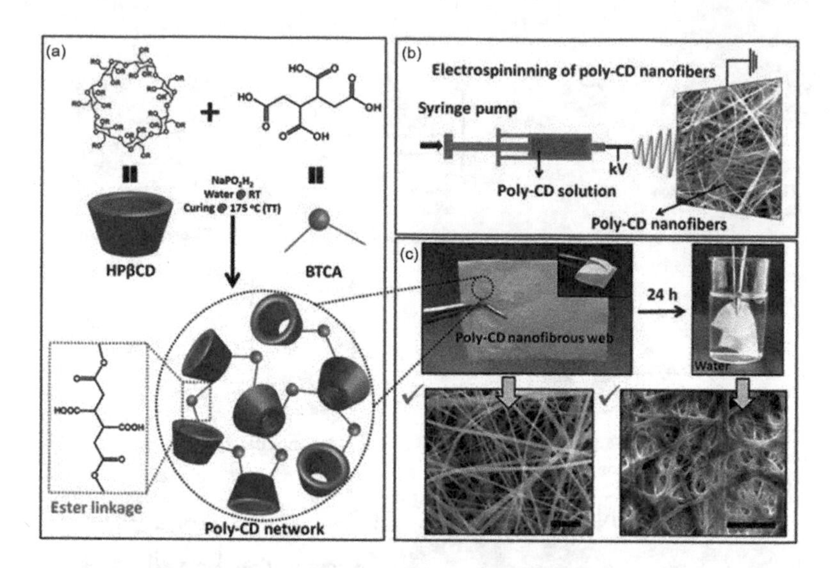

FIGURE 9.10 Fabrication of cross-linked poly-CD nanofibrous web. (a) Chemical structure and the schematic view of the HPβCD and BTCA. The schematic illustration of the cross-linked poly-CD network structure linked with ester linkage (randomly formed). (b) Schematic diagram of the electrospinning of the CD solutions containing BTCA and initiator as well. (c) Digital photograph of the self-standing feature and insoluble property of the poly-CD nanofibrous webs. The SEM images of poly-CD nanofibers before and after immersing in water for 24 h (scale bar: 10 μm). (From Celebioglu, A. et al., *Sci. Rep.*, 7, 7369, 2017.)

HPβCD solution, and then sodium hypophosphite hydrate (SHP) was added as an initiator to CD-BTCA solution. This was maintained at 50°C with continuous stirring and then electrospun to obtain fibers. The nanofibers were then subjected to heat at 175°C for 1 hour to obtain cross-linked and insoluble poly-CD nanofibers. The schematic illustration of the fabrication of cross-linked poly-CD nanofibrous web is depicted in Figure 9.10. Adsorption kinetics of the poly-CD nanofibers were evaluated using methylene blue (MB) dye at varied time intervals. A relatively faster adsorption process was observed within the first 5 min due to the existence of several active adsorption sites, and approximately 70% of the dye was removed during the first 15 min. The adsorptions reached saturation by 360 min with equilibrium uptake of 96.2% ± 0.8% [59].

Recently, Bai et al. (2018) reported the use of a simple, green, and effective air plasma etching treatment of electrospun PLLA nanofiber membrane, to be used as an adsorbent for cationic methylene blue dye [60]. PLLA nanofiber membrane was fabricated by electrospinning PLLA solution in chloroform and N, N-dimethylformamide. Then, the PLLA nanofiber membrane was treated with air plasma at different time intervals including 1, 3, 5, 8, 10, 15, and 20 min. This was followed by immersing the treated nanofibers in MB dye solution, to study their adsorption performances. The adsorption experiment showed that the removal efficiency of the MB dye increased with increasing plasma treatment time. The mechanism of the adsorption was attributed to the electrostatic interaction and increased surface area generated for binding sites after the plasma treatment. Also, the recycle tests revealed the ability of the membranes to achieve 86% after ten adsorption-regeneration cycles, implicating good reusability of the adsorbent [60].

9.2.2 Removal of Metal Ions in Water

Over several decades, numerous adsorbents have been developed for metal ion removal using bacteria, plant biomass, chitosan, cyclodextrin, and various nanomaterials. Chitosan is considered to be the largest naturally occurring biomass polysaccharide component obtained from chitin that is generally found in the exoskeleton of crustacean shells, such as crabs, insects, and shrimps. In recent years, chitosan has been extensively used as an effective sorbent of metal ions owing to cost effectiveness, high hydrophilicity, nontoxicity, high contents of amino, and hydroxyl functional groups, which act as adsorption sites for enhanced performance [61,62]. Generally, larger surface area of the adsorbent facilitates higher adsorption capacity; however, a common issue encountered in such materials is poor electrospinnability. Therefore, it is of utmost significance to find ways to make nanofibrous structure. In this regard, Min et al. (2015) demonstrated the addition of a compatible polymer, poly(ethylene oxide) (PEO), that enhances the spinnability of chitosan and produces defect-free nanofibrous structures [63]. The resulting chitosan based electrospun nanofiber has been applied for potential in arsenic removal. The attained results demonstrated an effective arsenic adsorption at acidic condition; however, their adsorption efficiency is limited when pH was around 7.0. Previous evidence has already shown enhanced adsorption capability for arsenic by of surface functionalized nanofibers [64]. Interestingly, iron-based materials have been reported as an effective adsorbent for arsenic removal in water under neutral pH; iron doped chitosan nanofiber (ICS-ENF) has been successfully produced and applied for As(V) removal [65]. The result shows that ICS-ENF possesses superior arsenic removal capacity under acid and weak basic condition (more than 90%). Notably, ICS-ENF has high adsorption capability at pH = 7, which is a significant outcome since most natural water bodies are at near neutral condition. Overall, the study suggests that the ICS-ENF demonstrated remarkable adsorption performance in terms of adsorption capacity at neutral pH as compared to chitosan-based electrospun nanofiber [65].

The electrospun nanofibrous matrix has received substantial attention as a carrier matrix to hold various functional groups. In addition, it is worth highlighting that the functional groups can be converted to active sites for adsorption of metal ions through special grafting reactions. For instance, plenty of nitrile groups present in polyacrylonitrile fiber matrices can be easily converted into carboxyl, amide, and amidoxime groups [66]. Polyacrylonitrile fiber (PANF) possesses many advantages including low-cost, high mechanical strength, easy availability, and excellent flexibility [67–69]. Xu et al. (2018) demonstrated simple preparation of aminophosphonic acid functionalized fiber ($PAN_{AP}F$) and used as a mercury ion adsorbent [70]. A chemical grafting approach has been adopted for immobilizing

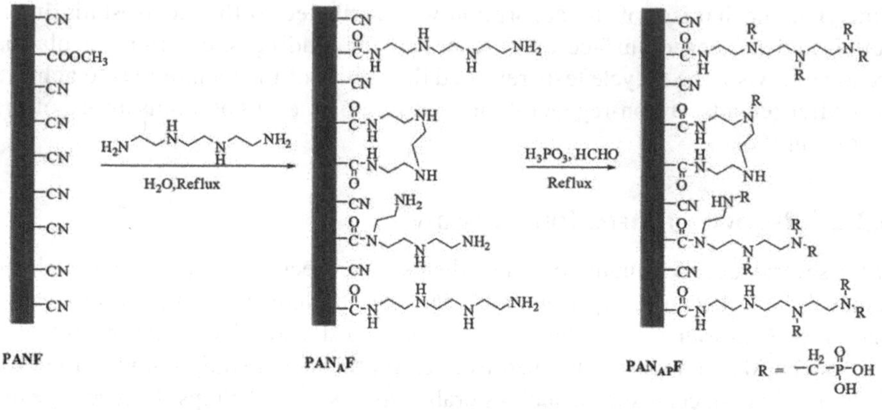

FIGURE 9.11 Schematic diagram of the synthesis of aminophosphonic acid functionalized polyacrylonitrile fiber. (Reprinted from *J. Hazard. Mater.*, Xu, G. et al., Highly selective and efficient adsorption of Hg^{2+} by a recyclable aminophosphonic acid functionalized polyacrylonitrile fiber, 344, 679–688, Copyright (2018), with permission from Elsevier.)

the aminophosphonic acid onto polyacrylonitrile fiber as shown in Figure 9.11. First, aminated fiber (PAN_AF) was prepared by modifying the PANF with triethylene tetramine and then aminophosphonic acid groups were grafted onto the PAN_AF using Mannich reaction. The effect of amine concentration, reaction time, and temperature has been extensively studied during the amination process since higher loading reduced mechanical strength of the functionalized fibers. Morphological examination showed that the grafting reaction didn't destroy fibrous structures. The adsorption capacities of the nanofibers were studied and the obtained results show that functionalized fiber ($PAN_{AP}F$) demonstrated higher adsorption efficiency and selectivity for Hg^{2+} when in the presence of other coexisting ions, viz. Pb^{2+}, Cd^{2+}, Ag^{+}, Zn^{2+}, Cu^{2+}, Ni^{2+}, Co^{2+}, Ca^{2+}, and Mg^{2+} [70].

Chelating fibrous adsorbents have gained significant attention for enhancing adsorption of metal ions for water because of their quick mass transfer velocity. Most of the existing synthesis methods involve oil-bath/water-bath heating and electrical heating that generally have drawbacks in terms of lower efficiency, adsorption capacity, low safety, and elongated synthesis time [71,72]. Microwave (MW) irradiation is an alternative and convenient method to overcome these issues in carrying out the surface modification process. Deng et al. (2017) demonstrated an MW-assisted process to prepare a thio-functionalized fibrous adsorbent (PANMW-Thio) for the enhanced and selective removal of Hg^{2+} and Cd^{2+} ions [73]. In this study, pH effect on adsorption performance, adsorption kinetics and isotherm, adsorption selectivity, recycling properties, and adsorption mechanism of surface modified fibers were carefully investigated and reported [73].

Notably, the adsorption properties of electrospun fibers can be enriched by regulating their composition and structure. Different types of electrospun nanofibers have been proposed for removal of metal ions including organic nanofibers, inorganic nanofibers, and inorganic/organic composite nanofibers [74,75]. In most cases, nanofibers serve as a flexible platform to hold the adsorbent materials. Zhao et al.

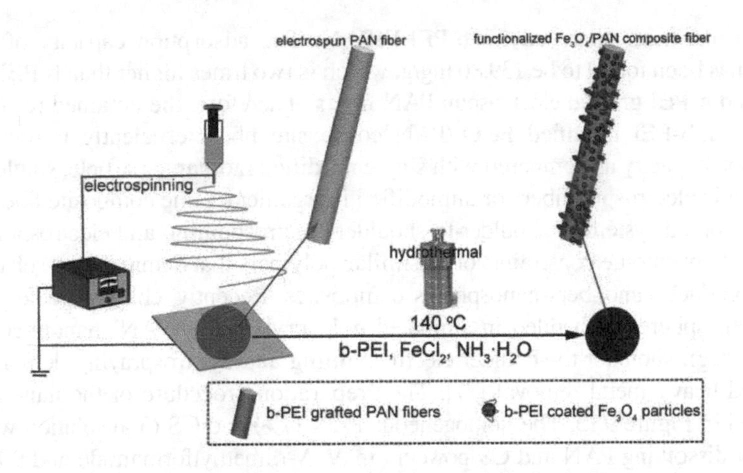

FIGURE 9.12 Schematic diagram of the synthesis of functionalized Fe_3O_4/PAN composite electrospun fibers. (Reprinted from *J. Colloid Interface Sci.*, Zhao, R. et al., Functionalized magnetic iron oxide/polyacrylonitrile composite electrospun fibers as effective chromium (VI) adsorbents for water purification, 505, 1018–1030, Copyright (2017), with permission from Elsevier.)

designed and demonstrated a b-PEI functionalized Fe_3O_4/polyacrylonitrile through combination of electrospinning and a one-step hydrothermal reaction [76]. The prepared electrospun PAN fibers served as substrate to load Fe_3O_4 and b-PEI, where Fe_3O_4 increased the adsorption capacity and utilization efficiency of the organic component. The preparation procedure is presented in Figure 9.12.

The morphology of the obtained fibers has been characterized by SEM and results showed that the pure PAN fibers were smooth and randomly distributed. The color of the fibers changed from white to black after hydrothermal processing as shown in Figure 9.12. Moreover, the fibers became thick, bent, and intertwined with each other and particles were formed on the outer surface of the fibers following the process. On another note, tiny particles were also formed uniformly on the surface of the fibers. Further, the stability of anchored particles was analyzed by ultrasonification of the prepared fibers in deionized water for 30 min. There was no precipitation observed and moreover the mass of the b-PEI-FePAN remained same indicating that the Fe_3O_4 particles were firmly immobilized on the surface. Additionally, the electrospun fiber displayed porous structure, which is beneficial for the adsorption process. The calculated porosity of electrospun PAN fibers was 82.6%, which increased to 88.7% after the hydrothermal reaction.

The enhanced adsorption capacity of the b-PEI modified Fe_3O_4/PAN composite fibers has been verified by comparing the adsorption capability of the related adsorbents (pure electrospun PAN fibers (EPAN), Fe_3O_4/PAN composite fibers (Fe_3O_4/PAN, the preparation procedure is similar to b-PEI-FePAN in the absence of b-PEI), b-PEI grafted electrospun PAN fibers (b-PEI-PAN, the preparation procedure is similar to b-PEI-FePAN in the absence of $FeCl_2$), and b-PEI coated Fe_3O_4 nanoparticles (b-PEI-Fe_3O_4), respectively) under the same adsorption conditions ($C_0 = 70$ mg/L, Volume = 100 mL, adsorbent dosage = 45 mg). The highest Cr(VI) removal efficiency

(91.34%) has been observed for b-PEI-FePAN. The adsorption capacity of b-PEI-FePAN has been found to be 139.86 mg/g, which is two times higher than b-PEI coated Fe_3O_4 and b-PEI grafted electrospun PAN fibers. Therefore, the obtained results suggested that b-PEI modified Fe_3O_4/PAN composite fibers efficiently improved the adsorption capacity as compared with single modified inorganic particle, single modified organic electrospun fibers or unmodified inorganic/organic composite fibers.

A combined system of shoulder-to-shoulder electrospinning and electrospraying is designed to overcome constraints of dissimilar polymers that demand a co-solvent system to produce nanofibers-nanospheres composites. Recently, chitosan-rectorite (CS/REC) nanospheres embedded in aminated polyacrylonitrile (PAN) nanofibers fabricated through shoulder-to-shoulder electrospinning and electrospraying demonstrated enhanced heavy metal removal [77]. The preparation procedure of the nanofibers is presented in Figure 9.13. The homogeneous PAN (7%) and CS (3%) solution was prepared by dissolving PAN and CS powders in *N*, *N*-dimethylformamide and 90% acetic acid, respectively. Then, CS/REC solution was prepared by blending REC and CS in the weight ratio of 5:1 in 90% acetic acid aqueous with stirring for 72 h at 60°C. The solution was filled into separate plastic syringes which were placed on both sides of the drum collector adjacent to one another. The optimized parameters for electrospinning of PAN were as follows: flow rate, 1 mL/h; voltage, 16 kV; and distance between the tip and collector, 15 cm. Meanwhile, the feed rate, high voltage, and the distance for electrospraying CS or CS/REC were 2 mL/h, 16 kV, and 15 cm, respectively. The speed of the rotating collector was fixed at 200 rpm. The PAN-CS samples were attained by shoulder-to-shoulder electrospinning of PAN and electrospraying of CS, and the same procedure has been followed for producing the PAN-CS/REC composites. All samples were dried under vacuum at 60°C for 24 h to remove trace solvent.

Generally, pH of solution is considered to be the most important factor affecting the adsorption performance. Therefore, pH was set in the range from 2.0 to

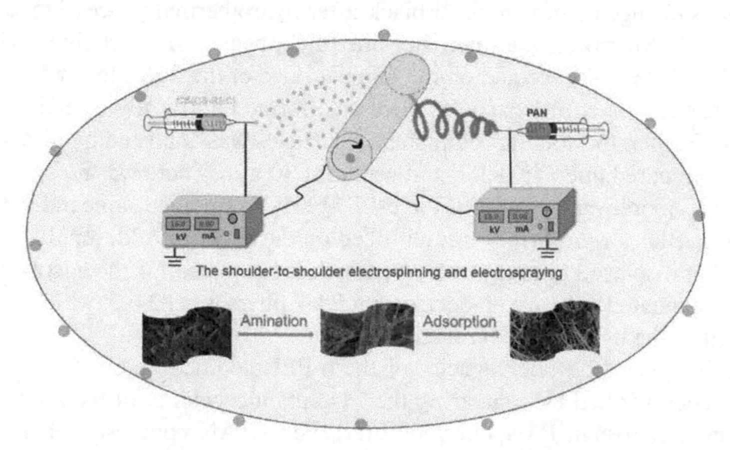

FIGURE 9.13 Fabrication process of composite mats via shoulder-to-shoulder electrospinning and electrospraying. (Reprinted from *Appl. Surf. Sci.* Huang, M. et al. Chitosan-rectorite nanospheres embedded aminated polyacrylonitrile nanofibers via shoulder-to-shoulder electrospinning and electrospraying for enhanced heavy metal removal, 437, 294–303, Copyright (2018), with permission from Elsevier.)

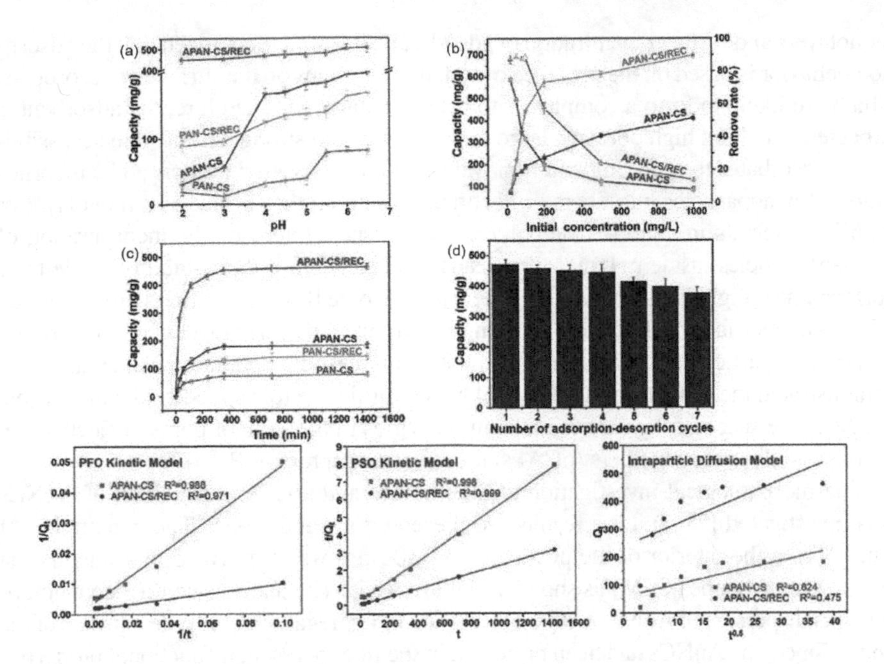

FIGURE 9.14 Effect of (a) pH values, (b) initial concentrations of Pb^{2+}, (c) contact time on the adsorption capacity of Pb^{2+} onto composite mats, (d) desorption performance and reusability of APAN-CS/REC and adsorption kinetics of Pb^{2+} onto the APAN-CS and APAN-CS/REC based on the PFO kinetic model, PSO kinetic model, and intraparticle diffusion model. (Reprinted from *Appl. Surf. Sci.* Huang, M. et al. Chitosan-rectorite nanospheres embedded aminated polyacrylonitrile nanofibers via shoulder-to-shoulder electrospinning and electrospraying for enhanced heavy metal removal, 437, 294–303, Copyright (2018), with permission from Elsevier.)

6.5 to understand the effect on adsorption at the PAN-CS/REC composites [77]. The observed results showed that the adsorption capacity of PAN-CS increased with increasing pH (Figure 9.14) and further, the adsorption capacity reached equilibrium with 77.40 ±4.5 mg/g when the pH reached 5.5. The attained adsorption capacity of APAN-CS increased from 29.80 ± 5.7 mg/g to 196.33 ± 7.3 mg/g when pH was varied from 2 to 6, as compared with PAN-CS. The combined results confirmed that amination could greatly improve the adsorption capacity due to the enhanced amine ($-NH_2$ and -NH) group content in the APAN-CS fibrous mats. The adsorption capacity of PAN-CS/REC was increased gradually from 88.4 ± 4.7 to 162.95 ± 8.6 mg/g with increasing pH. It was observed that the Pb^{2+} adsorption on PAN-CS/REC was insensitive to pH changes. After the amination process, the tendency of insensitivity to pH changes was more obvious in the APAN-CS/REC sample. The adsorption capacity had reached 449.05 ± 15.6 mg/g at pH = 2, and afterwards, the maximum adsorption capacity reached 500.95 ± 21.4 mg/g at pH = 6. The attained results predicted that the addition of REC not only allowed the membrane to extract the lead ions without being affected by the pH changes of the solution but also synergy with the amination increased the adsorption capacity, proving the validity of the previous hypothesis.

The most critical problem encountered with adsorbents, including nanofibers, is their limited adsorption capacity due to the adsorption of metal ions on their surface as monolayers and further denaturation of adsorbents. It is well recognized that the adsorption behavior is based on the presence of functional groups on the surface of adsorbents, which are likely to form a complex with heavy metal ions. Therefore, the adsorbent is expected to exhibit high porosity, large surface area, and strong binding-site accessibility for adsorbates to result in rapid removal and high adsorption capacity. The incorporation of nanoparticles into fibers or decorating them on the surface as a monolayer or multilayer results in nanoparticle/polymer composites. Interestingly, incorporation of the nanoparticles inside polymer fiber leads to protection of their stability while their performance might be decreased. In order to overcome this, the particles are decorated on the outer surface of the fibers, resulting in enhanced performance, while the protection of nanoparticles against the environment is in question. Senthamizhan et al. (2016) demonstrated the "nanotrap" assembly in which dithiothreitol capped gold nanoclusters (AuNCs) are successfully encapsulated into cavities in the form of pores in electrospun porous cellulose acetate fibers (pCAFs) for effective capture of Pb^{2+} [78].

The morphological investigation pCAFs before and after encapsulation of AuNCs was investigated [78] and the results are presented in Figure 9.15. The penetration of AuNCs into the interior of the pCAF up to ~300 nm was confirmed by transmission electron microscope (TEM) as shown in Figure 9.15a. The analysis confirmed a homogeneous dispersion of AuNCs without formation of aggregates on the outer surface of the fibers. Since, the AuNCs had been protected in the fiber pores their functional properties

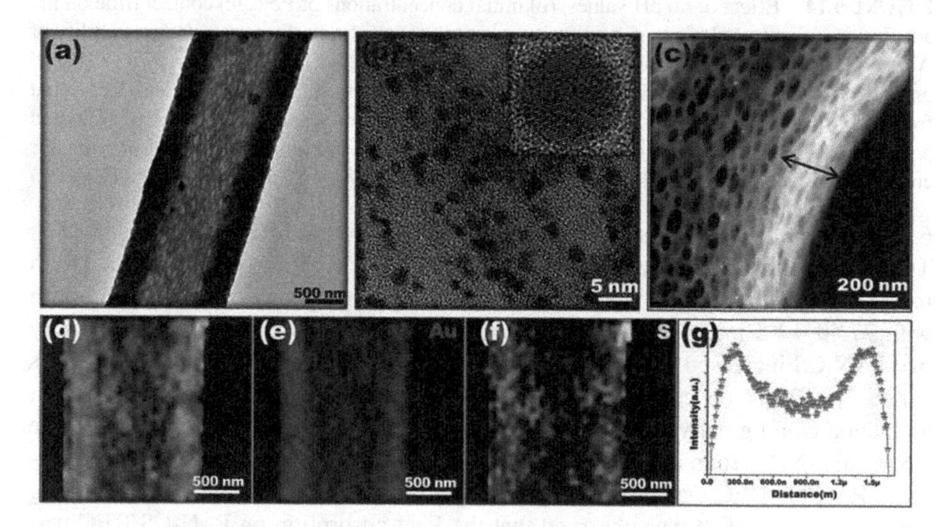

FIGURE 9.15 TEM image of the gold nanocluster (AuNC) encapsulated porous cellulose acetate fiber (pCAF/AuNC). The image clearly demonstrates that the penetration depth is almost uniform throughout the fiber, confirming the deep penetration of AuNCs into the fiber interior. (b) TEM image of AuNCs. The inset shows AuNCs having a lattice spacing of 0.235 nm which corresponds to the (111) lattice plane of the face-centered cubic (fcc) gold. (c and d) HAADF-STEM image and EDS elemental mapping of (e) Au and (f) S. The image confirms that the porous nature of the fiber is highly persistent and does not degrade following the incorporation of AuNCs. (g) EDS intensity line profile of Au taken across the pCAF/AuNC surface. (Senthamizhan, A. et al., *J. Mater. Chem. A*, 4, 2484–2493, 2016. Reproduced by permission of The Royal Society of Chemistry.)

were protected against the environment for a prolonged time period. The encapsulated AuNCs into a fibrous matrix firmly anchored on the fiber surfaces act as nanotraps for capturing toxic metal ions in water as demonstrated in Figure 9.15a and b. The scanning transmission electron microscopy (STEM) image of the pCAF/AuNC confirms that the porous nature of pCAFs was highly persistent and didn't degrade following incorporation of AuNCs, which is expected to enhance the rapid diffusion and fast response to heavy metal ions as seen in Figure 9.15c. The size of the AuNC and pores in pCAFs are considered to be important factors for incorporation. It is extremely important that the AuNCs are smaller than the pores for effective incorporation, to avoid blockage of the pore wall. The above discussed facts aided in optimizing the dipping time as 3 h for the successful incorporation of AuNCs into pCAFs with shaking. The uniform distribution of AuNCs in the pCAF was confirmed by STEM-EDX elemental mapping as depicted in Figure 9.15d–f. The line intensity profile in Figure 9.15g further confirms that the penetration of AuNCs is uniform along fibers.

Wetting features of the fibrous membrane play a significant role in the process of adsorption, but is not considered as a primary factor. If the fibrous membrane does not permit penetration of polluted water, the reaction takes place only on the surface of the fiber mat resulting in lower adsorption capacity. The authors paid more attention to study the interaction between fibrous mat and water and their transport behavior to improve the overall removal capacity of metal ions [78]. The enhanced capillary effect of the fibrous membrane is presented in Figure 9.16. Until now,

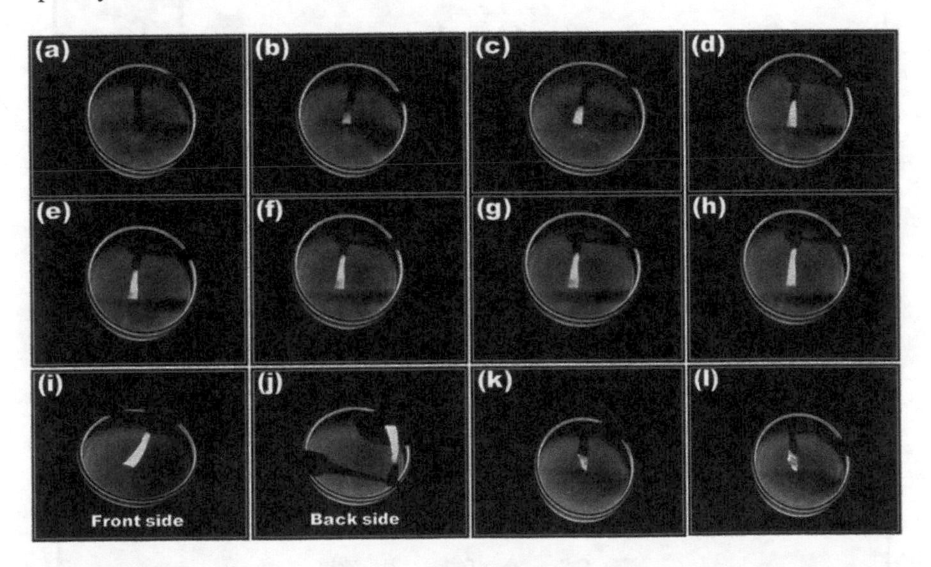

FIGURE 9.16 (a–h) A sequence of frames showing the propagation of water upward in the pCAFM/AuNC strip. The red and green color represents the characteristics of AuNCs and FITC dye. (i and j) Front and back sides of the pCAFM/AuNC after spread out of water. The water is efficiently diffused into the membrane, confirmed from the back side of the membrane emitting green fluorescence due to the presence of transported FITC dye. (k and l) Photographs of the nCAFM/AuNC showing the propagation of water after 30 s and 2 min, respectively. (Senthamizhan, A. et al., *J. Mater. Chem. A*, 4, 2484–2493, 2016. Reproduced by permission of The Royal Society of Chemistry.)

limited attention has been paid to liquid moisture transport behaviors of electrospun fibrous membranes especially when they are applied in removal of metal ions. It has been observed that various surface modification processes can be used to design the wetting behavior of fibrous membranes. The diffusion of water droplets is uniform in all directions at a fraction of seconds whereas the water droplet is stable on the pristine membrane. The overall results confirmed that the occurrence of pores greatly enhances the capillary effect of the pCAFM/AuNC.

Lu et al. (2014) developed a hydrophilic poly(vinyl alcohol-*co*-ethylene) (PVA-*co*-PE) nanofiber membrane for heavy metal ions removal. The PVA-*co*-PE nanofiber surface was functionalized with iminodiacetic acid (IDA) through solid phase synthesis [79]. The schematic representation of the functionalization procedure is presented in Figure 9.17. The desired amount of iminodiacetic acid (IDA) was dissolved in sodium hydroxide solution and then cyanuric chloride activated nanofiber membranes were immersed in the IDA solution with concentrations ranging from 0.05 to 4 mol/L for a time period of two hours at 30°C. The nanofiber membranes were then thoroughly rinsed with sodium hydroxide solution and deionized water, as demonstrated in Figure 9.17. The excellent wetting property of adsorbents plays an important role in determining the adsorption efficiency and capability of heavy metal ions. The wetting characteristics of pristine, cyanuric chloride activated, and

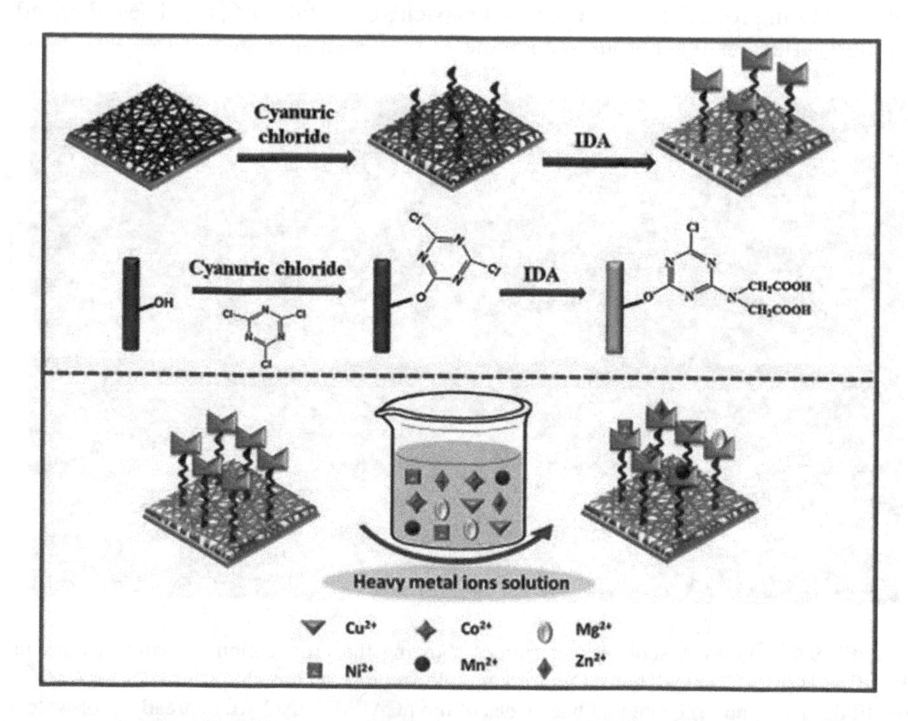

FIGURE 9.17 Fabrication of hydrophilic PVA-*co*-PE nanofiber membranes for heavy metal ions removal by cyanuric chloride activation and IDA functionalization. (Reprinted from *React. Funct. Polym.*, Lu, Y. et al., Hydrophilic PVA-co-PE nanofiber membrane functionalized with iminodiacetic acid by solid-phase synthesis for heavy metal ions removal, 82, 98–102, Copyright (2014), with permission from Elsevier.)

IDA functionalized PVA-*co*-PE nanofiber membranes were evaluated by measuring their water contact angle [79]. Due to the presence of hydroxyl groups, the pristine PVA-*co*-PE nanofiber membrane had a contact angle of 56.06°. The cyanuric chloride activation introduced triazine rings on surfaces of PVA-*co*-PE nanofiber membranes, leading to a higher contact angle of 62.84°. After the IDA was covalently linked to surfaces of PVA-*co*-PE nanofiber membranes, the contact angle significantly dropped to 0°, showing excellent hydrophilicity and wetting property. This could greatly facilitate the contact and chelation between diacetic acid groups from IDA on nanofiber membranes with heavy metal ions in the water. After adsorption of Cu^{2+}, the water contact angle was 49.17°, which might be attributed to the consumption of hydrophilic diacetic groups in IDA by chelating Cu^{2+}.

The impact of IDA amount that is linked with activated PVA-*co*-PE nanofiber membranes on the adsorption of heavy metal ions was taken into consideration, so different concentrations of IDA ranging from 0.05 to 4.0 mol/L were added for reacting with the cyanuric chloride on activated PVA-*co*-PE nanofiber membranes [79]. The adsorption capacity of Cu^{2+} on PVA-*co*-PE nanofiber membranes functionalized with diverse concentrations of IDA were evaluated. The outcome indicated a significant increase in the adsorption capacity upon increasing the concentration of IDA in PVA-*co*-PE nanofiber membranes. The adsorption amount of Cu^{2+} gradually increased from 45.7 to 101.87 mg/g, reaching the equilibrium when the IDA concentration was above 2 mol/L. The overall experimental results proved that the increase in the IDA concentration led to the higher adsorption amount of Cu^{2+}. The IDA functionalized nanofiber membranes exhibited equilibrium adsorption amounts of heavy metal ions in the order of $Cu^{2+} > Co^{2+} > Zn^{2+} > Ni^{2+}$ at 30°C. The excellent reusability of IDA functionalized nanofiber membranes for heavy metal ions removal was provided by the reversible conversion between carboxylate ions and acetic acid during repeated adsorption and desorption cycles.

9.3 CONCLUSION AND PERSPECTIVES

The development of fibrous composite membrane with a tailored surface for adsorption of water pollutants has been one of the most significant improvements during the past decade. This review comprehensively summarizes the challenges associated with the current methods for fabricating composite fibrous membrane and their efficiency in adsorption of pollutants. However, the effective control of a fibrous surface still remains a challenge. For example, in-depth attention is required to the elution of nanoparticles/functional groups from the fiber surfaces and their subsequent toxic impact on the environment. With these issues resolved, there looks to be a promising demand for surface functionalized nanofibers in treating polluted water.

REFERENCES

1. The World Health Organization, 2016. An estimated 12.6 million deaths each year are attributable to unhealthy environments. http://www.who.int/mediacentre/news/rcleases/2016/deaths-attributable-to-unhealthy-environments/en/ (Accessed April 10, 2018).
2. The World Health Organization. Water sanitation hygiene. http://www.who.int/water_sanitation_health/en/ (Accessed April 10, 2018).

3. Shimizu, Y., Okuno, Y. I., Uryu, K., Ohtsubo, S., and Watanabe, A. 1996. Filtration characteristics of hollow fiber microfiltration membranes used in membrane bioreactor for domestic wastewater treatment. *Water Res.* 30:2385–2392.
4. Radjenović, J., Petrović, M., Ventura, F., and Barceló, D. 2008. Rejection of pharmaceuticals in nanofiltration and reverse osmosis membrane drinking water treatment. *Water Res.* 42:3601–3610.
5. Matlock, M. M., Howerton, B. S., and Atwood, D. A. 2002. Chemical precipitation of heavy metals from acid mine drainage. *Water Res.* 36:4757–4764.
6. Moghaddam, S. S., Moghaddam, M. A., and Arami, M. Coagulation/flocculation process for dye removal using sludge from water treatment plant: optimization through response surface methodology. *J. Hazard. Mater.* 175:651–657.
7. Michalak, I., Chojnacka, K., and Witek-Krowiak, A. 2013. State of the art for the biosorption process: A review. *Appl. Biochem. Biotechnol.* 170:1389–1416.
8. Ali, I., and Gupta, V. K. 2006. Advances in water treatment by adsorption technology. *Nat. Protoc.* 1:2661–2667.
9. Ali, I. 2012. New generation adsorbents for water treatment. *Chem. Rev.* 112, 5073–5091.
10. Das, R., Vecitis, C. D., Schulze, A. et al. 2017. Recent advances in nanomaterials for water protection and monitoring. *Chem. Soc. Rev.* 46:6946–7020.
11. Bethi, B., Sonawane, S. H., Bhanvase, B. A., and Gumfekar, S. P. 2016. Nanomaterials-based advanced oxidation processes for wastewater treatment: A review. *Chem. Eng. Process.* 109:178–189.
12. Goh, K., Karahan, H. E., Wei, L. et al. 2016. Carbon nanomaterials for advancing separation membranes: A strategic perspective. *Carbon* 109: 694–710.
13. Daer, S., Kharraz, J., Giwa, A., and Hasan, S. W. 2015. Recent applications of nanomaterials in water desalination: A critical review and future opportunities. *Desalination* 367:37–48.
14. Cincinelli, A., Martellini, T., Coppini, E., Fibbi, D., and Katsoyiannis A. 2015. Nanotechnologies for removal of pharmaceuticals and personal care products from water and wastewater. A review. *J. Nanosci. Nanotechnol.* 15:3333–3347.
15. Kumar, S., Ahlawat, W., Bhanjana, G., Heydarifard, S., Nazhad, M. M, and Dilbaghi, N. 2014. Nanotechnology-based water treatment strategies. *J. Nanosci Nanotechnol.* 14:1838–1858.
16. Ghasemzadeh, G., Momenpour, M., Omidi, F., Hosseini, M. R., Ahani, M., and Barzegari, A. 2014. Applications of nanomaterials in water treatment and environmental remediation. Front. *Environ. Sci. Eng.* 8:471–482.
17. Santhosh, C., Velmurugan, V., Jacob, G. et al. 2016. Role of nanomaterials in water treatment applications: A review. *Chem. Eng. J.* 306:1116–1137.
18. Adeleye, A.S., Conway, J. R., Garner, K., Huang, Y., Su, Y., and Keller, A. A. 2016. Engineered nanomaterials for water treatment and remediation: Costs, benefits, and applicability. *Chem. Eng. J.* 286:640–662.
19. Xu, P., Zeng, G. M., Huang, D. L. et al. 2012. Use of iron oxide nanomaterials in wastewater treatment: A review. *Sci. Total Environ.* 424:1–10.
20. Ramakrishna, S., Fujihara, K., Teo, W. E., Yong, T., Ma, Z., and Ramaseshan, R. 2006. Electrospun nanofibers: Solving global issues. *Mater. Today* 9:40–50.
21. Thavasi, V., Singh, G., and Ramakrishna, S. 2008. Electrospun nanofibers in energy and environmental applications. *Energy Environ. Sci.* 1:205–221.
22. Thenmozhi, S., Dharmaraj, N., Kadirvelu, K., and Kim, H. Y. 2017. Electrospun nanofibers: New generation materials for advanced applications. *Mater. Sci. Eng. B* 217:36–48.
23. Kenry, and Lim, C. T. 2017. Nanofiber technology: Current status and emerging developments. *Prog. Polym. Sci.* 70:1–17.

24. Ahmed, F. E., Lalia, B. S., and Hashaikeh, R. 2015. A review on electrospinning for membrane fabrication: Challenges and applications. *Desalination* 356:15–30.
25. Uyar, T. and E. Kny. 2017. *Electrospun Materials for Tissue Engineering and Biomedical Applications: Research, Design and Commercialization*. Woodhead Publishing Series in Biomaterials, Elsevier.
26. Peng, S., Jin, G., Li, L. et al. 2016. Multi-functional electrospun nanofibres for advances in tissue regeneration, energy conversion & storage, and water treatment. *Chem. Soc. Rev.* 45: 1225–1241.
27. Gupta, V. K., and Suhas. 2009. Application of low-cost adsorbents for dye removal-a review. *J. Environ. Manage.* 90:2313–2342.
28. Moussavi, G., and Mahmoudi, M. 2009. Removal of azo and anthraquinone reactive dyes by using MgO nanoparticles. *J. Hazard. Mater.* 168:806–812.
29. Iqbal M. 2008. *Textile Dyes*. Rehbar Publishers, Karachi, Pakistan.
30. Sharma, P., Kaur, H., Sharma, M., and Sahore, V. 2011. A review on applicability of naturally available adsorbents for the removal of hazardous dyes from aqueous waste. *Environ. Monit. Assess.* 183:151–195.
31. Nidheesh, P. V., Gandhimathi, R., and Ramesh, S. T. 2013. Degradation of dyes from aqueous solution by Fenton processes: A review. *Environ. Sci. Pollut. Res.* 20:2099–2132.
32. Crini, G. 2006. Non-conventional low-cost adsorbents for dye removal: A review. *Bioresour. Technol.* 97:1061–1085.
33. Oller, I., Malato, S., and Sánchez-Pérez, J. A. 2011. Combination of advanced oxidation processes and biological treatments for wastewater decontamination: A review. *Sci. Total Environ.* 409:4141–4166.
34. Natarajan, S., Bajaj, H. C., and Tayade, R. J. 2018. Recent advances based on the synergetic effect of adsorption for removal of dyes from waste water using photocatalytic process. *J. Environ. Sci.* 65:201–222.
35. Martínez-Huitle, C. A., and Brillas, E. 2009. Decontamination of wastewaters containing synthetic organic dyes by electrochemical methods: A general review. *Appl. Catal., B* 87:105–145.
36. Fernández, C., Larrechi, M. S., and Callao, M. P. 2010. An analytical overview of processes for removing organic dyes from wastewater effluents. *TrAC, Trends Anal. Chem.* 29:1202–1211.
37. Sadegh, H., Ali, G. A. M., Gupta, V. K. et al. 2017. The role of nanomaterials as effective adsorbents and their applications in wastewater treatment. *J. Nanostruct. Chem.* 7:1–14.
38. Cai, Z., Sun, Y., Liu, W., Pan, F., Sun, P., and Fu, J. 2017. An overview of nanomaterials applied for removing dyes from wastewater. *Environ. Sci. Pollut. Res. Int.* 19:15882–15904.
39. Tan, K. B., Vakili, M., Horri, B. A., Poh, P. E., Abdullah, A. Z., and Salamatinia, B. 2015. Adsorption of dyes by nanomaterials: Recent developments and adsorption mechanisms. *Sep. Purif. Technol.* 150:229–242.
40. Shanker, U., Rani, M., and Jassal, V. 2017. Degradation of hazardous organic dyes in water by nanomaterials. *Environ Chem Lett.* 15:623–642.
41. Ahmad, A., Mohd-Setapar, S. H., Chuong, C. S. et al. 2015. Recent advances in new generation dye removal technologies: novel search for approaches to reprocess wastewater. *RSC Adv.* 5:30801–30818.
42. Satilmis, B., Budd, P. M., and Uyar, T. 2017. Systematic hydrolysis of PIM-1 and electrospinning of hydrolyzed PIM-1 ultrafine fibers for an efficient removal of dye from water. *React. Funct. Polym.* 121:67–75.
43. Qureshi, U. A., Khatri, Z., Ahmed, F., Khatri, M., and Kim, I. S. 2017. Electrospun zein nanofiber as a green and recyclable adsorbent for the removal of reactive black 5 from the aqueous phase. *ACS Sustainable Chem. Eng.* 5:4340–4351.

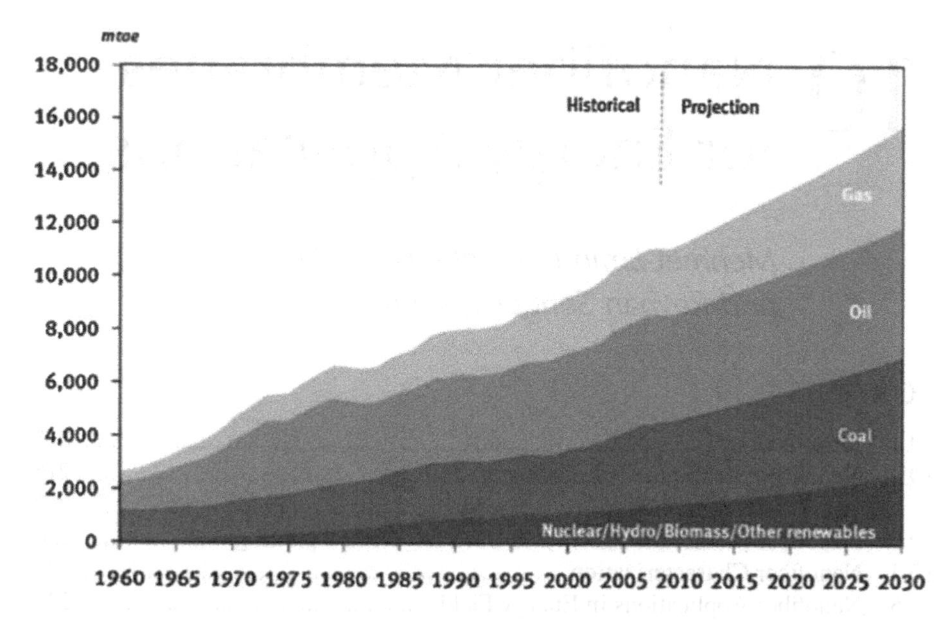

FIGURE 10.1 World energy supply according to fuel type. (Reprinted from *J. Power Sour.*, 196, Dong, Z. et al., Electrospinning materials for energy-related applications and devices, 4886–4904, Copyright 2011, with permission from Elsevier.)

Currently, these nanostructured materials are synthesized by methods such as chemical vapor deposition, wet chemical synthesis, sol-gel, self-assembly and electrospinning. Other than electrospinning, rest of the methods have some limitations in terms of material restrictions, high cost as well as highly complex to process (Dong et al., 2011). On the other hand, electrospinning is used to fabricate nanofibers that are extensively used for energy in terms of developing energy conversion and storage devices due to their high surface area, controllable porosity and ease of accessibility. Electrospinning is an efficient and low-cost method and offers fabrication of variety of different morphologies such as core-shell, hollow, multilayer and, porous (Sun et al., 2016). Proper selection of material, electrospinning parameters and characterization are particularly important for achieving these morphologies.

In this chapter, first materials used for nanofiber membrane fabrication are investigated, followed by deep insights into nanofiber membrane fabrication and characterization. Information related to nanofiber membrane applications for energy-related field are summarized in terms of fuel cells, supercapacitors, solar cells, lithium-ion batteries, hydrogen storage and pressure-retarded osmosis (PRO).

10.2 NANOFIBER MEMBRANE FABRICATION

Various engineering fields involved in nanoscale materials such as nanofibers in the fiber industry. Nanofibers are defined as fibers with a diameter of 100 nm or less (Ramakrishna, 2005) and remarkable for their characteristic features such as larger surface area to volume ratio, extremely small pore diameters and superior mechanical properties (Huang et al., 2003). Due to these characteristic features, nanofibers have very wide range of applications in fields such as filtration, battery separator, wound dressing, vascular grafts, enzyme immobilization, electrochemical sensing, composite materials, reinforcements, blood vessel engineering and tissue engineering (Bergshoef and Vancso, 1999; Bhattarai et al., 2004; Xu et al., 2004; Nayak et al., 2011).

Because of the process limitations, existing fiber technologies cannot produce strong fibers having a diameter smaller than 2 μm. The most commonly used process for nanofiber fabrication is electrospinning (Patanaik et al., 2007; Zhou and Gong, 2008). The reason for using electrospinning is due to its simplicity and suitability for allowance to use different polymers, ceramics and metals. There are other methods in literature including melt blowing, flash spinning, bicomponent spinning, phase-separation and drawing. Nanofiber mats known as nanowebs are used for collecting fibers having several nanometers to hundreds of nanometers scale (Nayak et al., 2011).

10.2.1 Electrospinning

Electrospinning is a technique originated from Formhals patent for the production of artificial filaments using high electric field in 1934. This technique was mainly based on the electrostatic charged liquid material which forms a cone shape near a droplet, if the charge density is very high, small jets may be ejected from the tip of the cone (Formhals, 1934). Electrospinning process mainly classified as solution electrospinning and melt electrospinning. Polymer properties affect type of electrospinning process.

Many researches have been reviewed the factors affecting characteristic features and different applications of the nanofiber web (Bognitzki et al., 2001; Duan et al., 2004; Koski et al., 2004; Mckee et al., 2004; Lin et al., 2004). Solution electrospinning method includes the use of toxic solvents. Therefore additional solvent extraction step is needed. Because of this reason, solution electrospinning process has some drawbacks such as low productivity (up to 300 mg/hr) and environmental concerns.

On the other hand, melt electrospinning has higher viscosity of molten polymer. The biggest challenge about application of polymeric melt is electrical discharge problem. This problem leads difficulties inherent in finer fiber formation.

In order to obtain higher productivity in solution electrospinning variable number of jets adopting different techniques such as: (a) multi-jets from single needle, (b) multi-jets from multiple needles and (c) needleless systems.

Multi-jets from single needle: Single needle electrospinning (SNE) defines as a single-jet initiated from the Taylor cone by applying an electric field. Yamashita et al. (2007) introduced the use of multi-jets for the first time. Polybutadiene (PB) beadless membranes were prepared by growing multi-jets at the needle tip of an SNE setup. Significant discrepancy in electric field distribution and degree of solution blockage at the needle tip were suggested as two possible formation mechanisms for the formation of multi-jets in an SNE. Vaseashta (2007) used another approach for the formation of multi-jets from multiple Taylor cones in an SNE system using a curved collector. Multi-jets can be applied to a single needle using separate polymer sub-jets on its path to the collector.

Multiple-jets from multiple needles: Multiple needle electrospinning (MNE) systems are less studied in literature (Regele et al., 2002; Kim et al., 2006; Lee et al., 2006; Teo et al., 2007). Needle configuration, numbers and dimensions are determining factors for MNE systems. Needle configurations for MNE systems are linear configuration and two-dimensional configurations (Figure 10.2). Linear and two-dimensional configurations have subdivisions such as elliptical, circular, triangular, square and hexagonal. While using MNE systems setup, large operating space and more carefully design between needles in order to beware of strong charge repulsion between the jets (Nayak et al., 2011).

Multi-jets from needleless system: One of the most popular systems is needleless electrospinning which is substantially improved by provoking numerous polymeric jets from free liquid surfaces. Principle of a needleless electrospinning system is as follows: when the applied electric field reaches above a critical point, electrically conductive liquid start to form jets. Yarin and Zussman (2004) first reported needleless electrospinning. At this work, two-layered system combination was used. Lower layer consists of a ferromagnetic suspension and the upper layer was a polyethylene oxide (PEO) solution. With this method numerous continuous spikes can be generated using magnetic field at the available surface of the magnetic fluid. Application of high voltage to polymer layer creates visible perturbations at the surface of the polymer layer. Electrode collected fibers from multiple jets after it reached a critical voltage (Nayak et al., 2011).

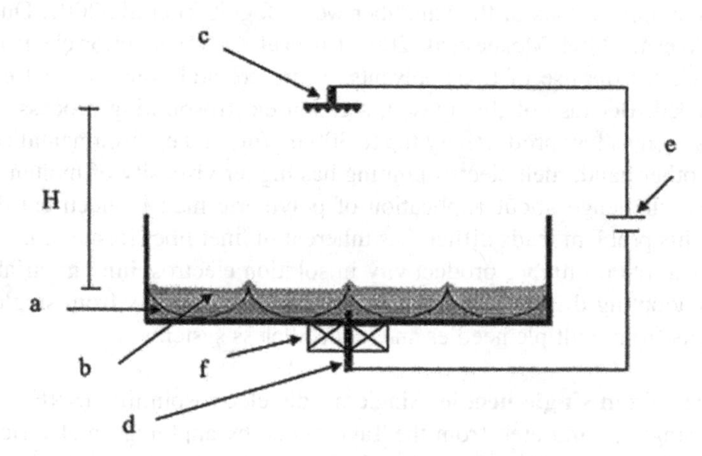

FIGURE 10.2 Schematic drawing of the experimental setup. (a) Layer of magnetic liquid, (b) layer of polymer solution, (c) counter-electrode located at a distance H from the free surface of the polymer, (d) electrode submerged into magnetic fluid, (e) high voltage source, and (f) strong permanent magnet or electromagnet. (Reprinted from *Polymer*, 45, Yarin, A. and Zussman, E., Upward needleless electrospinning of multiple nanofibers, 2977–2980, Copyright 2004, with permission from Elsevier.)

Jirsak et al. (2005) studied the formation of multi-jets from the free surface of a liquid uploaded in a slowly rotating horizontal cylinder. Elmarco Company (Liberec) commercialized this method under the brand name of Nanospider™. Dosunmu et al. (2006) presented a novel needless electrospinning method which includes a cylindrical porous polyethylene (PE) tube for the fabrication of nanofibers from multiple jets of nylon 6,6 solution. In this method, electricity applied from the bottom of the tube, pressurized air goes through the walls of the porous tube until it reaches top of the tube. Fiber diameter was similar single jet electrospinning with a broader distribution range.

Another needleless electrospinning system has been reported by Wang et al. (2009). In the study, a conical metal wire coil was used as the spinneret and polyvinyl alcohol (PVA) nanofibers were fabricated. The advantage of needleless system is the possibility of producing finer nanofibers on a larger scale compared to conventional electrospinning.

Other potential approaches in electrospinning: Alternative to needle including and needleless systems mentioned above, electrospinning may also be classified into a confined feed system (CFS) and unconfined feed system (UFS). In CFS system, the polymer solution or melt is injected by a syringe pump at a constant rate. The advantages of CFS are:

- Controlled flow rate
- Uniform fiber diameter
- Better fiber quality

Besides CFS has some disadvantages such as increased system complexity and prone to clogging.

In UFS, the polymer solution or melt flows unconstrained over the surface of another material. UFS includes bubble electrospinning; electro blowing; electrospinning by using a porous hollow tube, a micro fluidic manifold and roller electrospinning (Nayak et al., 2011).

Bubble electrospinning: Bubble electrospinning is explored by Liu and He (2007) for the mass production of nanofibers. System components are a high-voltage DC generator, a gas pump and a vertical liquid reservoir where opening in lid exist, a gas tube is centered at the bottom of the reservoir, a thin metal electrode and a grounded collector.

Electro blowing: Electro blowing is an electrospinning process which is conducted with air blowing. Under high voltage, polymer solution which dissolved in a solvent is pumped through a nozzle (Figure 10.3). Compressed air is injected through the lower end of the spinning nozzle and fibers are collected on a grounded collector. In electro blowing, simultaneously electrical force and air blowing shear force interact with each other to fabricate the nanofibers from the polymeric solution. Electro blowing nanofiber process is suitable both thermoplastic and thermosetting resins (Nayak et al., 2011).

of alternatively energy technologies such as solar, wind, biomass, hydropower, tidal power, ocean thermal energy conversion, etc. (Bui and McCutcheon, 2014).

Salinity gradient energy naturally occurs when freshwater encounter saline water all over the world. Osmosis process can be used for capturing the energy of mixing fresh and saline waters, which is ultimately lost when freshwater dilutes saline water. Estimated of all salinity gradients in the world is 1.4–2.6 TW from which approximately 980 GW (Ramon et al., 2011; Logan and Elimelech, 2012) can be effectively harnessed with an appropriate designed system (Bui and McCutcheon, 2014). An osmotic pressure difference occurs between two solutions from a dilute solution across a semipermeable membrane to a more concentrated draw solution. Hydrostatic pressure applied by the draw side, which is less than osmotic pressure, the osmotic flow is retarded. The Norwegian power company Statkraft estimates that osmotic power generation may be developed to be competitive with other energy sources without the drawbacks (Bui and McCutcheon, 2014).

Figure 10.5 describes osmotic processes probabilities between fresh and salty water separated by a semipermeable membrane (Cath et al., 2006). Power generation with

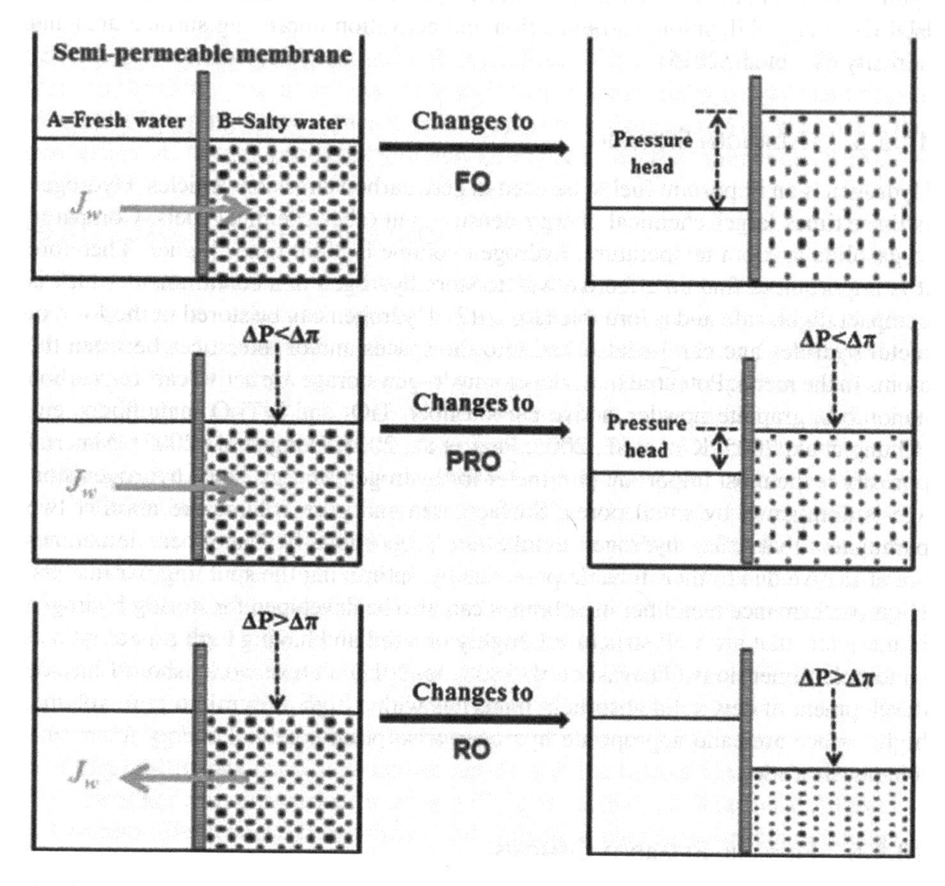

FIGURE 10.5 Schematic representation of osmotic processes. (Reprinted from Han, G. et al., *Prog. Polym. Sci.*, 51, 1–27, 2015. With permission.)

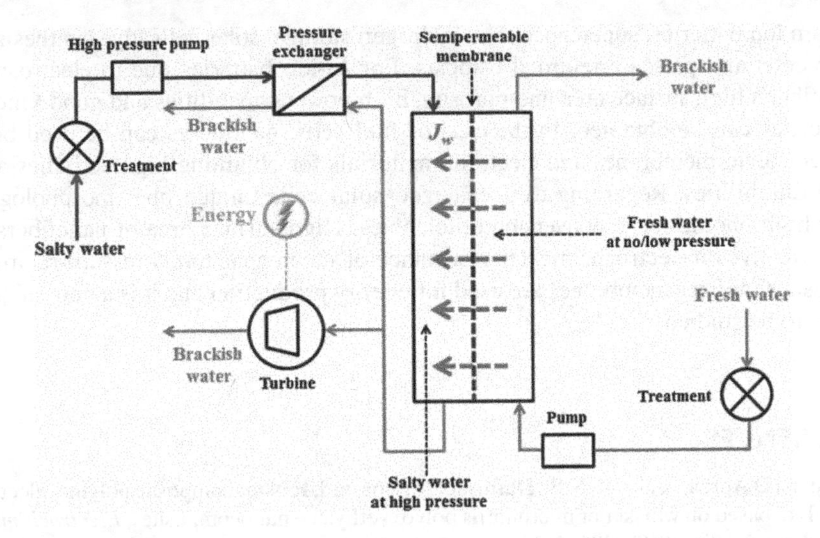

FIGURE 10.6 Schematic diagram of a typical PRO osmotic power plant with continuous and steady state flow. (Reprinted from Han, G. et al., *Prog. Polym. Sci.*, 51, 1–27, 2015. With permission.)

PRO process could be well explained via PRO osmosis phenomenon at Figure 10.6 shows a schematic diagram of a typical PRO osmotic power process at steady-state flow (Thorsen and Holt, 2003, 2009; Skilhagen et al., 2008). In order to remove impurities, feed streams have to be pretreated to reduce membrane fouling (Han et al., 2015).

A novel membrane preparation method found by Bui et al. (2011), flat-sheet polyamide (PA) composite membranes supported by a nonwoven web of electrospun nanofibers. Because of nanofiber's superior porosity, effective area of PA-layer increases. In the study, PSf and polyethersulfone (PES) electrospun nanofiber supports were coated by inter polymerization technique. PA layer shows stronger adhesion to PSf support rather than PES supports and fabricated membranes displayed five times higher water flux than a commercial osmotic membrane. Electrospinning method is a great promise for forward osmosis and PRO applications (Bui et al., 2011; Alsvik and Hägg, 2013).

10.6 CONCLUSIONS

Energy is one of the most important concerns for the sake of society in twenty-first century. Finding alternative devices to be used in energy applications is getting important. Electrospinning and electrospun nanofibers offer great advantages due to their high surface area, porosity and it is a clean energy alternative. Current advances in electrospun nanofibers deal with conversion and storage of energy. These advances in electrospun nanofibers enable the production of high value added products. Many different materials and different types of electrospinning can be used for the fabrication of nanofibers. Energy-related applications of nanofibers are lithium ion batteries, supercapacitors, hydrogen storage, solar cells, dye synthesized solar cells and pressure-retarded osmosis. For Li-ion batteries due to electrospun nanofiber's high surface area and porosity, high power capabilities and good kinetic properties can be obtained. In the case of fuel cells, nanofibers can be used both as electrolyte membranes and electrode materials for obtaining high activities and good durabilities. Regarding dye-sensitized solar cells unique fiber morphologies offer high photoelectric conversion efficiencies. High surface area of nanofibers is also effective on electrochemical performance of supercapacitors. Pressure-retarded osmosis nanofiber membranes are used for energy production and it is a new subject needs to be studied.

REFERENCES

Alloin, F., D'Aprea, A., Kissi, N.E., Dufresne, A., Bossard, F. Nanocomposite polymer electrolyte based on whisker or microfibrils polyoxyethylene nanocomposites, *Electrochimica Acta*, 55(18) (2010) 5186–5194.

Alsvik, I.L., Hägg, M.B. Review: Pressure retarded osmosis and forward osmosis membranes: Materials and methods. *Polymers*, 5 (2013) 303–327.

Angulakshmi, N., Thomas, S., Nair, J.R., Bongiovanni, R., Gerbaldi, C., Stephan, A.M. Cycling profile of innovative nanochitin-incorporated poly- (ethylene oxide) based electrolytes for lithium batteries, *Journal of Power Sources*, 228 (2013) 294–299.

Angulakshmi, N., Prem Kumar, T., Thomas, S., Manuel Stephan, A. Ionic conductivity and interfacial properties of nanochitin-incorporated polyethylene oxide-LiN(C2F5SO2)2 polymer electrolytes, *Electrochimica Acta*, 55(4) (2010) 1401–1406.

Angulakshmi, N., Thomas, S., Nahm, K.S., Stephan, A.M., Elizabeth, R.N. Electrochemical and mechanical properties of nanochitin-incorporated PVDF-HFP-based polymer electrolytes for lithium batteries, *Ionics*, 17(5) (2011) 407–414.

Azizi Samir, A., Chazeau, L., Alloin, F., Cavaillé, J.Y., Dufresne, A. Sanchez, J.Y. POE-based nanocomposite polymer electrolytes reinforced with cellulose whiskers, *Electrochimica Acta*, 50(19) (2005) 3897–3903, 75.

Azizi Samir, M.A.S., Alloin, F., Sanchez, J.-Y., El Kissi, N., Dufresne, A. Preparation of cellulose whiskers reinforced nanocomposites from an organic medium suspension, *Macromolecules*, 37(4) (2004a) 1386–1393.

Azizi Samir, M.A.S., Mateos, A.M., Alloin, F., Sanchez, J.-Y., Dufresne, A. Plasticized nanocomposite polymer electrolytes based on poly(oxyethylene) and cellulose whiskers, *Electrochimica Acta*, 49(26) (2004b) 4667–4677, 105.

Bai, Y., Huang, Z.H., Kang, F. Electrospun preparation of microporous carbon ultrafine fibers with tuned diameter, pore structure and hydrophobicity from phenolic resin, *Carbon*, 66 (2014) 705–712.

Bergshoef, M., Vancso, G. Transparent nanocomposites with ultrathin, electrospun nylon-4, 6 fiber reinforcement, *Advanced Materials*, 11 (1999) 1362–1365.

Bhattarai, S., Bhattarai, N., Yi, H., Hwang, P., Cha, D., Kim, H. Novel biodegradable electrospun membrane: Scaffold for tissue engineering, *Biomaterials*, 25 (2004) 2595–2602.

Bognitzki, M., Czado, W., Frese, T. et al. Nanostructured fibers via electrospinning, *Advanced Materials*, 13 (2001) 70–72.

Bui, N.N., Lind, M.L., Hoek, E.M.V., McCutcheon, J.R. Electrospun nanofiber supported thin film composite membranes for engineered osmosis, *Journal of Membrane Science*, 385–386 (2011) 10–19.

Bui, N.N., McCutcheon, J.R. Nanofiber supported thin-film composite membrane for pressure-retarded osmosis, *Environmental Science & Technology*, 48 (2014) 4129–4136.

Cath, T.Y., Childress, A.E., Elimelech, M. Forward osmosis: Principles, applications, and recent developments, *Journal of Membrane Science*, 281 (2006) 70–87.

Cavaliere, S., Subianto, S., Savych, I., Jones, D.J., Rozi`ere, J. Electrospinning: Designed architectures for energy conversion and storage devices, *Energy and Environmental Science*, 4(12) (2011) 4761–4785.

Chandrasekar, R., Zhang, L., Howe, J.Y., Hedin, N.E., Zhang, Y., Fong, H. Fabrication and characterization of electrospun titania nanofibers, *Journal of Material Science*, 44(5) (2009) 1198–1205.

Chen, Y., Liu, F., Qiu, F. et al. Cobalt-doped porous carbon nanosheets derived from 2D hypercrosslinked polymer with CoN4 for high performance electrochemical capacitors, *Polymers (Basel)*, 10(12) (2018) 1339.

Chen, Y.Q., Zheng, X.J., Feng, X. The fabrication of vanadium-doped ZnO piezoelectric nanofiber by electrospinning, *Nanotechnology*, 21 (2009) 055708.

Chen, Y.Q., Zheng, X.J., Mao, S.X., Li, W. Nanoscale mechanical behavior of vanadium doped ZnO piezoelectric nanofiber by nanoindentation technique, *Journal of Applied Physics*, 107(9) (2010) 094302.

Chiappone, A. Natural nanofibers in polymer membranes for energy applications, in Visakh P.M. and Olga Nazarenko (Eds.), *Nanostructured Polymer Membranes*. 2, (379–412) 2016. Beverly, MA: Scrivener Publishing LLC.

Chiappone, A., Nair, J.R., Gerbaldi, C., Jabbour, L., Bongiovanni, R., Zeno, E., Beneventi, D., Penazzi, N. Microfibrillated cellulose as reinforcement for Li-ion battery polymer electrolytes with excellent mechanical stability, *Journal of Power Sources*, 196(23) (2011) 10280–10288.

Choi, S.W., Fu, Y.Z., Ahn, Y.R., Jo, S.M., Manthiram, A. Nafion-impregnated electrospun polyvinylidene fluoride composite membranes for direct methanol fuel cells, *Journal of Power Sources*, 180 (2008) 167–171.

Chuangchote, S., Sagawa, T., Yoshikawa, S. Efficient dye-sensitized solar cells using electrospun TiO_2 nanofibers as a light harvesting layer, *Applied Physics Letters*, 93(3) (2008) 266.

Chung, H.J., Lee, D.W., Jo, S.M., Kim, D.Y., Lee, W.S. Electrospun poly(vinylidene fluoride)-based carbon nanofibers for hydrogen storage, *Materials Research Society Symposium Proceedings*, 837 (2005) 77–82.

Deng, L., Young, R.J., Kinloch, I.A., Zhu, Y., Eichhorn, S.J. Carbon nanofibres produced from electrospun cellulose nanofibers, *Carbon*, 58 (2013) 66–75.

Ding, Y., Zhang, P., Long, Z., Jiang, Y., Huang, J., Yan, W., Liu, G. Synthesis and electrochemical properties of Co_3O_4 nanofibers as anode materials for lithium-ion batteries, *Materials Letters*, 62 (2008) 3410–3412.

Dong, Z., Kennedy, S.J., Wu, J. Electrospinning materials for energy-related applications and devices, *Journal of Power Sources*, 196 (2011) 4886–4904.

Dosunmu, O., Chase, G., Kataphinan, W., Reneker, D. Electrospinning of polymer nanofibres from multiple jets on a porous tubular surface, *Nanotechnology*, 17 (2006) 1123.

Duan, B., Dong, C., Yuan, X., Yao, K. Electrospinning of chitosan solutions in acetic acid with poly (ethyleneoxide), *Journal of Biomaterials Science, Polymer Edition*, 15 (2004) 797–811.

Fan, Q., Whittingham, M.S. Electrospun manganese oxide nanofibers as anodes for lithium-ion batteries, *Electrochemical and Solid-State Letters*, 10(3) (2007) A48–A51.

Fatema, U.K., Uddin, A.J., Uemura, K., Gotoh, Y. Fabrication of carbon fibers from electrospun poly(vinyl alcohol) nanofibers, *Textile Research Journal*, 81(7) (2011) 659–672.

Formhals, A. United States Patent No US 1975504. Process and apparatus for preparing artificial threads. 1934. Retrieved from http://www.freepatentsonline.com/1975504.html.

Fujihara, K., Kumar, A., Jose, R., Ramakrishna, S., Uchida, S. Spray deposition of electrospun TiO_2 nanorods for dye-sensitized solar cell, *Nanotechnology*, 18(36) (2007) 365709.

Gu, Y., Chen, D., Jiao, X., Liu, F. LiCoO2-MgO coaxial fibers: Co-electrospun fabrication, characterization and electrochemical properties, *Journal of Materials Chemistry*, 17(8) (2007) 1769–1776.

Gu, Y., Jian, F., Wang, X. Synthesis and characterization of nanostructured Co_3O_4 fibers used as anode materials for lithium ion batteries, *Thin Solid Films*, 517 (2008) 652–655.

Han, G., Zhang, S., Li, X., Chung, T.S. Review: Progress in pressure-retarded osmosis (PRO) membranes for osmotic power generation, *Progress in Polymer Science*, 51 (2015) 1–27.

Hong, S.E., Kim, D.K., Jo, S. M., Kim, D.Y., Chin, B.D, Lee, D.W. Graphite nanofibers prepared from catalytic graphitization of electrospun poly(vinylidene fluoride) nanofibers and their hydrogen storage capacity, *Catalysis Today*, 120(3–4) (2007) 413–419.

Hossain, M., Kim, A. The effect of acetic acid on morphology of PZT nanofibers fabricated by electrospinning, *Materials Letters*, 63 (2009) 789–792.

Hu, G., Meng, X., Feng, X., Ding, Y., Zhang, S., Yang, M. Anatase TiO_2 nanoparticles/carbon nanotubes nanofibers: Preparation, characterization and photocatalytic properties, *Journal of Materials Science*, 42(17) (2007) 7162–7170.

Huang, Z., Zhang, Y., Kotaki, M., Ramakrishna S. A review on polymer nanofibers by electrospinning and their applications in nanocomposites, *Composites Science and Technology*, 63 (2003) 2223–2253.

Jeong, I., Lee, J., Joseph, K.V., Lee, H.I., Kim, J.K., Yoon, S. Low-cost electrospun WC/C composite nanofiber as a powerful platinum-free counter electrode for dye-sensitized solar cell, *Nano Energy*, 9 (2014) 392–400.

Ji, L., Jung, K.H., Medford, A.J., Zhang, X. Electrospun polyacrylonitrile fibers with dispersed Si nanoparticles and their electrochemical behaviors after carbonization, *Journal of Materials Chemistry*, 19(28) (2009) 4992–4997.

Ji, L., Lin, Z., Zhou, R., Shi, Q., Toprakci, O., Medford, A.J., Millns, C.R., Zhang, X. Formation and electrochemical performance of copper/carbon composite nanofibers, *Electrochimica Acta* 55 (2010) 1605–1611.

Ji, L., Zhang, X. Fabrication of porous carbon/Si composite nanofibers as high-capacity battery electrodes, *Electrochemistry Communications*, 11 (2009) 1146–1149.

Jirsak, O., Sanetrnik, F., Lukas, D., Kotek, V., Martinova, L., Chaloupek, J. Patent No WO2005024101. A method of nanofibres production from a polymer solution using electrostatic spinning and a device for carrying out the method. Retrieved from https://patentscope.wipo.int/search/en/detail.jsf?docId=WO2005024101&tab=PCTBIBLIO&maxRec=1000.

Jo, S.M. Electrospun nanofibrous materials and their hydrogen storage, hydrogen storage jianjun liu, *IntechOpen*, (2012) doi:10.5772/50521. Available from: https://www.intechopen.com/books/hydrogen-storage/electrospun-nanofibrous-materials-and-their-hydrogen-storage

Kim, D.K., Park, S.H., Kim, B.C., Chin, B.D., Jo, S.M., Kim, D.Y. Electrospun polyacrylonitrile based carbon nanofibers and their hydrogen storage, *Macromolecular Research*, 13(6) (2005) 521–528.

Kim, G.H., Cho, Y.S., Kim, W.D. Stability analysis for multi-jets electrospinning process modified with a cylindrical electrode. *European Polymer Journal*, 42 (2006) 2031–2038.

Kim, H.J., Kim, Y.S., Seo, M.H, Choi, S.M., Cho, J., Huber, G.W., Kim, W.B. Highly improved oxygen reduction performance over Pt/C-dispersed nanowire network catalysts, *Electrochemistry Communications*, 12 (2010) 32–35.

Kim, H.J., Kim, Y.S., Seo, M.H., Choi, S.M., Kim, W.B. Pt and PtRh nanowire electrocatalysts for cyclohexane-fueled polymer electrolyte membrane fuel cell, *Electrochemistry Communications*, 11 (2009a) 446–449.

Kim, J.U., Park, S.H., Choi, H.J., Lee, W.K., Lee, J.K., Kim, M.R. Effect of electrolyte in electrospun poly(vinylidene fluoride-cohexafluoropropylene) nanofibers on dye-sensitized solar cells, *Solar Energy Materials & Solar Cells*, 93 (2009b) 803–807.

Koski, A., Yim, K., Shivkumar, S. Effect of molecular weight on fibrous PVA produced by electrospinning, *Materials Letters*, 58 (2004) 493–497.

Lee, J.R., Jee, S.Y., Kim, H.J., Hong, Y.T. S SK. Patent No WO2006/123879A1. Filament bundle type nano fiber and manufacturing method thereof. 2006. Retrieved from https://patentimages.storage.googleapis.com/b3/16/a9/18b1f4cdf0033a/WO2006123879A1.pdf.

Lin, T., Wang, H., Wang, X. The charge effect of cationic surfactants on the elimination of fibre beads in the electrospinning of polystyrene, *Nanotechnol*, 15 (2004) 1375.

Liu, Y., He, J. Bubble electrospinning for mass production of nanofibers, *International Journal of Nonlinear Sciences and Numerical Simulation*, 8 (2007) 393.

Logan, B.E., Elimelech, M. Membrane-based processes for sustainable power generation using water. *Nature*, 4888 (2012) 313–319.

Lu, H-W., Li, D., Sun, K., Li, Y.S., Fu, W. Carbon nanotube reinforced NiO fibers for rechargeable lithium batteries, *Solid State Sciences*, 11 (2009) 982–987.

Maensiri, S., Nuansing, W. Thermoelectric oxide NaCo2O4 nanofibers fabricated by electrospinning, *Materials Chemistry and Physics*, 99 (2006) 104–108.

McKee, M., Wilkes, G., Colby, R., Long T. Correlations of solution rheology with electrospun fiber formation of linear and branched polyesters, *Macromolecules*, 37 (2004) 1760–1767.

Nayak, R., Padhye, R., Kyratzis, I.L., Truong, Y.B, Arnold, L. Recent advances in nanofibre fabrication techniques, *Textile Research Journal*, 82(2) (2011) 129–147.

Onozuka, K., Ding, B., Tsuge, Y., Naka, T., Yamazaki, M., Sugi, S., Ohno, S., Yoshikawa, M., Shirator, S. Electrospinning processed nanofibrous TiO_2 membranes for photovoltaic applications, *Nanotechnology*, 17(4) (2006) 1026–1031.

Park, S.H., Kim, B.C., Jo, S.M., Kim, D.Y., Lee, W.S. Carbon nanofibrous materials prepared from electrospun polyacrylonitrile nanofibers for hydrogen storage, *Materials Research Society Symposium Proceedings*, 837 (2005) 71–76.

Park, S.H., Kim, J.U., Lee, S.Y., Lee, W.K., Lee, J.K., Kim, M.R. Dye-sensitized solar cells using polymer electrolytes based on poly(vinylidene fluoride-hexafluoro propylene) nanofibers by electrospinning method, *Journal of Nanoscience and Nanotechnology*, 8(9) (2008) 4889–4894.

Patanaik, A., Anandjiwala, R., Rengasamy, R., Ghosh, A., Pal, H. Nanotechnology in fibrous materials–A new perspective, *Textile Progress*, 39 (2007) 67–120.

Peng, S., Li, L., Kong, J., Lee, Y., Tian, L., Srinivasan, M., Adams, S., Ramakrishna, S. Electrospun carbon nanofibers and their hybrid composites as advanced materials for energy conversion and storage, *Nano Energy*, 22 (2016) 361–395.

Qiao, H., Wei, Q. Chapter 10—*Functional Nanofibers in Lithium-Ion Batteries in Functional Nanofibers and Their Applications*, Oxford, UK: Woodhead Publishing Series in Textiles 2012, 197–208.

Ramakrishna, S. *An Introduction to Electrospinning and Nanofibers*. Hackensack, NJ: World Scientific Pub Co Inc, 2005.

Ramon, G.Z., Feinberg, B.J., Hoek, E.M.V. Membrane-based production of salinity-gradient power, *Energy & Environmental Science*, 4 (2011) 4423.

Regele, J, Papac, M., Rickard, M., Dunn-Rankin, D. Effects of capillary spacing on EHD spraying from an array of cone jets, *Journal of Aerosol Science*, 33 (2002) 1471–1479.

Rose, M., Kockrick, E., Senkovska, I., Kaskel, S. High surface area carbide-derived carbon fibers produced by electrospinning of polycarbosilane precursors, *Carbon*, 48(2) (2010) 403–407.

Shi, X., Zhou, W., Ma, D., Ma, Q., Bridges, D., Ma, Y., Hu, A. Review article: Electrospinning of nanofibers and their applications for energy devices, *Journal of Nanomaterials*, 16(1) (2015) 20.

Shim, H.S., Kim, J.W., Kim, W.B. Fabrication and optical properties of conjugated polymer composited multi-arrays of TiO 2 nanowires via sequential electrospinning, *Journal of Nanoscience and Nanotechnology*, 9(8) (2009) 4721–4726.

Shu, J., Li, J.C.M., Platinum nanowire produced by electrospinning, *NanoLetters*, 9(4) (2009) 1307–1314.

Skilhagen, S.E., Dugstad, J.E., Aaberg, R.J. Osmotic power—Power production based on the osmotic pressure difference between waters with varying salt gradients, *Desalination*, 220 (2008) 476–482.

Song, M.Y., Kim, D.K., Ihn, K.J., Jo, S.M., Kim, D.Y. Electrospun TiO_2 electrodes for dye-sensitized solar cells, *Nanotechnology*, 15(12) (2004) 1861–1865.

Song, M.Y., Kim, D.K., Jo, S.M., Kim, D.Y. Enhancement of the photocurrent generation in dye-sensitized solar cell based on electrospun TiO_2 electrode by surface treatment, *Synthetic Materials*, 155 (2005) 635–638.

Stephan, A.M., Kumar, T.P., Kulandainathan, M.A., Lakshmi, N.A., Chitin-incorporated poly(ethylene oxide)-based nanocomposite electrolytes for lithium batteries. *Journal of Physical Chemistry B*, 113(7), (2009) 1963–1971.

Sun, G., Sun, L., Xie, H., Liu, J. Electrospinning of nanofibers for energy applications, *Nanomaterials*, 6 (2016) 129. doi:10.3390/nano6070129.

Teo, W, Gopal, R., Ramaseshan, R., Fujihara, K., Ramakrishna, S. A dynamic liquid support system for continuous electrospun yarn fabrication, *Polymer*, 48 (2007) 3400–3405.

Thavasi, V., Singh, G., Ramakrishna, S. Electrospun nanofibers in energy and environmental applications, *Energy & Environmental Science*, 1 (2008) 205–221.

Thorsen, T., Holt, T. Hydrophile semipermeable membrane. WO Pat 047733 A1, 2003.

Thorsen, T., Holt, T. Semi permeable membrane for use in osmosis, and method and plant for providing elevated pressure by osmosis to create power. US Pat, 7,566,402 B2, 2009.

Vaseashta, A. Controlled formation of multiple Taylor cones in electrospinning process, *Applied Physics Letters*, 90 (2007) 093115.

Wang, L., Yu, Y., Chen, P.C., Zhang, D.W., Chen, C.H., Electrospinning synthesis of C/Fe3O4 composite nanofibers and their application for high performance lithium-ion batteries, *Journal of Power Sources*, 183 (2008) 717–723.

Wang, P., Zhang, D., Ma, F., Ou, Y., Chen, Q.N., Xie, S., Li, J. Mesoporous carbon nanofibers with a high surface area electrospun from thermoplastic polyvinylpyrrolidone, *Nanoscale*, 4(22) (2012) 7199–7204.

Wang, X, Niu, H., Lin, T. Needleless electrospinning of nanofibers with a conical wire coil, *Polymer Engineering & Science*, 49 (2009) 1582–1586.

Wang, X., Um, I.C., Fang, D., Okamoto, A., Hsiao, B.S., Chu, B. Formation of water-resistant hyaluronic acid nanofibers by blowing-assisted electro-spinning and non-toxic post treatments, *Polymer*, 46(13) (2005) 4853–4867.

Wang, Y., Serrano, S., Santiago-Avilés, J.J., Raman characterization of carbon nanofibers prepared using electrospinning, *Synthetic Metals*, 138(3) (2003) 423–427.

Xu, C., Inai, R., Kotaki, M., Ramakrishna, S. Aligned biodegradable nanofibrous structure: A potential scaffold for blood vessel engineering. *Biomaterials*, 25 (2004) 877–886.

Xu, S., Shi, Y., Kim, S.G. Fabrication and mechanical property of nano piezoelectric fibres, *Nanotechnology*, 17(17) (2006) 4497–4501.

Yamashita, Y., Ko, F., Tanaka, A., Miyake, H. Characteristics of elastomeric nanofiber membranes produced by electrospinning, *Journal of Textile Engineering*, 53 (2007) 137–142.

Yan, H., Mahanta, N.K., Wang, B., Wang, S., Abramson, A.R., Cakmak, M., Structural evolution in graphitization of nanofibers and mats from electrospun polyimide–mesophase pitch blends, *Carbon*, 71 (2014) 303–318.

Yang, Y., Centrone, A., Chen, L., Simeon, F., Hatton, T.A., Rutledge, G.C., Highly porous electrospun polyvinylidene fluoride (PVDF)-based carbon fiber, *Carbon*, 49 (2011) 3395–3403.

Yarin, A., Zussman, E. Upward needleless electrospinning of multiple nanofibers, *Polymer*, 45 (2004) 2977–2980.

Yin, T., Liu, D., Ou, Y., Ma, F., Xie, S., Li, J.F., Li, J., Nanocrystalline thermoelectric Ca3Co4O9 ceramics by sol-gel based electrospinning and spark plasma sintering, *Journal of Physics Chemistry C*, 114(21) (2010) 10061–10065.

Yu, Y., Gu, L., Wang, C., Dhanabalan, A., Van Aken, P.A., Maier, J., Encapsulation of Sn@carbon nanoparticles in bamboo-like hollow carbon nanofibers as an anode material in lithium-based batteries, *Angewandte Chemie: International Edition*, 48(35) (2009) 6485–6489.

Zhang, B., Kang, F., Tarascon, J-M., Kim, J-K. Recent advances in electrospun carbon nanofibers and their application in electrochemical energy storage. *Progress in Materials Science*, 76 (2016) 319–380.

Zhang, H., Wang, X., Liang, Y., Preparation and characterization of a Lithium-ion battery separator from cellulose nanofibers, *Heliyon*, (2015) e00032, doi:10.1016/j.heliyon.2015.e00032 2405-8440.

Zhang, X., Ji, L., Toprakci, O., Liang, Y., Alcoutlabi, M., Electrospun nanofiber-based anodes, cathodes, and separators for advanced lithium-ion batteries, *Polymer Reviews*, 51(3) (2011) 239–264, doi:10.1080/15583724.2011.593390.

Zhou, F., Gong, R. Manufacturing technologies of polymeric nanofibres and nanofibre yarns, *Polymer International*, 57 (2008) 837–845.

Zou, L., Gan, L., Kang, F., Wang, M., Shen, W., Huang, Z. Sn/C non-woven film prepared by electrospinning as anode materials for lithium ion batteries, *Journal of Power Sources*, 195 (2010) 1216–1220.

11 Electrospun Nanocomposite Polymer Electrolyte Membrane for Fuel Cell Applications

Nor Azureen Mohamad Nor and Juhana Jaafar

CONTENTS

11.1 INTRODUCTION

The scarcity of fossil fuels as an energy source together with the destruction of the environmental, due to the emission of toxic gases, has raised the need for clean and green technologies for power generation. Fuel cells have now become important technology that has been widely explored due to the promises of clean and efficient energy conversion. Despite their importance in modern technology, fuel cells have actually been known to science for more than 150 years. Though generally considered a curiosity in the 1800s, fuel cells became the subject of intense research and development during the 1900s (Kirubakaran et al. 2009). The fuel cell concept was demonstrated successfully in the early nineteenth century by Humphry Davy. It was in 21838 that the scientist Christian Friedrich Schönbein continued the work to invent fuel cells technology (Andujar and Segura 2009). Fuel cells are a promising technology as a source of heat and electricity for buildings, and as an electrical power source for electric motors propelling vehicles. Fuel cells cleanly and efficiently convert chemical energy from hydrogen-rich fuels into electrical power and usable high-quality heat in an electrochemical process while the crude oil prices surge (Kucernak and Toyoda 2018). A fuel cell system offers a platform of a clean, quiet, and highly efficient process in the conversion of fuel to energy via an electrochemical process. This technology is two to three times more efficient than fuel burning. No other energy generation technology offers the combination of benefits that fuel cells technology can offer. In addition to low or zero emission, the benefits include high efficiency and reliability, multifuel capability, flexibility, durability, scalability, and ease of maintenance (Miyake et al. 2017). Fuel cells operate silently, so they will reduce noise pollution as well as air pollution, and the waste heat can be used to provide hot water or space heating for a home or office.

The basic operation of fuel cell applications is to produce an electrical current that can be directed outside the cell to do work, such as powering an electric motor or illuminating a light bulb. The chemical reactions that produce the current are the key to how fuel cell works (Chaurasia et al. 2003). There are several kinds of fuel cells and each operates differently. A fuel cell system consists of electrolyte, electrodes, and gas diffusion layers. The electrolyte layer plays an important role in a fuel cell system and can be described as the heart of the fuel cell. Different types of fuel cell consist of different types of electrolyte. An electrolyte layer is a selective layer that permits only the appropriate ions to pass through, it separates the product gases, and provides electrical insulation to the electrodes (Femmer et al. 2015). If free electrons or other substances passed through the electrolyte, they would disrupt the chemical reaction (Wang et al. 2011). Meanwhile, the electrodes layer consists of anode and cathode on each side of the electrolyte. Both electrodes consist of a catalyst layer that facilitates the reaction of oxygen and fuel. It is usually made of platinum nanoparticles very thinly coated onto carbon paper or cloth. The catalyst is rough and porous so that a maximum surface area of the platinum can be exposed to the fuel sources and oxygen (Zielke et al. 2016).

The gas diffusion layer (GDL) is another layer that is important in a fuel cell system. It is located at the outside of both of the electrodes (anode and cathode) and facilitates transport of reactants into the catalyst layer as well removing the water product. Each GDL is typically composed of a sheet of carbon paper in which the carbon fibers are partially coated with polytetrafluoroethylene (PTFE) (El-kharouf et al. 2012). Gases diffuse rapidly through the pores in GDL. These pores are kept open by the hydrophobic PTFE, which prevents excessive water build-up. In many cases, the inner surface of the GDL is coated with a thin layer of a high surface area carbon mixed with PTFE to form what is called the microporous layer. The microporous layer can help adjust the balance between water retention that is needed to maintain the membrane conductivity and water release that is used to keep the pores open so that hydrogen and oxygen can diffuse into the electrodes (Park et al. 2016).

Fuel cell systems are classified primarily by the type of electrolyte that has been employed. The type of electrolyte used will determine the type of electrochemical reaction occurring in the cell, the catalyst needed, the operating temperature, and the reactant required in the cell (Thepkaew et al. 2008; Tada et al. 2015). These operating conditions will determine what kind of fuel cell type is suitable. In fuel cell technology, various types of fuel cells that are currently under development and they have advantages, limitations, and potential applications (U.S. Department of Energy, 2016). There are six different types of fuel cell that can be classified by the type of electrolyte that are summarized in Table 11.1, these are alkaline fuel cells (AFC), proton exchange membrane fuel cells (PEMFC), direct methanol fuel cells (DMFC), solid oxide fuel cells (SOFC), molten carbonate fuel cells (MCFC), and phosphoric acid fuel cells (PACFs).

TABLE 11.1
Fuel Cell Types

Type of Fuel Cell	Common Electrolyte	Operating Temperature	Anode/Cathode Gas	Applications
Alkaline fuel cells (AFC)	Aqueous KOH or alkaline polymer membrane	<100°C	H_2/O_2	Military Space Backup power Transportation
Proton exchange membrane fuel cells (PEMFC)	Polymer membrane	<120°C	H_2/O_2 or atmospheric O_2	Backup power Portable power Distributed generation Transportation Specialty vehicles
Direct methanol fuel cell (DMFC)	Polymer membrane	<120°C	Methanol/atmospheric O_2	Portable power Auxiliary power Distributed generation
Solid oxide fuel cells (SOFC)	Ceramic oxide	500°C–1000°C	H_2 or methane/atmospheric O_2	Auxiliary power Electric utility Distributed generation
Molten carbonate fuel cell (MCFC)	Alkali-carbonate	600°C–700°C	H_2 or methane/atmospheric O_2	Electric utility Distributed generation
Phosphoric acid fuel cells (PACFs)	Phosphorus	150°C–200°C	H_2/atmospheric O_2	Distributed generation

AFC, PEMFC, and DMFC lie under the same category in which they are using polymer membrane as an electrolyte which has been called as polymer electrolyte membrane. Polymer electrolyte membrane offers advantages over the other types of electrolyte. Polymer electrolyte membrane operates at a lower temperature, they are lighter and more compact, which makes them ideal for applications such as cars. However, the polymer electrolyte membrane must be water-saturated during operation (Dodds et al. 2015). Some fuel cell systems operate without humidifiers, relying on the rapid water generation and the high rate of back diffusion through the thin membrane layers to maintain membrane hydration. High-temperature polymer electrolyte membrane operates between 100°C and 200°C and potentially offers benefit to electrode kinetics, heat management, and better tolerance to fuel impurities are the advantage to be applied as polymer electrolyte membrane in fuel cell applications (Nomnqa et al. 2016). This improvement potentially could lead to higher overall system efficiencies. Unfortunately, higher cell operation temperature will results in polymer electrolyte membrane to lose its functionality rapidly and begins to deform through creep, resulting in localized thinning and overall lower system lifetimes. As a result, new anhydrous polymer materials that can withstand higher temperatures are actively being studied for the development of suitable polymer electrolyte membrane with an enhanced cell performance.

11.2 POLYMER ELECTROLYTE MEMBRANE FUEL CELL

The basic principle of a fuel cell system is similar to all kinds of fuel cell types. Herein, the principle of proton exchange membrane fuel cell (PEMFC) was discussed. In PEMFC, the fuel used is hydrogen and the charge carrier is the hydrogen ion (proton). As the hydrogen gases are supplied at the anode, the hydrogen molecules undergo chemical reaction and split into electrons and protons. The proton permeates through the polymer electrolyte membrane to the cathode, while the electrons travel along the external circuit and produce electrical power. The commonly used oxidant in this cell is oxygen supplied in air at the cathode and combines with the electrons and proton to produce water (Bakangura et al. 2016). The reactions at the electrodes are described in Figure 11.1.

Electrolyte is a substance that produces an electrically conducting solution when dissolved in a polar solvent such as water. The dissolved electrolyte separates into cations and anions, which disperse uniformly through the solvent. If an electric potential is applied to such a solution, the cations of the solution are drawn to the electrode that has an abundance of electrons while the anions are drawn to the electrode that has a deficit of electrons. The movement of anions and cations in the opposite directions within the solution amounts to a current (Arges et al. 2010). The electrolyte is the key to operating the fuel cell system. The electrolyte used is generally classifies the different types of fuel cell because it determines the operating temperature of the system and in part the kind of fuel that can be employed. Electrolytes must permit only the appropriate ions to pass between the anode and cathode. If free electrons or other substances pass through the electrolyte, they would disrupt the chemical reaction (Abdulkareem et al. 2012).

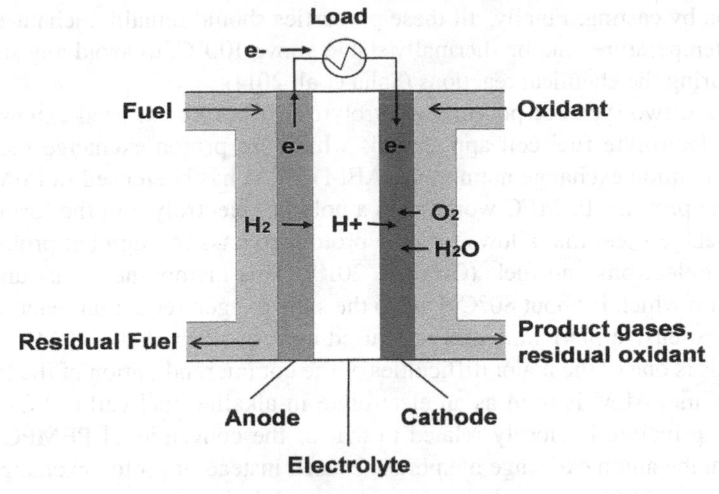

FIGURE 11.1 Basic principle of PEMFC.

11.2.1 Polymer Electrolyte Membrane

The polymer electrolyte membrane is the central component contributing to the efficiency of the fuel cell application. Polymer electrolyte membrane is a semipermeable membrane generally made from ionomers and designed to conduct protons while acting as a barrier to the reactant and electron insulator (Kyu and Nazir 2013). The essential function of the polymer electrolyte membrane in a fuel cell system is as the reactants separator and protons transportation medium, while blocking the passage of electrons through the membrane. Generally, there are several required properties of the polymer electrolyte membrane in order for it to be compatible with the operation of the various fuel cell types. One of them is that there is an efficient transfer of the ions from one electrode to the other, which requires a high conductivity (Couture et al. 2011). Furthermore, the membrane has to be a barrier to hydrogen and oxygen since their diffusion would make the fuel cell less efficient, with lower performances, and the mixture of these gases is explosive.

Polymer electrolyte membrane should be chemically stable as it is acting in strong acidic or basic environment (Espiritu et al. 2016). Ideally, the durability and mechanical strength properties of the polymer electrolyte membrane are desired regardless of the hydration state and temperature of the membrane. In fact, the quantity of water in the fuel cell is difficult to quantify as it is continuously produced. It has to be noted that standard membranes in the fuel cell systems are highly hydrophilic and their properties highly depend on the hydration state. However, they have to remain water-insoluble and if possible to be soluble in suitable organic solvents to allow membrane fabrication by casting. Finally, all these properties should remain unchanged at the working temperature, and be thermally stable above 100°C, to avoid any structural change during the chemical reactions (Sahu et al. 2014).

There are two types of polymer electrolyte that have been used extensively in polymer electrolyte fuel cell applications which are proton exchange membrane (PEM) and anion exchange membrane (AEM). PEM has been used in PEMFC for conducting protons. PEMFC works with a polymer electrolyte in the form of thin and permeable sheet that allows for only protons to pass through but prohibits the passage of electrons and fuels (Gu et al. 2010). This membrane works under low temperature which is about 80°C. Due to the slow oxygen reduction reaction at the cathode, expensive platinum catalyst is used at both sides of the PEM (Sui et al. 2017). This is one of the major difficulties of the commercialization of the PEMFC.

Meanwhile, AEM is used as an electrolyte in alkaline fuel cell (AFC) system. The basic principle is closely related to that of the conventional PEMFC system except that the anion exchange membrane is used instead of proton exchange membrane. In solid AFC, hydroxide ions are generated during the electrochemical oxygen reduction at the cathode. They are transported from the cathode to the anode through the anion conducting (but electronically insulating) polymer electrolyte, wherein they combine with hydrogen to form water. It is the opposite to the H^+ conduction reaction of PEM. Then, the electrons generated during H_2 oxidation pass through the external circuit to the cathode, where they participate in the electrochemical reduction of oxygen to produce OH^-. In the overall reaction, both PEM and AEM will produce H_2O as a byproduct. In contrast to PEM, water will be the reactant at the cathode and water will also be generated at the anode twice as much as in PEM per electron. The general comparisons between AEM and PEM are summarized in Table 11.2.

TABLE 11.2
Comparison of AEM and PEM

	AEM	PEM
Reactant	H_2	H_2
Oxidant	O_2/H_2O	O_2
Anode	$2H_2 + 4OH^- \rightarrow 4H_2O + 4e^-$	$2H_2 \rightarrow 4H^+ + 4e^-$
Cathode	$O_2 + 2H_2O + 4e^- \rightarrow 4OH^-$	$O_2 + 4H^+ + 4e^- \rightarrow 2H_2O$
Counter ion	OH^- conductive	H^+ conductive
Ion-exchange group	Quaternary ammonium group, hydrocarbon polymer backbone	Fluorinated polymer, sulfonic acid group
Features	Pt-free catalyst available, advantage for cathode O_2 reduction	High ion conductivity, excellent ionomer solution
Issues	High cost materials Low ion conductivity Low thermostability	High cost materials Fuel crossover Low oxygen reduction rate at cathode

11.3 CHALLENGES IN POLYMER ELECTROLYTE MEMBRANE FUEL CELL TECHNOLOGY

Cost, performance, and durability are the crucial challenges in the fuel cell industry. Industries are working closely to overcome the critical technical barriers to fuel cell development. The most crucial factor that hinders the development of the fuel cell industry is the cost. Platinum that been applied on the catalyst layer represents one of the largest cost components in a fuel cells operation. Modern research and development (R&D) focuses on the utilization of the current platinum group metal (PGM) as well as non-PGM catalyst approaches for long-term application.

The development of a fuel cell system with excellent performance is also a big issue for the fuel cell industry. R&D focuses on developing ion-exchange membrane electrolytes with an enhanced performance at a reduced cost. There are many factors that need to be taken into account to improve the efficiency of a fuel cells performance. Some of the factors that need to be taken seriously to enhance fuel cell performance including improving membrane electrode assemblies (MEAs) through the integration of crucial properties for MEA components, identifying their degradation mechanisms, developing approaches to mitigate these effects, and maintaining core activities of components tailored for power of stationary or portable devices.

Another serious issue related to the development of a fuel cell system is the durability of the cell system lifetime that meets the application expectations. The Department of Energy (DOE) is targeting the durability for stationary and transportation fuel cells to be about 40,000 and 5,000 h under realistic operating conditions, respectively. Currently, R&D focuses on understanding the fuel cell degradation mechanisms and developing fuel cell materials that will reduce the operating cost and at the same time contribute to excellent performance and higher stability.

11.4 NANOCOMPOSITE POLYMER ELECTROLYTE MEMBRANE

An ion-exchange membrane generally consists of a polymeric skeleton on which anionic and cationic functional groups are incorporated, allowing the transportation of selectively charged ions. Polymer electrolyte membrane are categorized according to the various types of polymer backbones and ionic charged group that are able to transport ions throughout the membrane. High ionic conductivity, high mechanical stabilities in terms of strength and flexibility, high chemical stability toward oxidation and reduction, zero electrical conductivity, low gas permeability, and low production cost are the most desirable properties for a polymer electrolyte membrane developed for fuel cell applications (Sebastian and Baglio 2017). Polymeric materials are currently the preferred choice of proton conducting membrane. The general requirement for good polymer electrolyte membrane is high proton conductivity of more than 10^{-2} S/cm that enables the proton transport between the electrodes. Polymer electrolyte membrane also requires humidification to conduct ions, and this sometimes causes membrane degradation due to the effects of hydration (Ji and Wei 2009). This condition contributes to losses in cell efficiency.

The problem of the polymer electrolyte membrane fuel cells based on perfluorosulfonic acid ionomers are high-cost, low-proton conductivity along with high temperature requirements, CO poisoning, and water management issues have limited the commercialization of the fuel cell system based polymeric membrane. Perfluorosulfonic acid ionomers type membrane or Nafion-based polymer electrolyte membrane by DuPont is the most common and commercially available PEM on the market (Kyu and Nazir 2013). It has been widely used for PEMFC and can be specified based on thickness and applications. Nafion 117, Nafion 115, Nafion 212, and Nafion 211 membranes are non-reinforced films and the last digits refer to the membrane thickness (e.g., Nafion 212 and Nafion 211 have the thickness of ~0.002″ and ~0.001″, respectively). Different conditions in operating the fuel cell system require different thicknesses of the Nafion membrane (Rangel-Cardenas and Koper 2017). A thicker membrane is needed when the fuel cell is running under high differential pressure operation and vice versa. For the past decades, conventional Nafion membrane has greatly and effectively influenced PEMFC performance due to its benefits such as high proton conductivity and mechanical and chemical stabilities (Wee 2007). Unfortunately, Nafion has a limited dehydration temperature below 100°C that can lead to slower kinetic reactions, poor cost efficiency, and CO poisoning of platinum electrode catalysts. An efficient Nafion membrane in fuel cell applications is obtained only at high hydration thus limiting the operating temperature of PEMFC at around 80°C with a relative humidity approaching 100% (Sahu et al. 2009).

To date, the contribution of the polymeric nature toward the mechanical properties of polymer electrolyte membrane in fuel cell systems is poor. Therefore, the development of stronger polymeric structure is still a focus study to resolve this problem (Schmitt et al. 2011; Eldin et al. 2017). In the past decades, nanocomposite membranes have been actively studied to succeed the conventional Nafion-based polymer electrolyte membranes in fuel cell applications. Nanometer-sized particles as filler with controllable structure and superior properties have been introduced to improve the polymer electrolyte membrane. This might not otherwise be achieved by conventional reinforcement of the polymeric materials. Various kinds of fillers, either organic or inorganic, have been studied to improve the membrane stability of conventional polymer electrolyte membrane. Organic–inorganic composite membrane has triggered a considerable interest due to its extraordinary properties arising from the synergy between the organic and inorganic material properties. It has been demonstrated that the inorganic materials within the membrane play an important role because they can interact with the polymer chain packing and create a more tortuous diffusion path.

The inorganic component in the polymeric material can enhance mechanical strength and stability of the membranes while suppressing fuel diffusion and swelling of membranes simultaneously. The organic component can not only combine the advantages of each component but also reduce their weaknesses. Also, it has been demonstrated that the addition of inorganic particles, such as titania, silica, alumina, zirconia, or clay into a polymer matrix, can improve the mechanical properties of a polymer electrolyte membrane (Jeon and Baek 2010; Chen et al. 2012). Organic–inorganic nanocomposite membrane can be broadly defined as molecular or nanocomposite with organic and inorganic components intimately mixed where at least one of the component domains has a dimension ranging in the nanoscale. Incorporating inorganic particles at the nanoscale range presents remarkable opportunities to produce unique material properties that contribute to excellent cell performance (Klaysom et al. 2011; Hong et al. 2015). For example, incorporating SiO_2 nanoparticles in perfluorosulfonic acid membrane showed good mechanical and thermal stabilities that are favorable in fuel cell systems (Wang et al. 2018).

11.5 NANOFIBERS

The incorporation of inorganic fillers into polymer membranes has drawn a significant amount of attention over the last few decades. The incorporation of inorganic nanomaterials influences the organic phase properties in terms of proton conductivity and membrane stability, in addition to decreased cost, improved water retention capacity, and improved mechanical strength (Staiti 2001). The preparation methods of the nanocomposite membrane and the intrinsic properties of the inorganic nanomaterials used as a filler such as size, type, shape, and the filler interactions with the polymer matrix can significantly affect the properties of the resultant matrix (Hanemann and Szabo 2010). In recent years, nanomaterials have received significant attention in the literature for both fundamental study and industrial applications. Recent studies on nanomaterials including nanoparticles, nanorods, nanotubes, nanosheets, nanowires, and nanofibers have provided very fascinating insight into nanomaterials properties and how nanomaterials will benefit fuel cell applications.

Nanofibers find potential applications in various areas such as tissue engineering scaffolds, blood vessel stents, wound dressings, batteries, gas, bio- and chemical sensors, catalysis, fuel cells, and water treatment (Kenry and Lim 2017). Inorganic materials in nanofiber form bring many advantages to fuel cell applications. This kind of nanostructure possesses many unique features including a large specific surface area, superior mechanical properties, and potential for use as nanoscale building blocks. The higher surface-to-volume ratio of the nanofibers has a significant influence increasing the mechanical properties of a polymeric membrane that they are incorporated into. Nanofibers have one dimension that is many times longer than those of the other types of nanomaterial where it provides a defined flow pattern along the longer dimension (Bjorge et al. 2009). The inorganic materials in the form of nanofiber surfaces will increase the hydrophilicity of the polymeric membrane thus increasing the water uptake for higher conductivity. These outstanding properties make the nanofibers to be key candidates for improvement of polymer electrolyte membrane and many important applications. A number of processing techniques such as electrospinning, drawing, template synthesis, phase separation, and self-assembly have been used to prepare nanofibers (Huang et al. 2003). There are only a few literature research examples that report on the properties of nanofibrous mats as polymer electrolyte membranes in the fuel cell applications. However, due to their promising features, nanocomposite-based polymer electrolyte membrane composed of nanofibers have been widely explored.

11.5.1 Nanofibers by Electrospinning

Electrospinning is one of the methods used to produce nanofibers with diameters ranging from 10 to 100 nm and it is a very simple and versatile technique. The technique can be easily employed in laboratories and can be scaled up to an industrial process. Nanofibers are produced from a variety of organic or inorganic polymers such as polyvinyl alcohol, polyurethanes, polyimides, copolymers, and composite polymers (Zhou and Gong 2008). The electrospun nanofiber properties are different from the other types of fibers produced from conventional spinning or melt-blown fibers as they exhibit a smaller fiber diameter. The potential applications based on the nanofibers structure, especially for their use as reinforcement in nanocomposite polymer electrolyte membrane development have been realized.

The basic setup for the electrospinning process requires a high voltage power supply, spinneret (metallic needle), syringe pump, and a nanofiber collector (Al or Si wafer). The prepared precursor solution is loaded into the syringe pump with a spinneret at the tip. The high voltage power (kV/cm) is supplied at the spinneret to yield an electrical field. The collector plate is placed several centimeters (cm) from the tip. The viscous solution is feed through the spinneret at a constant and slow rate, usually below 1 mL/h and flow rate is controlled by the syringe pump. The droplet from the spinneret immediately becomes highly electrified and the induced charges are evenly dispersed at the liquid surface (Zhang et al. 2009; Liu et al. 2012). The charged fibers are deposited on the collector plate because of the electrostatic attraction. The as-spun nanofibers are kept in the air to complete the hydrolysis reaction. The typical setup for electrospinning process is presented in Figure 11.2.

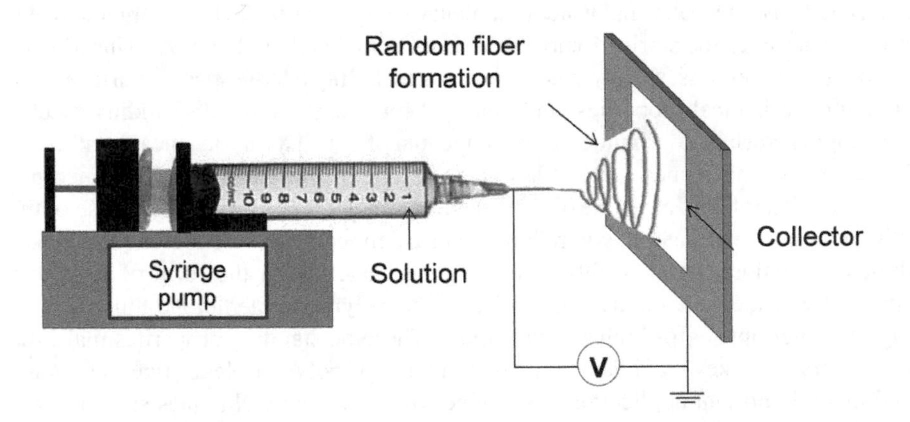

FIGURE 11.2 Typical setup for electrospinning process.

11.6 ELECTROSPINNING PROCESS PARAMETER

The unique internal structure and characteristics of nanofibers such as morphological structure, fiber diameter, phase separation, polymer chain orientation, and level of water uptake are crucially important to create effective ion transport pathways to achieve high ion conductivity in polymer electrolyte membrane. In addition to the high conductivity, nanofibers are known to have high thermal, mechanical, and chemical stabilities with controllable structure (Esfahani et al. 2017). The ideal target in forming nanofibers through an electrospinning process is to produce defect-free, consistent and controllable fiber diameters, defect-controllable fiber surfaces, and collectable continuous single nanofibers. Nanofibers fabrication can be optimized by controlling solution, process, and ambient conditions as the characteristics of the electrospun fibers are determined by these parameters (Homaeigohar and Elbahri 2014). It is very important to avoid the occurrence of beads, especially for smaller nanofibers.

11.6.1 Solution Parameters

To prepare nanofibers through the electrospinning technique, the most important thing is the preparation of the electrospinning solution; the characteristics of the electrospinning solution need to be taken into account. The electrospinning solution properties that are important include concentration, polymer molecular weight, viscosity, surface tension, and conductivity of the solution. The concentration of the polymer plays an important role in preparing the electrospinning solution as it will determine whether it can be electrospun into nanofibers and has an important effect on fiber morphology. To obtain the fibers, a feed solution is required with an optimum concentration for fibers without forming beads. Generally, the fiber diameter and uniformity will increase as the polymer concentration in the electrospinning solution increases. However, at a very low polymer solution concentration, electrospraying occurs instead of electrospinning owing to the low viscosity and high surface tension of the solution. When the concentration is a little bit higher, a mixture of beads and fiber will be formed (Liu et al. 2008). It is important to optimize the polymer solution concentration to form a very smooth and homogeneous fiber structure.

The polymeric material molecular weight also has an important effect of the morphologies of the electrospun fibers. Polymer with high molecular weight ensures that even at low concentration, the liquid droplets are able to stretch through the action of electric field (Wu et al. 2009). Furthermore, beads will form along the fibers even with the optimized concentration of the low molecular weight polymer solution. In principle, molecular weight reflects the entanglement of polymer chains in solution which is also referred to as solution viscosity. Suitable viscosity is required in the electrospinning process; a continuous and smooth fiber cannot be obtained under a very low viscosity whereas high viscosity solution results in the hard ejection of jets from the spinneret tip. It is important to note that viscosity of the solution is related to the polymer concentration and molecular weight. Under low viscosity solution, the surface tension of the electrospinning solution is the dominant factor for the formation of fibers (Amariei et al. 2017). The use of different solvents may contribute to different surface tensions for the electrospinning solution. Under fixed conditions such as viscosity of the solution, surface tension determines the upper and lower boundaries of the continuous fiber formation. Another factor that needs to be considered while using the electrospinning technique is the conductivity of the electrospinning solution under high voltage. Polymer is generally polyelectrolytic in nature where the ions increase the charge carrying ability of the polymer jet subjecting it to higher tension under the electrical field. Higher tension will contribute to the formation of poor fibers. To reduce the tension in the polymer jet and to reduce the fiber diameters, the electrical conductivity can be reduced by the addition of ionic salt (Yalcinkaya et al. 2015). Taking into account the conditions need to prepare the homogeneous and continuous fiber without defect, the electrospinning solution should be firstly optimized before it can be used in the electrospinning process.

11.6.2 Process Parameters

The setup for the electrospinning process is very simple and has been widely used. Applied voltage is the crucial factor in the electrospinning process. The applied voltage needs to be much higher than the threshold so that the charged jet can be ejected from the Taylor cone. Unfortunately, the effect of the applied voltage on the ejected fibers morphological structure is still under arguments; several groups reported different results on the relationship between the applied voltage and fiber diameters. Some studies reported that a high applied voltage creates a small fiber diameter and vice versa (Zhu et al. 2017). Based on the different arguments, it can be concluded that applied voltage does have an effect on the morphological structures of the fibers but the level of significance varies with another factor such as solution concentration or other process parameters.

Another factor that affects the fiber formation in the electrospinning process is the solution flow rate within the syringe. A low flow rate is recommended as the polymer solution will have enough time for polarization. A higher feed flow rate contributes to more beads in the fiber structure and fibers with larger diameters will form (Zargham et al. 2012). Tip-to-collector distance and type of collector used in the electrospinning process have less influence on the fiber properties compared to other factors. But, an optimum distance should be selected to provide enough time for solvent evaporation during the travel to the collector after the solution ejection from the polymer solution. The type of collector used will determine the final type of nanofibers that will be collected from the collector, either random or aligned nanofibers. Usually, a conductive substrate will be used as a fiber collector. Generally, aluminum foil is used but it is difficult to transfer the fiber formation to other substrates for different applications. Diverse collectors have been developed depending on the applications needed including wire mesh, membrane substrates, parallel or grid bars, and rotating rods.

11.6.3 Ambient Parameters

Ambient conditions are another parameter that affects the fiber formation; humidity and temperature of the ambient environment affect the morphological structure of the fiber. Increasing the ambient temperature will lead to a thinner fiber diameter while lower humidity may dry the solvent quickly. A high ambient humidity will lead to a larger fiber diameters, slower solvent evaporation, and smaller pore formation on the fibers surface (Nezarati et al. 2013). So, to produce the fibers with larger pores on the fibers surface, the ambient temperature should be controlled.

11.7 ELECTROSPUN NANOCOMPOSITE MEMBRANE

Nanofibers have promising applications in nanocomposite materials. The incorporation of nanofibers in composite membrane brings many advantages to fuel cell applications. Despite the excellent fuel cell performance from the integrated nanocomposite polymer electrolyte membrane, the techniques to develop the effective nanocomposite membrane are still challenging (Tripathi et al. 2010). There are a wide range of different modification methods available to develop nanocomposite membrane (Quartarone et al. 2017). There are three most widely used techniques to fabricate nanocomposite membrane which are physical blending, sol–gel, and infiltration methods.

11.7.1 Physical Blending

Recently, the incorporation of organic/inorganic electrospun nanofibers into the polymer matrix has been found to be an effective strategy to enhance properties of the polymer electrolyte membrane in fuel cell systems. In particular, the incorporation of organic/inorganic nanofibers of controlled size results in high performance for fuel cell applications. Physical blending is the most widely used and easiest way to develop the nanocomposite membrane consisting of organic/inorganic nanofibers. In this process, physical blending refers to the incorporation of a filler in any shape and size that is vigorously mixed with the polymer solution and subsequently cast. In a typical procedure, an appropriate amount of filler in nanofibers form is first mixed with the polymer solution under vigorous stirring. The membrane is then fabricated using film casting and dried until all the solvent is fully evaporated. Some studies used filler particles instead of nanofibers before mixing with the polymer solution to produce the electrospinning solution (Sethupathy et al. 2013). It is the simplest way to fabricate an electrospun nanocomposite membrane because the membrane thickness could be easily controlled. The physical blending technique to produce electrospun nanocomposite membrane is illustrated in Figure 11.3.

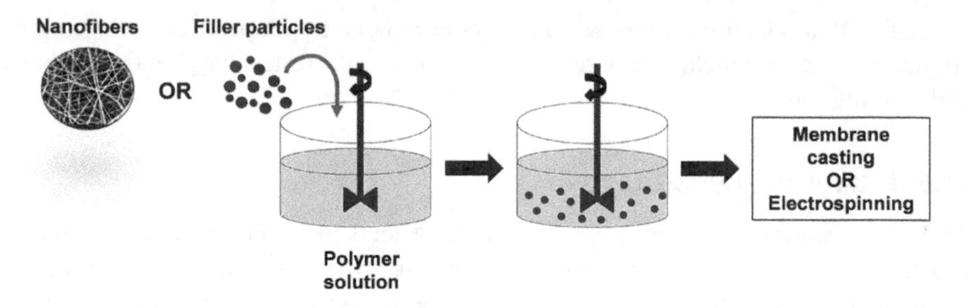

FIGURE 11.3 Physical blending method.

One of the limiting factors in blending nanofiber fillers in polymeric solution is that the aggregation/dispersion state of the nanoparticles needs to be controlled (Salarizadeh et al. 2017). The physical interaction between the nanofibers and polymer matrix such as inhomogeneity of the composite membrane represents difficulties. Larger nanoparticle sizes of more than 100 nm are difficult to disperse due to surface interactions and excessive loading of nanoparticles leads to agglomeration that reduces the active surface area for membrane water uptake (Schonvogel et al. 2017). Liu et al. (2017) employed this technique by blending electrospun sulfonated carbon nanofibers (SCNFs) into sulfonated poly (ether ether ketone) (SPEEK) solution to develop the electrospun SCNF/SPEEK nanocomposite membrane for DMFC. They reported that the incorporation of the SCNF into the SPEEK polymer matrix enhanced the mechanical strength and water uptake of the membrane. However, under high content of SCNF, the continuous proton conducting channel was terminated. This was related to dispersion of the filler into the polymer matrix where SCNF cannot disperse uniformly due to the slight aggregation of SCNF when the loading is too high.

11.7.2 Sol–Gel Method

Further work on the electrospun nanofibers used in polymer electrolyte membrane has been focused on through-plane ion conductivity and the improvement of the fuel cell performance. The sol–gel process is a versatile technique for the preparation of ion conducting nanocomposite materials (Donato et al. 2017). A direct mixing of the polymer and filler solution before casting or electrospinning leads to the formation of ionic conducting membranes. A nanocomposite structure is obtained at the molecular level with entirely different properties when using optimized proportions of the polymer and filler solutions. Figure 11.4 illustrates the preparation of nanocomposite membrane using a sol–gel method. The function of the electrospun nanofibers in the polymer matrix in the preparation of nanocomposite membrane has effectively increased the performance of the polymer electrolyte membrane in the fuel cell system. Park et al. (2017) studied the performance of electrospun Nafion/PVDF nanofiber membrane using the sol–gel technique before undergoing the hot pressing technique to develop a dense and defect-free membrane that is applicable in H_2/Br_2 fuel cells. The electrospun composite membrane fabricated was a very thin membrane of about 18 μm with a better cell membrane than MEA containing 55 μm Nafion membrane.

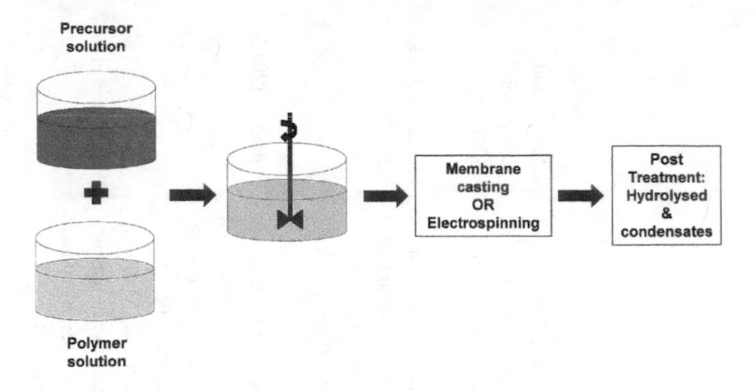

FIGURE 11.4 Sol–gel method.

11.7.3 Infiltration

The infiltration technique is another technique that has been used to prepare the electrospun nanocomposite membrane (Sood et al. 2016). The basic method using infiltration technique starts with the immersion of the polymer membrane or nanofiber mat in the filler or polymer solutions. After several hours, the membrane is taken out from the precursor solution followed by post-treatment to remove the excessive solvent and then dried. However, the impregnation of the polymer or nanofiber membrane experiences a difficulty in maintaining the concentration gradient of precursor solutions which limits the incorporation of filler or polymer concentration into the membrane to a certain degree (Ahlfield et al. 2017). This technique is illustrated in Figure 11.5.

The infiltration technique has also been applied to develop the electrospun nanocomposite membrane for the fabrication of polymer electrolyte membrane. Somehow, the ion conductive materials can be in the form of nanofibers or the polymer matrix itself. Abouzari-lotf et al. (2016) developed an electrospun nanocomposite membrane by immersing the Nylon-6,6 nanofibers mat into an ionic conductive solution which was phosphotungstic acid, $H_3PW_{12}O_{40}$. The prepared electrospun nanocomposite membrane offered comparable high ionic conductivity as well as mechanical and thermal stabilities compared to the commercial Nafion 112 (Abouzari-lotf et al. 2016). The summary on the development of the electrospun nanocomposite membrane using different techniques for fuel cell applications is shown in Table 11.3.

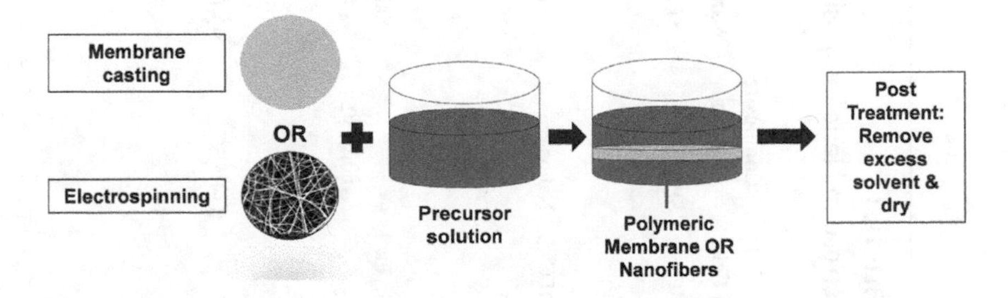

FIGURE 11.5 Infiltration method.

TABLE 11.3
Electrospun Nanocomposite Membrane for Fuel Cell Applications

Fuel Cells	Composite Material	Nanofiber Forms	Modification Method	T (°C)/RH (%)	Proton Conductivity (S/cm)	Remarks	Ref
PEMFC	SPI	SPI	Infiltration	80/98	0.37	A nanofibers composite consists of SPI nanofibers showed better fuel cell performance compared to that of the pristine SPI membranes.	Takemori and Kawakami (2010)
PEMFC	Nafion	SPI	Infiltration	27/24	0.0102	Sulfonic acid group in Nafion matrix attracted to the SPI nanofibers to construct acid-condensed layer that contributes to positive impact on proton conduction and water diffusion under low RH%.	Makinouchi et al. (2017)
H_2/Br_2 Fuel Cell	—	Nafion/PVDF	Sol–gel	27/-	0.043	Electrospun PVDF/Nafion membrane undergoes the hot pressing technique to build a dense and defect-free membrane. Single fiber membrane (18 μm) is significantly better than the MEA with commercial Nafion 212 (55 μm).	Park et al. (2017)
	Nafion/PVDF	—			0.033		
AFC	Q-PAES	Q-PAES	Infiltration	30/-	0.083	The nanofibers alone and within the composite membrane showed a very good performance as an anion exchange membrane.	Watanabe et al. (2017)

(Continued)

TABLE 11.3 (*Continued*)
Electrospun Nanocomposite Membrane for Fuel Cell Applications

Fuel Cells	Composite Material	Nanofiber Forms	Modification Method	T (°C)/RH (%)	Proton Conductivity (S/cm)	Remarks	Ref
DMFC	SPEEK	SCNF	Physical Blending	60/-	0.128	Increasing amount of SCNF improved the mechanical strength and water uptake.	Liu et al. (2017)
PEMFC	Nafion	Phy-PBI	Infiltration	80/40	0.0031	Nanofibers framework exhibited high gas barrier properties and distinguished membrane stabilities compared to recast Nafion membrane.	Tanaka et al. (2017)
AFC	Q-PAES	—	—	90/95	0.034	The aligned nanofibers had 10–15 times higher anion conductivity than the recast membrane.	Watanabe et al. (2016)
	—	Q-PAES			0.16		
DMFC	SPEEK	Titanium Oxides	Sol–gel	70/100	0.103	The SPEEK/TiNF composite membrane has higher proton conductivities than pristine SPEEK membrane due to the nature of hygroscopic substance of Ti that promotes water absorption for proton conductivity.	Dong et al. (2017)
PEMFC	Chitosan	SPEEK	Infiltration	25/100	0.481	Higher loading of the porous SPEEK nanofibers increased the proton conductivity of the composite membrane due to the presence of channels for proton transport.	Hou et al. (2018)

(*Continued*)

TABLE 11.3 (*Continued*)
Electrospun Nanocomposite Membrane for Fuel Cell Applications

Fuel Cells	Composite Material	Nanofiber Forms	Modification Method	T (°C)/RH (%)	Proton Conductivity (S/cm)	Remarks	Ref
PEMFC	$H_3PW_{12}O_{40}$ PTA	Nylon-6,6	Infiltration	30/100	0.1	Comparable with Nafion 112 under the same conditions.	Abouzari-lotf et al. (2016)
DMFC	$H_3PW_{12}O_{40}$ PTA/ Chitosan	PVDF	Infiltration	70/-	0.023	Chitosan acts as a filler to fill the porous structure of the PVDF/PWA nanofibers mat to get a dense and compact structure for PEMFC.	Gong et al. (2017)

11.8 SUMMARY

Fuel cells offer several advantages over conventional power sources including reduced dependence on fossil fuels, long useful life, high efficiency, relatively safe, essential zero toxicity, minimal maintenance costs, and carbon-free emission. The research with regard to the fuel cell system using polymer electrolyte membrane as an ion carrier in the fuel cell device in the past two decades has generated interest in improving the properties of polymer electrolyte membrane. Polymer electrolyte membrane-based electrospun nanocomposite membrane has triggered considerable interest due to excellent properties that arise from the both components of the composite material. Electrospun nanofibers are an excellent filler that have been used in nanocomposite-based polymer electrolyte membrane. The continuous fiber structure of the electrospun nanofibers offers a platform for the ion transportation channel in the nanocomposite membrane matrix. Thus, they increase the ion conductivity of the polymer electrolyte membrane that contributes to efficient fuel cell applications. Furthermore, the surface functional flexibilities of the electrospun nanofibers membrane is believed to increase the mechanical strength of the nanocomposite membrane. However, the technique to develop the electrospun nanocomposite membrane needs to be chosen carefully so as to minimize the loss of nanofiber functionalities in the nanocomposite membrane. It can be concluded that the incorporation of electrospun nanofiber membrane as part of nanocomposite-based polymer electrolyte membrane is a promising and interesting development path for the improvement of fuel cells via nanotechnology.

REFERENCES

Abdulkareem, A. S., Afolabi, A. S., Idibie, C. A., Iyuke, S. E., and Pienaar, H. C. v. Z. 2012. Development of Composite Proton Exchange Membrane from Polystyrene Butadiene Rubber and Carbon Nanoballs for Fuel Cell Application. *Energy Procedia* 14: 2026–2037.

Abouzari-lotf, E., Jacob, M. V., Ghassemi, H., Ahmad, A., Nasef, M. M., Zaker, M., and Mehdipour-Ataei, S. 2016. Enhancement of Fuel Cell Performance with Less-Water Dependant Composite Membranes Having Polyoxometalate Anchored Nanofibrous Interlayer. *Journal of Power Sources* 326: 482–489.

Ahlfield, J., Huang, G., Liu, L., Kaburagi, Y., Kim, Y., and Kohl, P. A. 2017. Anion Conducting Ionomers for Fuel Cell and Electrolyzers. *Journal of the Electrochemical Society* 164: 1648–1653.

Amariei, N., Manea, L. R., Bertea, A P., Bertea, A., and Popa, A. 2017. The Influence of Polymer Solution on the Properties of Electrospun 3D Nanostructures. *Materials Science and Engineering* 209: 012092.

Andujar, J. M., and Segura, F. 2009. Fuel Cells: History and Updating. A Walk along Two Centuries. *Renewable and Sustainable Energy Reviews* 13(9): 2309–2322.

Arges, C. G., Ramani, V., and Pintauro, P. N. 2010. Anion Exchange Membrane Fuel Cells. *The Electrochemical Society Interface, Summer* 2010: 31–35.

Bakangura, E., Wu, L., Ge, L., Yang, Z., and Xu, T. 2016. Mixed Matrix Proton Exchange Membranes for Fuel Cells: State of the Art and Perspectives. *Progress in Polymer Science* 57: 103–152.

Bjorge, D., Daels, N., Vrieze, S. D., Dejans, P., Camp, T. V., Audenaert, W., Hogie, J., Westbroek, P., de Clerck, K., and Hulle, S. W. H. 2009. Performance Assessment of Electrospun Nanofibers for Filter Applications. *Desalination* 249:942–948.

Chaurasia, P. B. L., Ando, Y., and Tanaka, T. 2003. Regenerative Fuel Cell with Chemical Reactions. *Energy Conversion & Management* 44: 611–628.

Chen, L., Chai, S., Liu, K., Ning, N., Gao, J., Liu, Q., Chen, F., and Fu, Q. 2012. Enhanced Epoxy/Silica Composites Mechanical Properties by Introducing Graphene Oxide to the Interface. *Applied Materials & Interfaces* 4(8): 4398–4404.

Couture, G., Alaaeddine, A., Boschet, F., and Ameduri, B. 2011. Polymeric Materials as Anion Exchange Membranes for Alkaline Fuel Cells. *Progress in Polymer Science* 36(11): 1521–1557.

Dodds, P. E., Staffell, L., Hawkes, A. D., Li, F., Grunewald, P., McDowall, W., and Ekins, P. 2015. Hydrogen and Fuel Cell Technologies for Heating: A Review. *International Journal of Hydrogen Energy* 40(5): 2065–2083.

Donato, K. Z., Matejka, L., Mauler, R. S., and Donato, R. K. 2017. Recent Applications of Ionic Liquids in the Sol-Gel Process for Polymer-Silica Nanocomposites with Ionic Interfaces. *Colloids and Interfaces* 1(5): 25.

Dong, C., Zhimin, H., Wang, Q., Zhu, B., Cong, C., Meng, X., and Zhou, Q. 2017. Facile Synthesis of Metal Oxide Nanofibers and Construction of Continuous Proton-Conducting Pathways in SPEEK Composite Membrane. *International Journal of Hydrogen Energy* 42: 25388–25400.

Eldin, M. S. M., Hashem, A. E., Tamer, T. M., Omer, A. M., Yossuf, M. E., and Saber, M. M. 2017. Development of Cross Linked Chitosan/Alginate Polyelectrolyte Proton Exchanger Membranes for Fuel Cell Applications. *International Journal of Electrochemical Science* 12: 3840–3858.

El-kharouf, A., Mason, T. J., Brett, D. L., and Pollet, B. G. 2012. Ex-situ Characterization of Gas Diffusion Layers for Proton Exchange Membrane Fuel Cells. *Journal of Power Sources* 218: 393–404.

Esfahani, H., Jose, R., and Ramakrishna, S. 2017. Electrospun Ceramic Nanofiber Mats Today: Synthesis, Properties and Applications. *Materials* 10: 1238.

Espiritu, R., Mamlouk, M., and Scott, K. 2016. Study on the Effect of the Degree of Grafting on the Performance of Polyethylene-Based Anion Exchange Membrane for Fuel Cell Application. *International Journal of Hydrogen Energy* 41(2): 1120–1133.

Femmer, R., Mani, A., and Wessling, M. 2015. Ion Transport through Electrolyte/polyelectrolyte Multi-layers. *Scientific Reports* 5: 11583. doi:10.1038/srep11583.

Gong, C., Liu, H., Zhang, B., Wang, G., Cheng, F., Zheng, G., Wen, S., Xue, Z., and Xie, X. 2017. High Level of Solid Superacid Coated Poly (Vinylidene Fluoride) Electrospun Nanofiber Composite Polymer Electrolyte Membranes. *Journal of Membrane Science* 535: 113–121.

Gu, W., Baker, D. R., Liu, Y., and Gasteiger, H. A. 2010. Proton Exchange Membrane Fuel Cell (PEMFC) Down-the-Channel Performance Model. In *Handbook of Fuel Cells: Fundamentals, Technology and Applications* (Vielstich, W., Ed.), John Wiley & Sons, Ltd, Hoboken, NJ.

Hanemann, T., and Szabo, D. V. 2010. Polymer-Nanoparticle Composites: From Synthesis to Modern Applications. *Materials* 3: 3468–3517.

Homaeigohar, S., and Elbahri, M. 2014. Nanocomposite Electrospun Nanofiber Membranes for Environmental Remediation. *Materials* 7: 1017–1045.

Hong, J. G., Glabman, S., and Chen, Y. 2015. Effect of Inorganic Filler Size on Electrochemical Performance of Nanocomposite Cation Exchange Membranes for Salinity Gradient Power Generation. *Journal of Membrane Science* 482: 33–41.

Hou, C., Zhang, X., Li, Y., Zhou, G., and Wang, J. 2018. Porous Nanofibers Composite Membrane for Unparalleled Proton Conduction. *Journal of Membrane Science* 550: 136–144.

Huang, Z. M., Zhang, Y. Z., Kotaki, M., and Ramakrishna, S. 2003. A Review on Polymer Nanofibers by Electrospinning and Their Applications in Nanocomposites. *Composites Science and Technology* 63: 2223–2253.

Jeon, I. Y., and Baek, J. B. 2010. Nanocomposites Derived from Polymers and Inorganic Nanoparticles. *Materials* 3(6): 3654–3674.

Ji, M., and Wei, Z. 2009. A Review of Water Management in Polymer Electrolyte Membrane Fuel Cells. *Energies* 2: 1057–1106.

Kenry and Lim, C. T. 2017. Nanofiber Technology: Current Status and Emerging Developments. *Progress in Polymer Science* 70: 1–17.

Kirubakaran, A., Jain, S., and Nema, R. K. 2009. A Review on Fuel Cell Technologies and Power Electronic Interface. *Renewable and Sustainable Energy Reviews* 13: 2430–2440.

Klaysom, C., Moon, S. H., Ladewig, B. P., Lu, G. Q. M, and Wang, L. 2011. The Influences of Inorganic Filler Particle Size on Composite Ion-Exchange Membranes for Desalination. *The Journal of Physical Chemistry C* 115(31): 15124–15132.

Kucernak, A. R., and Toyoda, E. 2018. Studying the Oxygen Reduction and Hydrogen Oxidation Reactions under Realistic Fuel Cell Conditions. *Electrochemistry Communications* 10(11): 1728–1732.

Kyu, T., and Nazir, N. A. 2013. Supramolecules Impregnated Proton Electrolyte Membranes. *Current Opinion in Chemical Engineering* 2(1): 132–138.

Liu, S., Liu, B., Nakata, K., Ochiai, T., Murakami, T., and Fujishima, A. 2012. Electrospinning Preparation and Photocatalytic Activity of Porous TiO_2 Nanofibers. *Journal of Nanomaterials* 2012:1–5.

Liu, X., Yang, Z., Zhang, Y., Li, C., Dong, J., Liu, Y., and Cheng, H. 2017. Electrospun Multifunctional Sulfonated Carbon Nanofibers for Design and Fabrication of SPEEK Composite Proton Exchange Membranes for Direct Methanol Fuel Cell Applications. *International Journal of Hydrogen Energy* 42: 10275–10284.

Liu, Y., He, J. H., Yu, J. Y., and Zeng, H. M. 2008. Controlling Numbers and Sizes of Beads in Electrospun Nanofibers. *Polymer International* 57: 632–636.

Makinouchi, T., Tanaka, M., and Kawakami, H. 2017. Improvement in Characteristics of a Nafion Membrane by Proton Conductive Nanofibers for Fuel Cell Applications. *Journal of Membrane Science* 530: 65–72.

Miyake, J., Taki, R., Mochizuki, T., Shimizu, R., Akiyama, R., Uchida, M., and Miyatake, K. 2017. Design of Flexible Polyphenylene Proton Conducting Membrane for Next Generation Fuel Cells. *Science Advances* 3(10): eaao0476.

Nezarati, R., Eifert, M. B., and Hernandez, E., C. 2013. Effects of Humidity and Solution Viscosity on Electrospun Fiber Morphology. *Tissue Engineering: Part C* 19(10): 810–819.

Nomnqa, M., Omoregbe, D. I., and Rabiu, A. 2016. Parametric Analysis of a High Temperature PEM Fuel Cell Based Microcogeneration System. *International Journal of Chemical Engineering* 4596251: 14.

Park, J., Pasaogullari, U., and Bonville, L. J. 2016. A New Membrane Electrode Assembly Structure with Novel Flow Fields for Polymer Electrolyte Fuel Cells. *Advancing Solid State and Electrochemical Science and Technology* 75(14): 55–62.

Park, J. W., Wycisk, R., Lin, G., Chong, P. Y., Powers, D., Nguyen, T. V., Dowd Jr., R. P., and Pintauro, P. N. 2017. Electrospun Nafion/PVDF Single-Fiber Blended Membranes for Regenerative H_2/Br_2 Fuel Cells. *Journal of Membrane Science* 541: 85–92.

Quartarone, E., Angioni, S., and Mustarelli, P. 2017. Polymer and Composite Membranes for Proton Conducting, High Temperature Fuel Cells: A Critical Review. *Materials* 10(7): 687.

Rangel-Cardenas, A. L., and Koper, G. J. M. 2017. Transport in Proton Exchange Membranes for Fuel Cell Applications—A systematic Non-equilibrium Approach. *Materials* 10: 576. doi:10.3390/ma10060576.

Sahu, A. K., Pitchumani, S., Sridhar, P., and Shukla, A. K. 2009. Nafion and Modified-Nafion Membranes for Polymer Electrolyte Fuel Cells: An Overview. *Bulletin of Material Science* 32(3): 285–294.

Sahu, I. P., Krishna, G., Biswas, M., and Das, M. K. 2014. Performance Study of PEM Fuel Cell under Different Loading Conditions. *Energy Procedia* 54: 468–478.

Salarizadeh, P., Javanbakht, M., and Pourmahdian, S. 2017. Enhancing the Performance of SPEEK Polymer Electrolyte Membranes Using Functionalized TiO_2 Nanoparticles with Proton Hoping Sites. *RSC Advances* 7: 8303–8313.

Schmitt, F., Granet, R., Sarrazin, C., Mackinzie, G., and Krausz, P. 2011. Synthesis of Anion Exchange Membranes from Cellulose: Crosslinking with Diiodobutane. *Carbohydrate Polymers* 86(1): 362–366.

Schonvogel, D., Hulstede, J., Wagner, P., Ktuusenberg, I., Tammeveski, K., Dyck, A., Agert, C., and Wark, M. 2017. Stability of Pt Nanoparticles on Alternative Carbon Supports for Oxygen Reduction Reaction. *Journal of the Electrochemical Society* 164(9): 995–1004.

Sebastian, D., and Baglio, V. 2017. Advanced Materials in Polymer Electrolyte Fuel Cells. *Materials* 10: 1163. doi:10.3390/ma10101163.

Sethupathy, M., Sethuraman, V., and Manisankar, P. 2013. Preparation of PVDF/SiO_2 Composite Nanofiber Membrane Using Electrospinning for Polymer Electrolyte Analysis. *Soft Nanoscience Letters* 3: 37–43.

Sood, R., Cavaliere, S., Jones, D. J., and Roziere, J. 2016. Electropaun Nanofiber Composite Polymer Electrolyte Fuel Cell and Electrolysis Membranes. *Nano Energy* 26: 729–745.

Staiti, P. 2001. Proton Conductive Membranes Based on Silicotungstic Acid/Silica and Polybenzimidazole. *Materials Letters* 47(4–5): 241–246.

Sui, S., Wang, X., Zhou, X., Su, Y., Riffat, S., and Liu, C. J. 2017. A Comprehensive Review of Pt Electrocatalysts for the Oxygen Reduction Reaction: Nanostructure, Activity, Mechanism and Carbon Support in PEM Fuel Cells. *Journal of Materials Chemistry A* 5: 1808–1825.

Tada, M., Uruga, T., and Iwasawa, Y. 2015. Key Factors Affecting the Performance and Durability of Cathode Electrocatalysts in Polymer Electrolyte Fuel Cells Characterized by In Situ Real Time and Spatially Resolved XAFS Techniques. *Catalysis Letters* 145: 58–70.

Takemori, R., and Kawakami, H. 2010. Electrospun Nanofibrous Blend Membranes for Fuel Cell Electrolytes. *Journal of Power Sources* 195: 5957–5961.

Tanaka, M., Takeda, Y., Wakiya, T., Wakamoto, Y., Harigaya, K., Ito, T., Tarao, T., and Kawakami, H. 2017. Acid-Doped Polymer Nanofiber Framework: Three-Dimensional Proton Conductive Network for High Performance Fuel Cells. *Journal of Power Source* 342: 125–134.

Thepkaew, J., Therdthianwong, A., and Therdthianwong, S. 2008. Key Parameters of Active Layers Affecting Proton Exchange Membrane (PEM) Fuel Cell Performance. *Energy* 33(12): 1794–1800.

Tripathi, B. P., Kumar, M., and Shahi, V. K. 2010. Organic-Inorganic Hybrid Alkaline Membranes by Epoxide Ring Opening for Direct Methanol Fuel Cell Applications. *Journal of Membrane Science* 360(1–2):90–101.

U.S Department of Energy, Comparisons of Fuel Cell Technologies. 2016. October 30, 2017. Available from:https://energy.gov/sites/prod/files/2016/06/f32/fcto_fuel_cells_comparison_chart_apr2016.pdf.

Wang, M., Liu, G., Cui, X., Feng, Y., Zhang, H., Wang, G., Zhong, S., and Luo, Y. 2018. Self-Crosslinked Organic-Inorganic Nanocomposite Membranes with Good Methanol Barrier for Direct Methanol Fuel Cell Applications. *Solid State Ionics* 315: 71–76.

Wang, Y., Chen, K. S., Mishler, J., Cho, S. C., and Adroher, X. C. 2011. A Review of Polymer Electrolyte Membrane Fuel Cells: Technology, Applications, and Needs on Fundamental Research. *Applied Energy* 88: 981–1007.

Watanabe, T., Tanaka, M., and Kawakami, H. 2016. Fabrication and Electrolyte Characterization of Uniaxially-Aligned Anion Conductive Polymer Nanofibers. *Nanoscale* 8: 19614–19619.

Watanabe, T., Tanaka, M., and Kawakami, H. 2017. Anion Conductive Polymer Nanofiber Composite Membrane: Effects of Nanofibers on Polymer Electrolyte Characteristics. *Polymer International* 66(3): 382–387.

Wee, J. H. 2007. Applications of Proton Exchange Membrane Fuel Cell Systems. *Renewable & Sustainable Energy Reviews* 11: 1720–1738.

Wu, N., Shao, D., Wei, Q., Cai, Y., and Gao, W. 2009. Characterization of PVAc/TiO_2 Hybrid Nanofibers: From Fibrous Morphologies to Molecular Structures. *Journal of Applied Polymer Science* 112(2008):1481–1485.

Yalcinkaya, F., Yalcinkaya, B., and Jirsak, O. 2015. Influence of Salts on Electrospinning of Aqueous and Nanoaqueous Polymer Solutions. *Journal of Nanomaterials* 2015, Article ID 134251: 12.

Zargham, S., Bazgir, S., Tavakoli, A., Rashidi, A. S., and Damerchley, R. 2012. The Effect of Flow Rate on Morphology and Deposition Area of Electrospun Nylon 6 Nanofiber. *Journal of Engineered Fibers and Fabrics* 7(4): 42–49.

Zhang, X., Xu, S., and Han, G. 2009. Fabrication and Photocatalytic Activity of TiO_2 Nanofiber Membrane. *Materials Letters* 63(21):1761–1763.

Zhou, F. L., and Gong, R. H. 2008. Review Manufacturing Technologies of Polymeric Nanofibers and Nanofiber Yarns. *Polymer International* 57:837–845.

Zhu, G., Zhao, L. Y., Zhu, L. T., Deng, X. Y., and Chen, W. L. 2017. Effect of Experimental Parameters on Nanofiber Diameter from Electrospinning with Wire Electrodes. *Materials Science and Engineering* 230: 012043.

Zielke, L., Vierrath, S., Moroni, R., Mondon, A., Zengerle, R., and Thiele, S. 2016. Three-Dimensional Morphology of the Interface between Micro Porous Layer and Catalyst Layer in a Polymer Electrolyte Membrane Fuel Cell. *RSC Advances* 84: 80700–80705.

12 Proton Transport Mechanisms in Nanofibers Ion Exchange Membrane

State of the Art and Perspectives

Nuha Awang, Ahmad Fauzi Ismail, Juhana Jaafar, Mohd Hafiz Dzarfan Othman, and Mukhlis A. Rahman

CONTENTS

12.1 INTRODUCTION

Proton conductivity is a crucial mechanism in many processes for example the photosynthesis and the production of electricity in hydrogen fuel cells. Since a proton has no electron shell of its own, it is able to interact strongly with the electron density around itself, which at that point takes some of the H (1s) characteristics Jannasch (2003), Murali and Eisenberg (1988). Nevertheless, in non-metallic compounds, proton interacts strongly with the closest valence electron. In the case of single oxygen, being separated from other electronegative species would cause the formulation of an O–H bond which is under 100 pm long contrasted with ~140 pm for the ionic "radius" of the oxide ion (Gilpa and Hogarth, 2001, Kerres et al., 1998; Kreuer et al., 1988).

The proton acquires its equilibrium position deeply embedded in the oxygen valence electron (Figure 12.1a). For medium distance between oxygen and oxygen

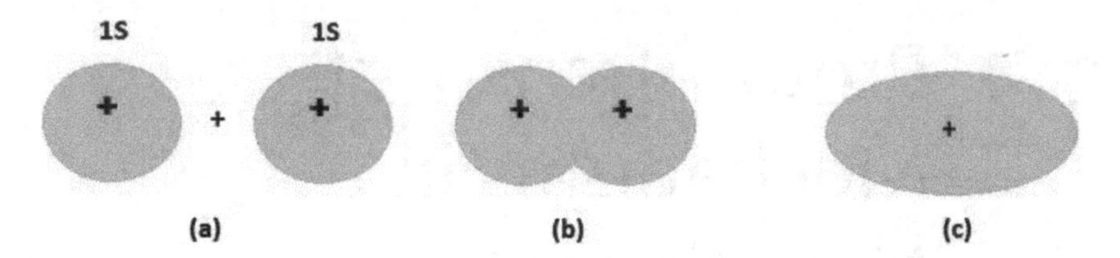

FIGURE 12.1 Proton binding cases, where the proton is interacted to one or two basic species: (a) Long oxygen separation of ~140 pm, (b) medium oxygen separation of ~250–280 pm, (c) short oxygen separation of ~240 pm.

(~250–280 pm) in a case of asymmetrical hydrogen bond (O–H•••O), the proton might be engaged with two bonds: a short, solid bond with the supposed proton donor, and a more drawn out, weaker bond with a proton acceptor (Figure 12.1b). For short oxygen separations (~240 pm), a symmetrical hydrogen bond might be formed, for instance the proton is involved in two equivalent bonds (Figure 12.1c) (Park, 2005, Park and Yamazaki, 2005).

Hydrogen bonds are an interaction type which is a typical element of proton conductors. In the presence of other stronger covalent or ionic bonds, hydrogen bonds tend to have less fluctuation. The proton acceptor or donor distance decides the hydrogen-bond interaction, and is then more confined by the stronger bonds of both proton acceptor and donor (Park, 2005). The assistant hydrogen bonds need to adjust to a structure, which is dominated by stronger interactions.

The materials with the highest proton diffusivity are hydrogen-bonded liquids or solids in which weak or medium hydrogen-bond interactions are not or just possibly restricted by the region of different type of bonds. A vital part of hydrogen bonding is to give a way for proton exchange from a proton donor to a proton acceptor, and the part of proton conductor is played by materials that can be a proton conductor especially for ion exchange membrane (Park and Yamazaki, 2005).

Ion exchange membranes are thin sheets of ion exchange materials that can be utilized to permit the transportation of cations or anions and separate the ions (Juda and McRae, 1950). Ion selective membranes in sheet shape with low electrical resistance, high selectivity, good mechanical and chemical properties were developed in 1950s (Moilanen et al., 2008). In the 1970s, DuPont created a cation exchange membrane from sulfonated polytetrafluoroethylene which has good chemical stability known as Nafion®. The Nafion® membrane is utilized for vital energy storage in the chlor-alkali generation industry and energy conversion industry (fuel cell) (Park et al., 2008).

To date, Nafion® membrane has been widely applied as a fundamental component in fuel cells due to its good chemical resistance and electrochemical properties (Awang et al., 2015). Perfluorinated vinyl ethers pendant side chains which are terminated by a sulfonate ionic group is an essential feature of the polytetrafluoroethylene backbone in Nafion® (Smitha et al., 2005). Nanophase separation occurs between hydrophilic ionic and hydrophobic matrix domains in the hydrated Nafion® membrane due to the amphiphilic composition (Peeters et al., 1998). The membrane must be completely hydrated for good proton conductivity to work at temperatures lower than 80°C. However, the working temperature must be increased over 100°C for transportation and to restrict the harming of anode catalysts by the existence of CO, and subsequently empower high-energy utilization applications (Xu, 2005).

Nafion® membrane demonstrates disadvantages and drastic modification which is caused by water dissipation. This leads Nafion® membrane to have low conductivity and the impact can be identified with the dehydration of initial ionic domains. Hence, economically accessible Nafion® membranes currently do not satisfy the criteria of a high performance fuel cell that can be operated over 100°C (Kariduraganavar et al., 2012).

There are two main mechanisms which portray proton diffusion in a manner that the proton stays protected by some electron density along the whole diffusion path. One of the well-known principal mechanism is the principal situation where "vehicles" show to claim local dynamics yet dwell on their destinations, the protons being exchanged inside of the hydrogen bonds from one "vehicle" to another. Extra redesign of proton environment involves reorientation of individual species or significantly more expanded gatherings that result in the formation of continuous trajectory for proton migration. This mechanism is often termed as the Grotthuss mechanism (Besse et al., 2002, Genies et al., 2001a, 2001b).

12.2 TRANSPORT OF PROTONS IN CATION EXCHANGE MEMBRANE

The transport of protons in cation exchange membrane is significantly influenced by the existence of water molecules. The water molecules provide the hydrogen bond network in which water is kept in the hydrophilic region. In this manner, water gives a superior medium for protons to move freely compared to other common ions. This is the consequence of the way that proton movement in water does not happen through normal diffusion, but rather by means of a procedure where the hydrogen bonds between water molecules are changed over into covalent bonds and the other way around (Ohya et al., 1995).

The movement of protons in water happens in two distinct processes. The first is the free solution-diffusion process as presented in Figure 12.2a, in which the protons and related water of hydration diffuse through the water phase. The second process is designated "proton hopping" in which the protons move in successive advances including the development and breakage of hydrogen holding of a water molecule series. The procedure is also known as the Grotthuss mechanism, Figure 12.2b. The procedure includes proton hopping from H_3O^+ to a neighboring H_2O atom, which thus discharges one of the protons to form a hydrogen bond with a neighboring proton. Along these lines, a proton "hops" starting with one water molecule then onto the next. Subsequently, the progressive proton hops prompt a compelling proton transport in water. In the Grotthuss mechanism, the first proton that enters the membrane is not the proton that leaves the membrane. This is due to the fact that it is not simply the proton that is transported; just the charge of the proton is transported, and not the mass. The charge is fundamentally passed on from one water molecule to another water molecule (Baradie et al., 1998, Antonucci et al., 1999).

Proton transport through a hydrated membrane can happen by means of the two processes specified previously. For proton transportation in a membrane, the water in the membrane itself can be comprehended as bound (or "non-freezable,"

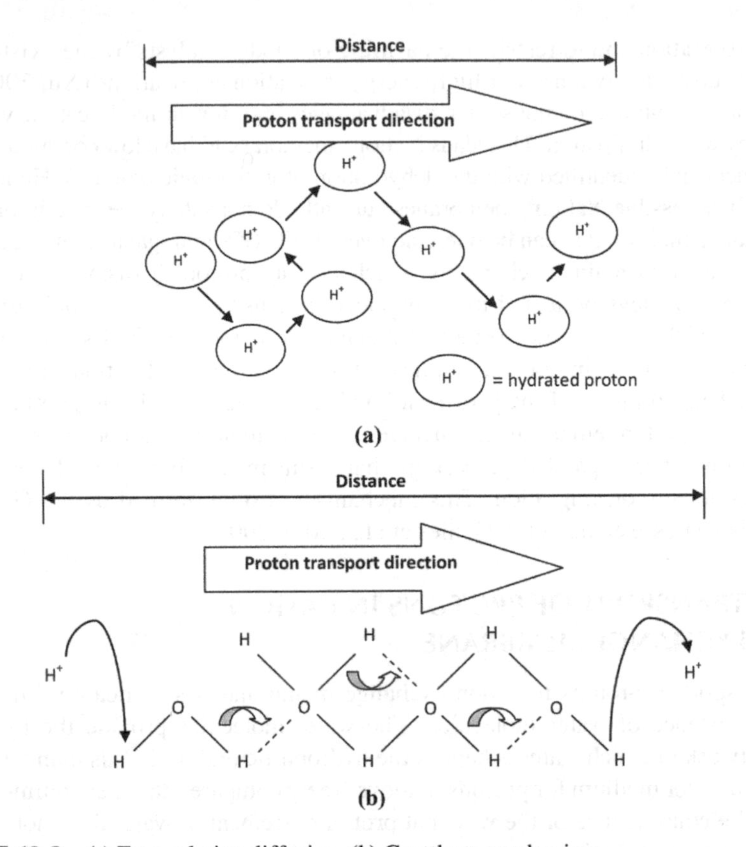

FIGURE 12.2 (a) Free solution diffusion, (b) Grotthuss mechanism.

"associated") water and free ("freezable," "bulk like") water. The free water takes after bulk water with hydrogen bonds as the major interactions.

Bound water molecules are in the region of the materials while the free water is further away. The bonding strength between the bound water and the materials is more prominent than that of hydrogen bonding in free water. The free water is bulk-like; however, the bound water is believed to be part of the materials. The transportation of protons through a hydrated membrane can be related to both free and bound water. However, the free diffusion is much slower in bound water. On the other hand, the proton hopping or the Grotthuss mechanism can occur in both free and bound water. This type of transportation in ion exchange membrane explains the ion transport in polyelectrolyte membranes (Baradie et al., 1998).

12.3 ENHANCING PROTON CONDUCTIVITY BY MODIFYING THE MEMBRANE MORPHOLOGY BY ELECTROSPINNING

Electrospinning is one of the most straightforward, adaptable, and advanced processing technique for fabricating nanofibers (Teo and Ramakrishna, 2006). The process is grouped into a couple of strategies, for instance magneto-electrospinning vibration-electrospinning, bubble electrospinning, and siro-electrospinning. The principal patent of electrospinning was issued by Formhals in October 1934 (Formhals, 1934). In spite of the fact that the procedure has been generally connected to fuel cell technology for about 70 years, the utilization of electrospun nanofibers as a part of delivering membrane for fuel cells is still new (Jaafar et al., 2017).

The electrospinning process became widely known in the sixteenth century when William Gilbert recorded the first electrostatic attraction of a liquid. In 1846, nitrate cellulose was developed by Christian Friedrich Schonbein. In 1900, the first electrospinning patent was filled by John Francis Cooley (Awang et al., 2017b).

In the work presented by Formhals and Zenely on polymers, compared to conventional techniques, for example, melt spinning, wet spinning, and dry spinning, electrospinning can create smaller pore size and larger specific surface area with the diameter of fiber ranging from 10 to 1000 nm; whilst conventional systems can only deliver fibers with a diameter in the range of 5–500 μm (Awang et al., 2015).

In the electrospinning technique for polymers, high voltage electrostatic field is associated with the charged surface of the polymer solution bead and in this manner will incite the launch of a liquid stream through a spinneret. Because of the present conditions, the electrostatic force overcomes the surface tension of the bead and the development of the Taylor cone happens when the solution is leaving the tip of the spinneret forming the charged jet (Awang et al., 2017a). The development of the Taylor cone is influenced by the associated voltage. The voltage will keep on increasing until the point when the balance condition is achieved between the surface tension and the electrostatic force (Figure 12.3). The electric field controls the course of charge stream and the solidified spun fibers are accumulated on the stationary or turning conductive collector (Tazi and Savadago, 2001). A setup for electrospinning is presented in Figure 12.4.

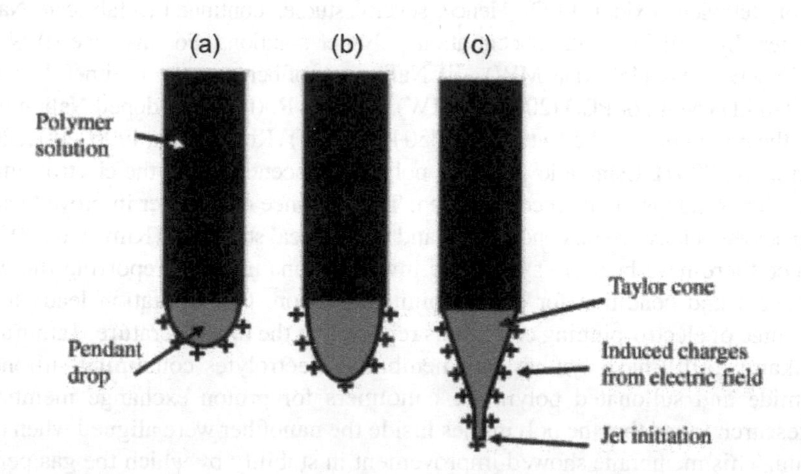

FIGURE 12.3 Development of the Taylor cone: (a) Pendant drop at initial stage, (b) pendant drop developed to become a Taylor cone, (c) Pendant drop is completely turned to a Taylor cone.

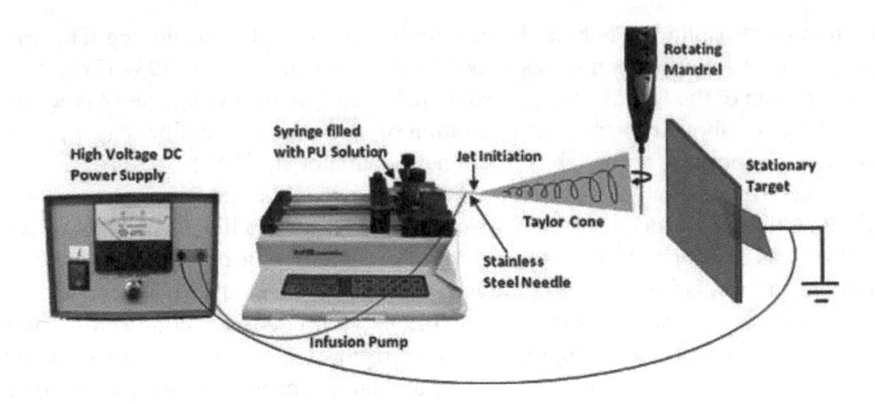

FIGURE 12.4 Electrospinning setup.

12.4 SPUN POLYMER BASED NANOFIBER MEMBRANES

Nanofibers have been made from hundreds of different polymers (Rikukawa and Sanui, 2000); however, only a few studies focus on the effect of polymer-based membrane electrospun fibers in fuel cell applications.

Electrospinning of Nafion perfluorosulfonic acid (PFSA) polymer has gained interest recently. Unfortunately, Nafion polymer electrospinning is difficult to achieve; Nafion does not totally dissolve in water or alcohol solution, and in common organic solvents, for example dimethylacetamide and dimethylformamide, forms micellar dispersion. Earlier attempts at electrospinning Nafion fibers from commercial solutions failed due to deficient polymer chain entanglements, in which case Nafion just electrosprayed as beads (Kolde et al., 1995).

Studies found that Nafion must be electrospun by including a high molecular weight carrier polymer, for example, poly (vinyl alcohol) (PVA), poly (acrylic acid) (PAA), and poly (ethylene oxide) (PEO). Hence, several studies continued to fabricate Nafion nanofibers by utilizing high-concentration polymer solutions. For instance (i) Nafion with 25 wt% PAA (450 kDa MW), (ii) Nafion nanofiber mats with either 16.7 wt% PVA (110 kDa MW) or PEO (200 kDa MW), and (iii) Ru(bpy)3 2^{+}-doped Nafion mats, where the fibers contained 29 wt% PAA (450 kDa MW) (Kerres et al., 1996, 2002, 2003, Arnett et al., 2007). Using a low relative polymer concentration in the electrospinning process for Nafion is highly recommended. The presence of polymer improved Nafion properties especially proton conductivity and mechanical stabilities (Kim et al., 2004).

Since there have been no systematic investigations made in reporting the exact parameters and condition for electrospinning Nafion, this limitation leads to the wide range of electrospinning conditions reported in the open literature. Tamura and Kawakami (2010) have synthesized membrane electrolytes containing sulfonated polyimide and sulfonated polyimide nanofibers for proton exchange membrane. The research found that the polyimides inside the nanofiber were aligned when electrospun. This membrane showed improvement in stability by which the gas permeability decreased while the aligned fibers increased. The proton conductivity of the membrane also showed a higher value in contrast to the membrane with nanofibers in perpendicular alignment (which was prepared through a casting technique).

Pan et al. (2008) concentrated on adding nanowire-based structures in the application of smaller scale fuel cells. Nafion/poly (vinyl pyrrolidone) (PVP) nanowires (NPNWs) were blended by electrospinning. 1.27 g of Nafion (E.I.DuPont Company, identical weight (EW) = 1100), 0.26 g of PVP (Sigma, MW~1300 000), and 2.1 mg tetramethylammonium chloride (Sigma) in 0.7 g ethanol were prepared as the antecedent polymer for electrospinning. Plastic syringes with stainless-steel needles were used with a 10 cm separation between the needle tip and the collector. The voltage connected was 16 kV. The use of PVP as a part of this study was devoted to high sub-atomic weight PVP. It was kept in mind that the end goal was to effectively electrospin the Nafion with no commitment on proton conductivity.

Electrospun Nafion itself is difficult to create alone because it is not soluble in most normal solvents and thus will inevitably lead to the arrangement of micelles, which diminish the chain entrapment (Li et al., 2003). Pan et al. (2008) utilized two conductive substrates (silicon) isolated by a void gap. The study found that the transportation of protons to reach the cathode in NPNWs turned out to be more effective, hence the proton conductivity of NPNWs was found to be bigger due to the Taylor cone effects which affected the structure of the electrospun fibers. They proposed that the proton conductivity can be improved by adjusting the distance across the NPNWs to below 2.3 μm. The other findings on the electrospun fiber polymer-based membrane for fuel cell application are outlined in Table 12.1.

Most of the previous studies have focused on polymer composite proton exchange membranes and the resulting membrane performance is dominant for proton conductivity rather than fuel barrier properties. A direct methanol fuel cell has shown a significant drawback on power density and efficiency if it is compared to a polymer exchange membrane fuel cell which operates with hydrogen due to the methanol crossover from anode to cathode (Genies et al., 2001b, Guo et al., 2002).

Another approach (Table 12.2) that concerns the preparation of new electrolyte composite membrane based on proton conducting materials has been investigated in the past few years for its good proton conductivity as well as its methanol barrier properties. These new electrolyte composite membranes consist of dispersion fillers, such as silica, heteropolyacid, zirconium phosphate, etc. within the polymer.

According to Kreuer (1995), sulfonated polyetherketones (PEEKK and PEEK) are durable under fuel cell operating conditions which are applied for several thousand hours. A hydrophilic characteristic is needed since the presence of water molecules can facilitate proton movement as well as increase the conductivity of the solid electrolytes, which is crucial in fuel cell applications (Lassegues et al., 2001). In addition, clay minerals have been used as a reinforcer in polymer membranes to improve the methanol crossover and water permeability in direct methanol fuel cells (DMFC) (Pu et al., 2001). Recently, sulfonated poly (ether ether ketone) has been electrospun and incorporated with SiO_2 as a supported proton exchange membrane for fuel cell applications (Liang et al., 2007). In this study, researchers impregnated SiO_2/SPEEK (sulfonated poly (ether ether ketone) nanofiber mat with Nafion solution in order to get a dense membrane for PEMFC applications. The potential of the impregnated SiO_2/SPEEK nanofiber mat with Nafion had showed the good properties as PEM for DMFC application.

Jaafar (2006) produced polymer-inorganic membrane consisting of SPEEK and Cloisite for DMFC application without the contribution of electrospinning process.

TABLE 12.1
Findings on Electrospun Fiber Polymer-Based Membrane for Fuel Cell Application

Based Polymer	Filler/Carrier Polymer	Proton Conductivity (mS cm^{-1})
Poly(vinyl alcohol)	Nafion®	22
Polyvinylidene fluoride	Nafion®	2
Sulfonated polyethersulfone	Nafion®	~85
Sulfonated random copolyimide	Sulfonated polyimide	Up to 370
Bromomethylated sulfonated polyphenylene oxide (BPPO)	Sulfonated poly(2,6-dimethyl-1,4-phenylene oxide) (SPPO)	30–80
3M perfluorosulfonc acid polymer	PEO	55
Nafion®	5wt% PVA or PEO	8.7–16
3M perfluorosulfonc acid polymer	PAA	498
Polymerized ionic liquid	Poly(MEBIm-BF4), PAA	7.1 × 10^{-4}
Sulfonated poly(ether ether ketone)	None	37 (solvent DMF) 41 (solvent DMAc)
Sulfonated poly(arylene ether sulfone)	None	86
	sulfonated polyhedral oligomeric silsesquioxane (sPOSS)	94
Sulfonated copolyimide		~100
Polyvinylidene	Phosphotungstic acid (PWA, up to 12.8 wt%)	~0.4
AquivionTM	PEO (Mw 1 × 106)	66
Sulfonated Zro2	PVP, poly(2-acrylamido-2-methylpropanesulfonic acid) (pAMPS)	240

Source: Li, L. et al., *J. Membr. Sci.*, 226, 159–167, 2003.

TABLE 12.2
Studies on Modification of PEM Based on Proton Conducting Materials

Approach	Purpose
1. Modifying perfluorinated ionomer membrane/preparing acid-base blends	To improve water retention properties at high temperature (>100°C)
2. Modifying ionomer membrane	To improve conductivity
3. Preparing new electrolyte composite membranes based on proton conducting materials	To improve the properties of polymer electrolyte membrane as desired properties especially the barrier properties and mechanical stability of the two components which can be combined into one composite

Source: Bozkurt, A. and Meyer, W.H., *Solid State Ion.*, 138, 259–265, 2001.

The membrane was found to reduce the methanol crossover problem as compared to pristine SPEEK and Nafion 112 membranes. The research demonstrated the advantages of clay composite over other nano-sized filler like nanofibers. They found that the surface area to volume ratio (A/V) for nanofibers was two times higher than that of a clay layer; the nanofiber would have a higher surface area of filler exposed to the polymer matrix, and thus a higher reinforcing ability compared to the clay layers. However, when nanofibers are oriented strongly in a particular direction in the composite, the clay layers dominate the reinforcement of the composite biaxially. In preventing the transport of gasses or fluid through the composite in a particular direction, a large amount of filler surface area is required; and for this highly oriented filler, the area would be significantly larger in a clay composite. This gave a rationale the research to prepare a polyethylene-clay nanocomposite. Depending on the clay quantity and the dispersion state, clay can act as a nucleating agent as well as an obstacle to the polymer mobility. However, electrospinning process favors the elongation of the chains and the ordering of the polymer (Doğan et al., 2011).

Smectites, which are a family of both montmorillonite (MMT) and hectorites, are a valuable mineral for industrial applications due to their high cation exchange capacity, high surface area, high surface reactivity, and high barrier property compared to others natural clays (Doğan et al., 2011). Organoclays, Cloisite 15 A® is a material prepared from montmorillonite (MMT) and cation di-tallow as mentioned. Tallow is the mixture of octadecyl (>60%), hexadecyl, and tetradecyl. Cloisite is an important additive in polymer nanocomposite membrane, as it has been prepared from MMT, processing its beneficial properties, thus and it is expected that the compatibility with the organic polymers will be enhanced (Jaafar et al., 2011). In order to produce an exfoliated polymer-clay nanocomposite membrane, the electrospinning process will be employed. SPEEK and Cloisite 15 A® will be used as the base materials in dope solution. The Cloisite 15 A® is expected to exfoliate well within the polymer matrix via electrospinning process; the crystalline structure of the clay will hinder the polymer movement as well as acting as a barrier for methanol crossover (Elabd et al., 2003).

12.5 CURRENT STUDIES AND FUTURE DIRECTIONS

A study by Jaafar et al. (2007) investigated SPEEK membrane prepared using a mixture of fuming sulfuric acid as the sulfonating agent. It exhibited higher proton conductivity and lower methanol permeability as compared to commercial ion exchange membrane particularly for direct methanol fuel cell (DMFC) applications. Further study on SPEEK has been carried out by Jaafar et al. (2009) by incorporating 2,4,6-triaminopyrimidine (TAP) as a compatibilizer to improve the compatibility between SPEEK and Cloisite 15 A® (Jaafar et al., 2009). The intercalated SPEEK/Cloisite 15 A®/TAP successfully improved the membrane barrier properties due to the unique features of Cloisite 15 A® which contributed to the formation of a longer pathway for methanol across the membrane. The methanol permeability also was significantly reduced.

The observation showed that non-fluorinated polymer membranes such as SPEEK have the highest potential and viable strategy to overcome the difficulties of the currently used Nafion® membrane. However, swelling at high degree of sulfonation may reduce the performance of DMFC. Several modifications need to be studied to curb the swelling aspects in SPEEK such as:

1. Offering bridging links to the reactive sulfonic groups via thermal activation and cross-linking of polymer chains by polyols (Mikhailenko et al., 2000, Manea and Mulder, 2002, Zhang et al., 2008).
2. Providing hydrophobic-hydrophilic block for SPEEK in different ratios (Karthikeyan et al., 2005).
3. Adjusting SPEEK by blending techniques. The compatible blends occur between SPEEK-polyether sulfone (PES), SPEEK- polyether imide (PEI), and polybenzimidazole (PBI) (Mikhailendo et al., 2004, Zhu et al., 2006, Tang et al., 2007, Zhu et al., 2007, Awang et al., 2015).

A significant amount of study on electrospun SPEEK/Cloisite fibers has successfully proven that electrospinning is one of the most suitable fabrication methods to reduce the size of filler; it is easier for the filler to disperse in the polymer matrix without agglomeration. A filler can act as an obstacle to the polymer mobility as well as a nucleating agent depending on the dispersion and loading of the clay itself.

Membrane morphology is very important in providing a new configuration of PEM with addition of electrospun fibers also responsible for improved proton conductivity. The Grotthus mechanism (proton hopping) plays an important role in bringing the proton from one site to another (Zhao et al., 2006). The protons are delivered through ionic cluster channels (free water and SO_3 that linked with non-freezing bound water) (Qin et al., 2009, Wu et al., 2009). It is desirable for a membrane to have high quantity of bound water (so called non-freezing bound water) since it is a crucial element in delivering protons. The study proved that the presence of Cloisite was an important factor that helped reduce the hydration in membrane by the Cloisite retaining water (Awang et al., 2018). Water retention capability of Cloisite has successfully enhanced the proton conductivity of the electrospun SPEEK/Cloisite nanofibers membranes. The increase in the proton conductivity of the nanocomposite membrane can be interpreted in the following way (Figure 12.5).

The absorbed water molecules are present mostly in the ionic cluster domains and ionic cluster channels. In particular, in the ionic cluster channels, the water molecules exist in two different forms. One is the protonated water (mostly non-freezing bound water) that is bound strongly to the ionic site. The other is free water that occupies the central space free from the influence of the ionic sites. The proton transfer through the ionic cluster channel occurs by two different mechanisms: (1) near the channel wall via the bound water, in which proton is transported by the Grotthuss mechanism, hopping from one ionic site to the other; or (2) via free water by vehicle mechanism, in which proton is facilitated by the water molecules moving through the interconnected central channel space. Water retention capacity of the nanocomposite membrane was enhanced due to the presence of electrospun fibers with good dispersion of Cloisite, which was expected to reduce the dehydration of membrane. Therefore, it can be said that the enhanced proton conductivity of the electrospun SPEEK/Cloisite nanocomposite membrane is attributed to the water retention capability. From the study, it is clear that the incorporation of appropriate

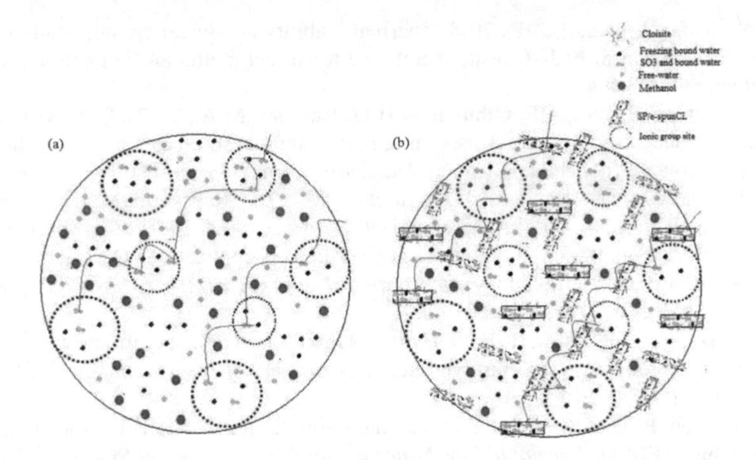

FIGURE 12.5 Transport Model of proton in (a) parent SPEEK and (b) electrospun SPEEK/Cloisite membranes. (From Awang, N. et al., *Polymers*, 10, 194, 2018.)

amounts of electrospun SPEEK/Cloisite fibers has synergistic effect to improve proton conductivity [62].

12.6 CONCLUSIONS

Functionalized polymeric material selection is essential to improve the properties of PEM. A significant amount of research has been carried out to solve two of PEM main problems, methanol crossover and low proton conductivity. As a consequence, SPEEK has drawn attention due to its capability to curb these problems. Most of the studies carried out have been focusing on optimizing and preparing SPEEK membranes for fuel cell application and modifying SPEEK using different techniques. Nevertheless, though SPEEK membranes have proven their potential and so challenge the dominance of Nafion membrane application within the industry, there are still a few weaknesses that need to be improved, notably morphology. The exfoliated morphological structure is important in providing winding methanol routes to alleviate the methanol crossover. Hence, fabrication of the functionalized SPEEK by electrospinning is suggested since it is believed to provide exfoliated morphology. In addition, electrospinning is the best solution to develop high-performance membrane for ion exchange membranes.

REFERENCES

Antonucci, P.L., Arico, A.S., Creti, P., Ramunni, E., Antonucci, V., 1999, Investigation of a direct methanol fuel cell based on a composite Nafion-silica electrolyte for high temperature operation, *Solid State Ionics* 125, 431–437.

Arnett, N.Y., Harrison, W.L., Badamia, A.S., Roy, A., Lane, O., Cromer, F., Dong, L., McGrath, J.E., 2007, Hydrocarbon and partially fluorinated sulfonated copolymer blends as functional membranes for proton exchange membrane fuel cells, *Journal of Power Sources* 172, 20–29.

Awang, N., Jaafar, J., Ismail, A.F., Matsuura, T., Junoh, H., Othman, M.H.D., Rahman, M.A., 2015, Functionalization of polymeric materials as a high performance membrane for direct methanol fuel cell: A review, *Reactive & Functional Polymers* 86, 248–258.

Awang, N., Jaafar, J., Ismail, A.F., 2018, Thermal stability and water content study of void-free electrospun SPEEK/Cloisite membrane for direct methanol fuel cell application, *Polymers* 10(2), 194.

Awang, N., Jaafar, J., Ismail, A.F., Othman, M.H.D., Rahman, M.A., 2017a, Effects of SPEEK/Cloisite concentration as electrospinning parameter on proton exchange membrane for direct methanol fuel cell application, *Materials Science Forum* 890.

Awang, N., Jaafar, J., Ismail, A.F., Othman, M.H.D., Rahman, M.A., Yusof, N., Azman, W.W.M.N., 2017b, Development of dense void-free electrospun SPEEK-Cloisite15A membrane for direct methanol fuel cell application: Optimization using response surface methodology, *International Journal of Hydrogen Energy* 42(42), 26496–26510.

Baradie, B., Dodelet, J.P., Guay, P., 1998, Hybrid Nafion®-inorganic membrane with potential applications for polymer electrolyte fuel cells, *Journal of Electroanalytical Chemistry* 489, 209–214.

Besse, S., Capron, P., Diat, O., 2002, Sulfonated polyimides for fuel cell electrode membrane assemblies (EMA), *Journal of New Materials for Electrochemical Systems* 5, 109–112.

Bozkurt, A., Meyer, W.H., 2001, Proton conducting blends of poly(4-vinylimidazole) with phosphoric acid, *Solid State Ionics* 138, 259–265.

Doğan, H., Inan, T.Y., Koral, M., Kaya, M., 2011, Organo-montmorillonites and sulfonated PEEK nanocomposite membranes for fuel cell applications, *Applied Clay Science* 52, 285–294.

Elabd, Y.A., Napadensky, E., Sloan, J.M., Crawford, D.M., Walker, C.W., 2003, Triblock copolymer ionomer membranes. Part I. Methanol and proton transport, *Journal of Membrane Science* 217(1–2), 227–242.

Formhals, A., 1934, U.S. Patent No. 1, 975,504. United States Patent Office.

Genies, C., Mercier, R., Sillion, B., 2001a, Stability study of sul-fonated phthalic and naphthalenic polyimide structures in aqueous medium, *Polymer* 42, 5097–5105.

Genies, C., Mercier, R., Sillion, B., Cornet, N., Gebel, G., Pineri, M., 2001b, Soluble sulfonated naphthalenic polyimides as materials for proton exchange membranes, *Polymer* 42, 359–373.

Gilpa, X., Hogarth, M., Department of Trade and Indus-try (UK) Homepage, 2001. Available on: http://www.dti-gov.uk/renewable/pdf/f0200189.pdf. Last retrieved August, 2013.

Guo, X., Fang, J., Watari, T., Tanaka, K., Kita, H., Okamoto, K., 2002, Novel sulfonated polyimides as polyelectrolytes for fuel cell application. Synthesis and proton conductivity of polyimides from 9,9-bis (4-aminophenyl)fluorene-2,7-disulfonic acid, *Macromolecules* 35, 6707–6713.

Jaafar, J., 2006, Development and characterization of sulfonated poly (ether ether ketone) membrane for direct methanol fuel cell. Universiti Teknologi Malaysia. MSc. Thesis.

Jaafar, J., Ismail, A., Matsuura, T., 2009, Preparation and barrier properties of SPEEK/Cloisite 15A®/TAP nanocomposite membrane for DMFC application, *Journal of Membrane Science* 345(1), 119–127.

Jaafar, J., Ismail, A.F., Matsuura, T., Nagai, K., 2011, *Journal of Membrane Science*, 382, 202–211.

Jaafar, J., Ismail, A.F., Othman, M.H.D., Rahman, M.A., Aziz, F., 2017, Fabrication of nanocomposite membrane via combined electrospinning and casting technique for direct methanol fuel cell, *Journal of Membrane Science and Research* 4, 146–157.

Jannasch, P., 2003, Recent developments in high-temperature proton conducting polymer electrolyte membranes, *Current Opinion in Colloid and Interface Science* 8(1), 96–102.

Juda, W., McRae, W.A., 1950, Coherent ion-exchange gels and membranes, *Journal of the American Chemical Society* 72(2), 1044–1044.

Kariduraganavar, M.Y., Kittur A.A., Kulkarni, S.S., 2012, *Ion Exchange Membranes: Preparation, Properties, and Applications Ion Exchange Technology I* (pp. 233–276). Springer, Dordrecht, the Netherlands.

Karthikeyan, C.S., Nunes, S.P., Prado, S.A., Ponce, M.L., Silva, H., Ruffmann, B., Schulte, K., 2005, Polymer nanocomposite membranes for DMFC application, *Journal of Membrane Science* 254, 139.

Kerres, J., Cui, W., Reichle, S., 1996, New sulfonated engineering polymers via the metalation route. I. Sulfonated poly(ethersulfone) PSU Udel via metalation–sulfination–oxidation, *Journal of Polymer Science Part A* 34, 2421–2438.

Kerres, J., Hein, M., Zhang, W., Graf, S., Nicoloso, N., 2003, Development of new blend membranes for polymer electrolyte fuel cell applications, *Journal of New Materials for Electrochemical Systems* 6, 223–229.

Kerres, J., Zhang, W., Cui, W., 1998, New sulfonated engineering polymers via the metalation route. II. Sulfinated/sulfonated poly(ether sulfone) PSU Udel and it crosslinking, *Journal of Polymer Science Part A* 36, 1441–1448.

Kerres, J., Zhang, W., Jörissen, L., Gogel, V., 2002, Application of different types of polyaryl-blend-membranes in DMFC, *Journal of New Materials for Electrochemical Systems* 5, 97–107.

Kim, Y.S., Harrison, W.L., McGrath, J.E., Pivovar, B.S., 2004, Effect of interfacial resistance on long term performance of direct methanol fuel cells, In: *205th Meeting, Electrochemical Society Inc.* (Abs. 334).

Kolde, J.A., Bahar, B., Wilson, M.S., Zawodzinski, T.A., Gottesfeld, S., 1995, Advanced composite polymer electrolyte fuel cell membranes, In: *Proceedings of the 1st International Symposium on Proton Conducting Membrane Fuel Cell*, Chicago, IL, pp. 193–201.

Kreuer, K.D., Hampele, M., Dolde, K., Rabenau, A., 1988, Proton transport in some heteropolyacidhydrates a single crystal PFG-NMR and conductivity study, *Solid State Ionics* 28, 589–593.

Kreuer, K., 1995, On the development of proton conducting materials for technological applications, *Solid State Ionics* 97(1), 1–15.

Lassegues, J.C., Grondin, J., Hernandez, M., Maree, B., 2001, Proton con-ducting polymer blends and hybrid organic inorganic materials, Proton transport in some heteropolyacidhydrates a single crystal PFG-NMR and conductivity study, *Solid State Ionics* 145, 37–45.

Li, L., Zhang, J., Wang, Y., 2003, Sulfonated poly (ether ether ketone) membranes for direct methanol fuel cell, *Journal of Membrane Science* 226, 159–167.

Liang, Y.F., Pan, H.Y., Zhu, X.L., Zhang, Y.X., Jian, X.G., 2007, Studies on synthesis and property of novel acid-base proton exchange membranes, *Chinese Chemical Letters* 18, 609–612.

Manea, C., Mulder, M., 2002, Characterization of polymer blends of polyethersulfone/sulfonated polysulfone and polyethersulfone/sulfonated polyetheretherketone for direct methanol fuel cell applications, *Journal of Membrane Science* 206, 443.

Mikhailenko, S.D., Zaidi, S.M.J., Kaliaguine, S., 2000, Electrical properties of sulfonated polyether ether ketone/polyetherimide blend membranes doped with inorganic acids, *Journal of Polymer Science B: Polymer Physics* 38, 1386.

Mikhailenko, S.D., Wang, K., Kaliaguine, S., Xing, P., Robertson, G.P., Guiver, M.D., 2004, Proton conducting membranes based on cross-linked sulfonated poly(ether ether ketone) (SPEEK), *Journal of Membrane Science* 233, 93.

Moilanen, D.E., Spry, D.B., Fayer, M.D., 2008, Water dynamics and proton transfer in nafion fuel cell membranes, *Journal of the American Chemical Society* 24(8), 3690–3698.

Murali, R., Eisenberg, A., 1988, Ionic miscibility enhancement in poly (tetrafluoroethylene)/poly (ethyl acrylate) blends I. Dynamic mechanical studies, *Journal of Polymer Science Part B: Polymer Physics* 26(7), 1385–1396.

Ohya, H., Paterson, R., Nomura, T., McFadzean, S., Suzuki, T., Kogure, M., 1995, Properties of new inorganic membranes prepared by metal alkoxide methods Part I: A new permselective cation exchange membrane based on Si/Ta oxides, *Journal of Membrane Science* 105(1), 103–112.

Pan, C., Wu, H., Wang, C., Wang, B., Zhang, L., Cheng, Z., Hu, P. et al., 2008. *Advanced Materials* 20, 1644–1648.

Park, H., Kim, Y., Hong, W.H., Choi, Y.S., Lee, H., 2005, Influence of morphology on the transport properties of perfluorosulfonate ionomers/polypyrrole composite membrane, *Macromolecules* 38(6), 2289–2295.

Park, S.H., Park, J.S., Yim, S.D., Lee, Y.M., Kim, C.S., 2008, Preparation of organic/inorganic composite membranes using two types of polymer matrix via a Sol–Gel process, *Journal of Power Sources* 181, 259–266.

Park, Y.S., Yamazaki, Y., 2005, Novel Nafion/Hydroxyapatite composite membrane with high crystallinity and low methanol crossover for DMFCs, *Polymer Bulletin* 53(3), 181192.

Peeters, J., Boom, J., Mulder, M., Strathmann, H., 1998, Retention measurements of nanofiltration membranes with electrolyte solutions, *Journal of Membrane Science* 145(2), 199–209.

Pu, H., Meyer, W.H., Wegner, G., 2001, Proton conductivity in acid blended poly-(4-vinylimidazole), *Macromolecular Chemistry and Physics* 202, 1478–1482.

Qin, L., Jens, O., Robert, F.S., Niels, J.B., 2009, High temperature proton exchange membranes based on polybenzimidazoles for fuel cells, *Progress in Polymer Science* 34, 449–477.

Rikukawa, M., Sanui, K., 2000, Proton-conducting polymer electrolyte membranes based on hydrocarbon polymers, *Progress in Polymer Science* 25, 1463–502.

Smitha, B., Sridhar, S., Khan, A., 2005, Solid polymer electrolyte membranes for fuel cell applications: A review, *Journal of Membrane Science* 259(1), 10–26.

Tamura, T., Kawakami, H., 2010. Aligned electrospun nanofiber composite membranes for fuel cell electrolytes, *Nano Letters* 10, 1324–1328.

Tang, H., Pan, M., Jiang, S.P., Wang, X., Ruan, Y., 2007, Fabrication and characterization of PFSI/ePTFE composite proton exchange membranes of polymer electrolyte fuel cells, *Electrochimica Acta* 52, 5304–5311.

Tazi, B., Savadago, O., 2001, New cation exchange membranes based on Nafion, Silicotungstic acid and thiophene, *Journal of New Materials for Electrochemical Systems* (in press) 185, 3–27.

Teo, W., Ramakrishna, S., 2006, A review on electrospinning design and nanofibre assemblies, *Nanotechnology* 17(14), R89.

Wu, D., Xu, T., Wu, L., Wu, Y., 2009, Hybrid acid–base polymer membranes prepared for application in fuel cells. *Journal of Power Sources* 186, 286–292.

Xu, T., 2005, Ion exchange membranes: State of their development and perspective. *Journal of Membrane Science* 263(1), 1–29.

Zhang, H., Li, X., Zhao, C., Fu, T., Shi, Y., 2008, Composite membranes based on highly sulfonated PEEK and PBI: Morphology characteristics and performance, *Journal of Membrane Science* 308, 66.

Zhao, C., Li, X., Wang, Z., Dou, Z., Zhong, S., 2006, Synthesis of the block sulfonated poly(ether ether ketone)s (S-PEEKs) materials for proton exchange membrane, *Journal of Membrane Science* 280, 643.

Zhu, X., Zhang, H., Liang, Y., Zhang, Y., Luo, Q., Bi, C., Yi, B., 2007, Challenging reinforced composite polymer electrolyte membranes based on disulfonated poly(arylene ether sulfone)-impregnated expanded PTFE for fuel cell applications, *Journal of Materials Chemistry* 17, 386–397.

Zhu, X., Zhang, H., Zhang, Y., Liang, Y., Wang, X., Yi, B., 2006, An ultrathin self-humidifying membrane for PEM fuel cell application: Fabrication, characterization, and experimental analysis, *Journal of Physical Chemistry B* 110, 14240–14248.

13 Fabrication of Mesoporous Nanofiber Networks by Phase Separation–Based Methods

Sadaki Samitsu

CONTENTS

13.1 INTRODUCTION

The unique morphology of nanofiber membranes (NMs) makes them well suited for a broad range of medical, environmental, and energy-related applications (Notario et al. 2016). For example, such membranes allow gas molecules to pass through while effectively removing particulate matter and viruses by adsorption, and can therefore, function as high-performance filters for gas-phase media (Sundarrajan et al. 2014). Moreover, NMs can also be used for liquid purification (enabling the fast separation of harmful biomolecules or colloidal contaminants such as oil droplets (Yoon et al. 2008)), as cathode–anode separators in high-capacity batteries (Lee et al. 2005; Zhang 2007; Cho et al. 2008), and as sound-absorbing materials for aircraft and luxury automobiles (Bell 2011; Kalinova 2017). Thus, the enhanced permeability and separation/adsorption/absorption capacities of these membranes are expected to result in a variety of future applications, thus contributing to making our living more comfortable.

NMs can be fabricated by several methods that are broadly categorized into two groups. The first group corresponds to techniques developed for the conventional manufacturing of fibers, nonwoven mats, and papers and featuring the assembly of nanofibers into thin sheets, as exemplified by melt-blowing (Ellison et al. 2007; Hassan et al. 2013), electrospinning (Ramakrishna et al. 2006), and filtration of nanofibers (Karan et al. 2013; Zhang et al. 2016). The second group corresponds to methods that rely on phase separation phenomena and feature the spontaneous formation of nanofibrous network morphology. Consequently, these techniques are commonly used for the production of commercial membranes and films (Mulder 1996; Witte et al. 1996). Importantly, the porosity features (pore size, distribution, volume, shape, and connectivity) of NMs can be optimized for target applications by careful choice of a suitable fabrication method. For example, although first-group methods afford NMs with large pores and high porosity and hence, with high fluid permeability, these membranes suffer from low mechanical strength and stability. Conversely, second-group methods afford NMs with smaller pores and good mechanical properties that can be used for liquid purification. On the other hand, the low porosity and flux of these membranes limit their industrial applications. Since other chapters in this book comprehensively describe the spinning technique-based fabrication of NMs and their versatile applications, we herein introduce the use of this method based on phase separation phenomena.

13.2 MEMBRANE FABRICATION BY MEANS OF PHASE SEPARATION IN POLYMERS

Phase separation is a universal phenomenon observed in various materials, including metals, ceramics, and polymers (Papon et al. 2006). Specifically, phase separation in polymer solutions is commonly characterized by two descriptors, with the first one representing the factor causing phase separation (thermally induced, non-solvent-induced, evaporation-induced, and reaction-induced phase separations), and the second one illustrating the types of generated phases. In particular, the separation of a homogeneous phase into two liquid phases with different concentrations is denoted as a liquid-liquid phase separation (Lloyd et al. 1991). Similarly, the formation of a solid phase (usually as a result of crystallization of a certain component) in a uniform liquid phase is denoted as a solid-liquid phase separation (Lloyd et al. 1990). Since polymer solutions contain two components (i.e., polymer and solvent molecules), their phase separation features the processes of polymer and solvent molecule crystallization, which both correspond to solid-liquid phase separation.

Since phase separations in polymer solutions afford two phases: a polymer-rich phase and a solvent-rich (polymer-lean) phase, fixation of phase separation morphology via solvent exchange or drying allows the fabrication of porous materials in the forms of films, sheets, particles, fibers, and monoliths. The general phase separation-based procedure used for the fabrication of NMs is shown in Figure 13.1. In polymer solutions, the solid polymers are completely dissolved and form a uniform fluidic phase, which can spontaneously separate into two (polymer-rich and solvent-rich) phases with different polymer concentrations under suitable conditions (solvent, temperature, evaporation, and polymerization (step 1)). Commonly, samples are immersed into an excess of a poorly polymer-solubilizing solvent, and the heterogeneous morphology resulting from phase separation is fixed by solvent exchange (step 2). Finally, the removal of the poorly polymer-solubilizing solvent by drying (step 3) results in the conversion of the solvent-rich phase into voids to afford NMs.

Liquid-liquid phase separation in polymer solutions has been widely used for the fabrication of polymer membranes with different pore sizes from a wide variety of polymers (Lloyd et al. 1991). However, the above technique usually affords asymmetric macroporous membranes with small pores present only on the membrane surface, which implies that a number of technological challenges still need to be solved. One of these challenges is the fabrication of mesoporous commodity polymers, particularly those with high surface-to-volume ratios. To address these challenges, we have developed two methods of NM fabrication based on solid-liquid phase separation in polymer solutions, namely thermally induced gelation driven by polymer crystallization (Samitsu 2018) and flash-freezing nanocrystallization of solvent molecules (Samitsu et al. 2013).

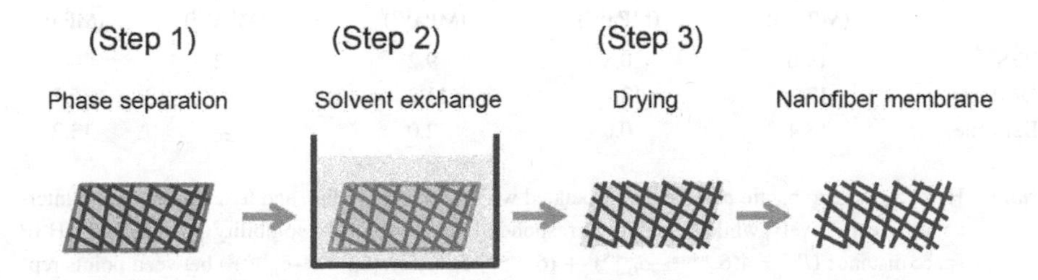

FIGURE 13.1 Schematic illustration of phase separation-based membrane fabrication.

13.3 FABRICATION OF MESOPOROUS NMs BY THERMALLY INDUCED GELATION

13.3.1 Membrane Materials

Thermally induced gelation in polymer solutions is driven by polymer crystallization in the solution state and is thus particularly well suited for crystalline polymers with high crystallization capabilities. The application of this method has been extensively documented for a number of crystalline polymers, e.g., polypropylene, high-density polyethylene, poly(4-methyl-l-pentene), and poly(vinylidene fluoride) (Lloyd et al. 1990), syndio- and isotactic polystyrene (Aubert 1988; Daniel et al. 2013), poly(lactic acid) (Ma and Zhang 1999), polycarbonate (Xin et al. 2012), and others (Samitsu et al. 2008a, 2008b; Liu and Ma 2009). Recently, we have also reported the fabrication of NMs based on solvent-induced crystallization of poly(ether sulfone) (PES) (Samitsu 2018).

13.3.2 Membrane Fabrication: Key Factors of Influence

The crystallization of polymers is induced by the reduction of their solubility, which is typically driven by cooling and/or addition of poor solvents and results in the formation of (super)saturated polymer solutions and the generation and growth of polymer nuclei. Therefore, membrane fabrication is affected by factors determining the solubility of polymers, i.e., type of solvent, quantity of poorly polymer-solubilizing solvent, temperature time profile, and polymer concentration.

Solvent composition is known to significantly influence the physical gelation of linear polymers, the solubilities of which can be quantitatively evaluated utilizing Hansen (2007) solubility parameters (HSPs), i.e., good solubility is observed when the HSP of a given solvent is similar to that of the polymer. Strictly speaking, in HSP space, dissolution is observed when the solvent is located inside a solubility sphere that is centered at the solubility parameter of the polymer and has a radius primarily dependent on polymer type. Table 13.1 summarizes the HSPs of PES, *N,*

TABLE 13.1
HSPs of PES, DMF, and Benzene

	δ_D ($MPa^{1/2}$)	δ_P ($MPa^{1/2}$)	δ_H ($MPa^{1/2}$)	R_0 ($MPa^{1/2}$)	R_a ($MPa^{1/2}$)
PES	19.6	10.8	9.2	6.2	—
DMF	17.4	13.7	11.3	—	5.7
Benzene	18.4	0.0	2.0	—	13.2

Note: Here, δ_D, δ_P, and δ_H are parameters associated with dispersive, polar, and hydrogen bonding interactions, respectively, while R_0 and R_a correspond to the radius of the solubility sphere and the HSP space distance ($R_a^2 = 4(\delta_D^{solvent} - \delta_D^{PES})^2 + (\delta_P^{solvent} - \delta_P^{PES})^2 + (\delta_H^{solvent} - \delta_H^{PES})^2$) between points representing PES and the chosen solvent (Hansen 2007). PES is dissolved in the case of $R_a/R_0 < 1$, with the reverse being true for $R_a/R_0 > 1$.

N-dimethylformamide (DMF), and benzene (Abbott and Hansen 2018), revealing that DMF is located inside the PES solubility sphere, whereas benzene is located outside of this sphere, which agrees with the high/poor solubilities of PES in DMF/benzene, respectively. Notably, pure solvents exhibit discrete HSP values, whereas the continuously variable HSP values of their mixtures allow polymer solubility to be controlled in the desired manner.

In contrast to that of small molecules, the solubility of polymers exhibits a considerable temperature dependence due to the large conformational entropy of polymer chains, generally increasing with temperature and thus allowing homogeneous transparent solutions to be obtained by complete dissolution of polymers at high temperatures. Cooling of such solutions results in the spontaneous assembly of crystalline polymer molecules into a periodic arrangement and the formation of crystalline domains in the solution. At low polymer concentrations, this process results in the generation of isolated objects with polymer crystal structure-determined shapes, e.g., particles, fibers, ribbons, and sheets (Guenet 2008). With increasing polymer concentration, these primary building blocks are progressively interconnected to form a three-dimensional network structure in solution, which finally results in gelation and is used to prepare NMs. The nucleation and growth rates of polymer crystals exhibit different temperature dependences, and crystallization behavior is therefore affected by temperature sequences, e.g., by cooling/heating rates and holding temperatures/times. However, the effects of temperature are much weaker than those of solvents, i.e., polymers form crystals at low temperatures only when solvents with moderate polymer solubilizing power are used.

Generally, NMs are prepared using relatively high concentration of polymers ($\geq$10 wt.%) due to the need for constructing three-dimensional polymer nanofiber frameworks. Specifically, high polymer concentrations markedly enhance the mechanical properties of NMs by reinforcing the network structure of nanofibers. On the other hand, polymer concentrations significantly affect the total pore volume of NMs, as solvent molecules act as porogens. Therefore, low polymer concentrations (i.e., high amounts of solvent) are expected to increase the total pore volume and pore size. However, when these concentrations are overly low, the produced NMs experience significant contraction and serious damage during solvent extraction and drying processes due to the low mechanical strength of the corresponding gels. As a result, optimal polymer concentration is determined by the trade-off between pore volume and mechanical stability. Herein, we describe the preparation of PES NMs as a specific example of membrane fabrication from gel precursors obtained by polymer crystallization.

13.3.3 Membrane Fabrication and Characterization

A solution of PES in DMF was prepared by dissolving PES in the above solvent a temperature exceeding 100°C overnight. PES solutions in DMF/benzene mixtures were prepared by adding small amounts of benzene to the solution of PES in DMF at $\leq$50°C. After stirring for 5–10 min, the obtained solutions were poured into a glass dish and subjected to 24-h incubation at 25°C for gelation. The thus prepared gels were subjected to room-temperature solvent exchange with excess methanol and vacuum-dried to afford porous PES membranes or monoliths. Sample thickness could be roughly controlled by varying the quantity of the PES solution used for gelation.

The gel/monolith crystal structures were analyzed using a benchtop X-ray diffractometer (MiniFlex600, Rigaku), and NM morphology was observed by field-emission scanning electron microscopy (SEM) (S-4800, Hitachi High-Technologies). The geometric specific surface area, S_{calc}, was determined as $S_{calc} = \sum_i \left(4f(d_i)/(\rho_r d_i)\right)$, where d_i is the nanofiber diameter obtained by SEM imaging, $f(d_i)$ is the probability of finding diameter d_i, and ρ_r is the true density of bulk PES. Material porosity was quantitatively assessed by nitrogen adsorption measurements performed utilizing a gas adsorption analyzer (BELSORP-max, MicrotracBEL). The specific surface area (S_{BET}) and pore size distribution were calculated using Brunauer-Emmett-Teller (BET) and Barrett-Joyner-Halenda (BJH) methods, respectively. The volumes of pores with radii of less than 25 nm and with radii of 25–100 nm were denoted as V_{meso} and V_{macro}, respectively.

13.3.4 Physical Gelation Driven by Polymer Crystallization

As mentioned above, DMF is a good solvent for PES, whereas benzene is a poor solvent. The physical gelation of PES solutions after 24-h incubation at 25°C was examined at different PES concentrations and DMF/benzene contents. For pure DMF, PES solutions with concentrations of ≤30 wt.% did not gel at room temperature. In contrast, the addition of suitable amounts of benzene to PES solutions caused gelation in a large PES concentration range. Although 20 wt.% PES solutions remained transparent sols at benzene contents of ≤10%, gelation was observed for benzene contents of ≥20%. Notably, gelation was observed for all solutions with PES concentrations of ≥10 wt.%, and the transparency of the obtained gels decreased with increasing PES concentration. Figure 13.2a displays a sol-gel phase diagram for PES solutions in PES concentration-benzene content space, revealing that the above solutions remain in the sol state in the lower left region of the diagram, whereas gels are formed in the upper right region corresponding to high PES concentrations and benzene contents. Transparent gels were obtained at the boundary between the sol and gel regions. To investigate the origin of gelation, we recorded X-ray diffraction (XRD) spectra of a PES gel and monolith (Figure 13.2b). The spectrum of PES gel demonstrated several peaks on a broad halo pattern, clearly indicating the presence of crystalline PES. An intense peak at $2\theta = 19.0°$ corresponded to a *d*-spacing of 4.6 Å, which was close to that of PES crystallized in dichloromethane (4.5 Å) (Blackadder et al. 1979). The other six peaks also agreed with previously reported values within a *d*-spacing range of 0.2 Å, even though these values were obtained for a gel containing a smaller amount of a different solvent. On the other hand, the XRD spectrum of the PES monolith exhibited a significantly reduced intensity compared to that of the gel sample and missed some of the peaks observed for the latter. Since the monolith contained less than 0.6 wt.% solvent (as confirmed by thermogravimetric analysis), the almost complete removal of solvent molecules was concluded to seriously deteriorate crystallinity, in agreement with a previous report (Blackadder et al. 1979).

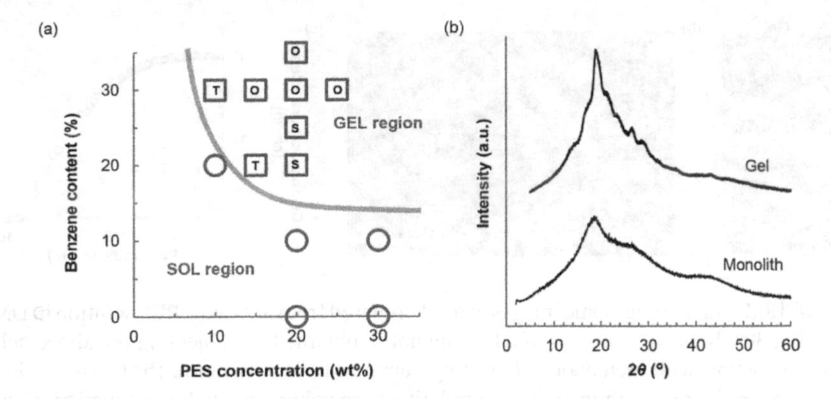

FIGURE 13.2 (a) Sol-gel phase diagram of PES solutions subjected to 24-h incubation at 25°C. PES concentration and benzene content are expressed as weight percentage relative to total solution weight and weight percentage relative to total solvent weight, respectively. Circles and squares represent the sol and gel states, respectively, and capital letters in squares indicate gel appearance (T: transparent, S: semitransparent, O: opaque). (b) XRD spectra of PES gel precursor and monolith. (Adapted with permission from Samitsu, S., *Macromolecules*, 51, 151–160, 2018. Copyright 2018 American Chemical Society.)

Although the physical gelation of crystalline polymers has been extensively studied, that of amorphous polymers has been underexplored. In particular, very few reports on the gelation behavior of engineering plastics are available. The high glass transition temperatures (T_gs) of engineering plastics limit their accessible crystallization temperatures to a narrow range, and the variety of possible chain conformations provides numerous potential crystalline structures and complicates crystallization in a well-defined arrangement. Solvent molecules sometimes support the crystallization of polymers by enhancing their chain mobility via the plasticizing effect. Since the formation of crystal complexes with solvent molecules, i.e., crystallosolvates, results in the spontaneous selection of energetically favorable crystal structures, solvent-induced crystallization is considered to be a promising way of producing crystalline engineering plastics despite lowering polymer concentrations.

13.3.5 Fabrication and Characterization of Mesoporous NMs Made of PES

Compared to conventional solution-phase casting, the use of a gel precursor is an effective means of improving the surface-to-volume ratio of a porous nanofiber monolith. To demonstrate the advantages of gel precursor usage, we prepared porous monoliths from both a PES sol (benzene content = 10%) and a gel (benzene content = 30%) and compared their structures (Figure 13.3a). The sol-derived monolith exhibited exclusively spherical macropores with radii of 2–10 μm, whereas the gel-derived one comprised nanofibers with an average diameter of 18 nm (Figure 13.3b) and featured an S_{calc} of 152 m^2/g determined from the nanofiber diameter distribution obtained by digital image analysis. The sparse reticular structure of the above nanofibers featured pores with radii of 20–200 nm, and the corresponding

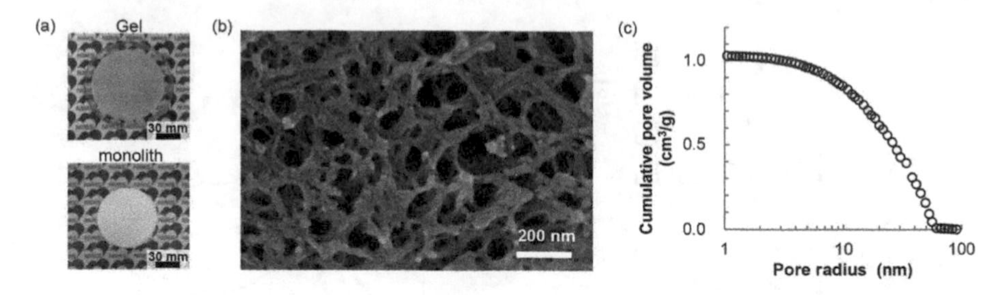

FIGURE 13.3 (a) Photographic images of a gel prepared from a 20 wt.% PES solution in DMF containing 30% benzene (top) and of a PES monolith obtained by subjecting the above gel to solvent extraction with methanol followed by vacuum-drying (bottom). (b) Cross-sectional SEM image of the monolith in (a). (c) Cumulative pore volume plotted as a function of pore radius for the monolith in (a). (Adapted with permission from Samitsu, S., *Macromolecules*, 51, 151–160, 2018. Copyright 2018 American Chemical Society.)

S_{BET} value of 180 m²/g was close to that of S_{calc}. Figure 13.3c shows that the pore volume gradually increased over the radius range of 2–50 nm, allowing V_{meso} and V_{macro} to be both estimated as 0.52 cm³/g. The broad pore size distribution exhibited a maximum at a radius of ~8 nm. Considering the network morphology of nanofibers, the gelation of PES solutions was thought to involve nanofiber network growth in the whole solution. When the solubility of PES was reduced by the addition of benzene, isolated PES chains first crystallized into nanofibers that were subsequently interconnected to form bundles and reticular structures, and the solution turned into a gel at the point of complete nanofiber network percolation throughout the sample. Owing to the high surface-to-volume ratio of nanofibers, the solvent molecules were retained in their network, which allowed the formation of a PES nanofiber–solvent gel. Notably, compared to the sol-derived monolith, the gel-derived one had a completely different structure in terms of building block (i.e., nanofiber) size, shape, and connectivity. Thus, a mesoporous monolith featuring a dense network of PES nanofibers could be successfully prepared using the gel precursor method.

13.3.6 Porous Structures Prepared Using Benzene Content/PES Concentration Changes

Benzene content and PES concentration were the key factors influencing the phase separation morphology of PES gels and hence, the porous structures of the corresponding monoliths. At a fixed benzene content of 30%, the monolith obtained at a low PES concentration of 10 wt.% featured a sparse network structure comprising straight nanofiber bundles with lengths of ~1 μm (Figure 13.4a), which was in stark contrast to the dense reticular nanofiber structures found in the monolith obtained at a high PES concentration (Figure 13.3b). S_{BET} values increased with PES concentration and reached a maximum of 206 m²/g at a PES concentration of 25 wt.%. Monoliths obtained from 15 to 25 wt.% gels exhibited similar pore-size distributions peaking at a radius of 7 nm, which indicated the presence of a large number of mesopores. Specifically, V_{meso} equaled 0.49–0.58 cm³/g at PES concentrations of 15–25 wt.%.

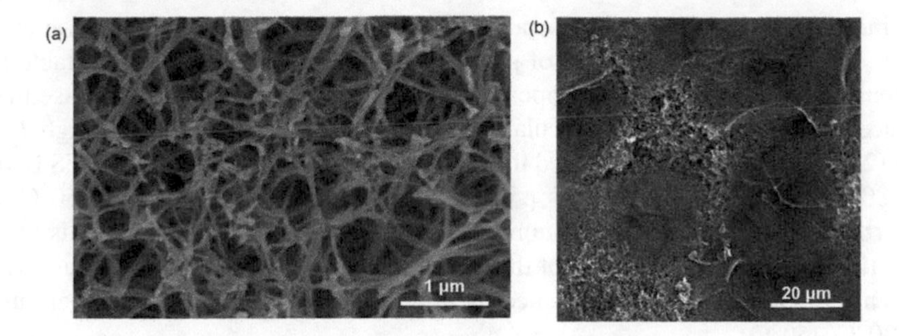

FIGURE 13.4 Cross-sectional SEM images of PES monoliths. (a) Sparse network morphology featuring straight nanofiber bundles. (b) Heterogeneous domain structures comprising dense and course reticular networks. The monoliths were fabricated using a gel precursor containing 10 wt.% PES and 30% benzene (a) and a gel containing 20 wt.% PES and 35% benzene (b). (Adapted with permission from Samitsu, S., *Macromolecules*, 51, 151–160, 2018. Copyright 2018 American Chemical Society.)

At a fixed PES concentration of 20 wt.%, S_{BET} values drastically increased at the benzene content corresponding to gelation onset and then plateaued at 170–180 m^2/g at benzene contents of 25%–35%. The above increase originated from the formation of nanofibers in precursor gels, and constant S_{BET} was indicative of similar nanofiber diameters, as confirmed by SEM. Higher benzene contents tended to result in the formation of smaller pores due to the generation of a dense nanofiber network, significantly reducing the solubility of PES and therefore promoting the nucleation of PES crystals. In fact, high benzene contents markedly increased the number of nanofibers and thus induced the formation of dense reticular structures with abundant mesopores. The pore size distribution peaked at a radius of 5 nm at a benzene content of 35%, and two reticular structures with different mesh sizes were found (Figure 13.4b). Specifically, the dense domains comprised small mesopores and enclosed other less dense regions that included both macro- and mesopores. This heterogeneous domain structure probably resulted from the successive crystallization of PES, since the number of isolated PES chains gradually decreased with progressing crystallization. Initially, isolated dense domains were nucleated in the homogeneous solution and began to connect with each other, which culminated in complete percolation throughout the sample and resulted in solution gelation. Due to the decreased PES concentration at the late stage, less dense domains surrounded by the percolating network of denser domains were subsequently generated in solution. The pore size distribution of the PES monolith exhibited a weak dependence on PES concentration above 10 wt.%, but strongly depended on the benzene content due to its strong influence on PES solubility in benzene/DMF mixtures.

13.3.7 Thermal Conductivity of Mesoporous PES NMs

Mesopores in polymeric materials are intrinsically unstable due to possessing a large surface energy and often degrade upon aging at high temperatures or as a result of capillary compression during drying. Therefore, the improvement of thermal and mechanical stabilities of mesopores in mesoporous polymers presents a challenge. The nanofiber frameworks of porous polymers are thought to be suitable for mechanical strengthening of mesoporous structures and significantly increase their surface-to-volume ratios. In particular, such nanofiber structures and the high T_g of PES (225°C) significantly improved the thermal stability of mesopores in PES NMs, i.e., 80% of these mesopores was preserved even after 600-h aging at 150°C, as confirmed by nitrogen gas adsorption characterization of aged samples. This result is particularly important in view of the fact that thermally stable mesoporous structures are promising high-performance thermal insulators at moderate temperatures of 100°C–150°C.

The thermal conductivity, $\lambda(T)$, of the NM was calculated as $\lambda(T) = C_p(T) \times \rho_{ap} \times \alpha(T)$, where C_p is the specific heat capacity, ρ_{ap} is the apparent density, and α is the thermal diffusivity. Assuming $C_p = 1.40$ J/g·K, $\rho_{ap} = 0.43$ g/cm^3, and $\alpha = 1.29 \times 10^{-7}$ m^2/s, the average thermal conductivity over the temperature range of 180°C–200°C was determined as 0.078 W/m·K. For the sake of simplicity, we neglected the temperature dependence of ρ_{ap} in these calculations. The thermal conductivity of the bulk was calculated as 0.287 W/m·K assuming $C_p = 1.42$ J/g·K, $\rho_{ap} = 1.39$ g/cm^3, and $\alpha = 1.45 \times 10^{-7}$ m^2/s. Since the C_p, ρ_{ap}, and α values of the NM were 99%, 31%, and 89% of those of the bulk sample, respectively, the low $\lambda(T)$ of the NM was primarily attributed to its low apparent density. Comparison with the $\lambda(T)$ values of other aerogels with similar ρ_{ap} showed that the $\lambda(T)$ of the NM was similar to those of methacrylate-based aerogels (Notario et al. 2015) but was four times larger than the best values reported for polyurethane-based (Biesmans et al. 1998) and resorcinol-formaldehyde aerogels (Lu et al. 1992). Thus, the low $\lambda(T)$ at high temperatures observed for the PES NM implied that it was well suited for thermal insulation applications.

13.3.8 Thermally Induced Gelation: Conclusion

Physical gelation is a promising method of fabricating mesoporous NMs, being well documented for crystalline polymers but relatively unexplored in the case of amorphous polymers. Herein, we demonstrated the first-time physical gelation of PES solutions in DMF induced by cooling and/or benzene addition and driven by the partial crystallization of PES, as confirmed by XRD measurements. Since PES is considered to be a typical amorphous polymer with high T_g, the solvent-induced crystallization of engineering plastics demonstrated herein presents a new way of fabricating NMs. The removal of residual solvent from gel precursors was found to be an effective means of fabricating porous PES NMs, and optimization of PES concentration and benzene content allowed us to realize mesoporous PES NMs comprising a fine network of PES nanofibers with an average diameter of 18 nm and exhibiting a high specific surface area of 206 m^2/g and a large mesopore volume of 0.52 cm^3/g. The obtained mesopores were thermally stable due to the high T_g of PES, and nearly 80% of them was preserved after 600-h aging at 150°C. At high temperatures (180°C–200°C), the thermal conductivity of the NM equaled 31% of that of the bulk polymer due to its low apparent density, suggesting that this material may function as a good thermal insulator in the range of 100°C–200°C.

13.4 FABRICATION OF MESOPOROUS NMs BY FLASH-FREEZING NANOCRYSTALLIZATION

Here, we consider the crystallization of solvent molecules in a polymer solution at a temperature lower than the solvent melting point. Since solvent crystals progressively grow by accumulating only solvent molecules, polymer molecules are selectively expelled and concentrated in regions between solvent crystals, which results in separation into solvent crystal and concentrated polymer phases. The removal of solvent crystals via freeze-drying affords pores replicating their shape, and the above technique is therefore denoted as the ice template method (Mahler and Bechtold 1980; Gutiérrez et al. 2008; Mukai et al. 2008) or the freeze-casting method (Deville 2008; Li et al. 2012) and is used for the fabrication of macroporous materials. However, despite the simplicity of this process, it could not previously be applied to the formation of NMs.

To reduce the pore size and obtain NMs by solvent crystallization, we have developed a new methodology of fabricating mesoporous polymers, namely flash-freezing nanocrystallization (Samitsu et al. 2013). From a classical viewpoint, crystallization comprises two steps, namely initial nucleation and subsequent nucleus growth. In order to form solvent nanocrystals, a high nucleation density and extremely slow growth of nuclei are desired. The classical theory predicts that a crystallization temperature decrease should result in an exponential increase of the nucleation rate and a significant decrease of the growth rate (Debenedetti 1997). Therefore, we expected that crystallization at very low temperatures should allow the formation of solvent nanocrystals in polymer solutions.

13.4.1 MEMBRANE FABRICATION

Figure 13.5 illustrates the principle of flash-freezing nanocrystallization. Solutions of polystyrene in DMF with concentration of 15 wt.% and above were sealed in a glass vial and quickly vitrified in liquid nitrogen. The vitrified solution was stored at −80°C in a cryogenic refrigerator to form solvent nanocrystals in the supercooled liquid, which finally resulted in phase separation to afford DMF nanocrystals and

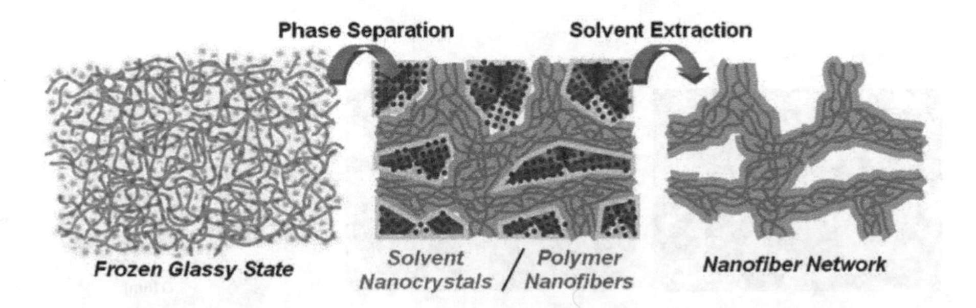

FIGURE 13.5 Schematic mechanism of nanofiber network formation via solvent nanocrystallization and subsequent solvent exchange and drying. (Adapted from Samitsu, S. et al., *Nat. Commun.*, 4, 1–7, 2013. With permission from CC-BY license.)

a vitrified highly concentrated polystyrene phase. The thus obtained structure was fixed by solvent exchange with excess methanol at −80°C, and subsequent freeze-drying from *t*-butanol furnished mesoporous polystyrene. Although freeze-drying was not essential for mesoporous structure formation, it was performed to reproducibly obtain mesopores in large quantities by minimizing the undesirable shrinkage of small pores during drying.

13.4.2 Membrane Characterization

Cross-sectional SEM imaging showed that the obtained NMs comprised highly branched interconnected nanofibers uniformly distributed in the whole imaging region without macrovoids. The nanofiber diameter distribution was determined by digital image analysis and allowed the average nanofiber diameter and S_{calc} to be estimated as 18 nm and 155 m^2/g, respectively. The membrane obtained at a low polystyrene concentration featured a coarse reticular structure with a pore size of 100 nm (Figure 13.6a), whereas the use of high concentrations resulted in the generation of smaller 15-nm-size mesopores due to the formation of a densely connected network of thin nanofibers (Figure 13.6b). S_{BET} values were determined to exceed 180 m^2/g in the concentration range of 5–40 wt.%, with a maximum of 282 m^2/g found at 40 wt.%. At low concentration, the pore size distribution (Figure 13.6c) peaked at a diameter of ~85 nm, and the total pore volume V_{total} was estimated as 2.23 cm^3/g. At high concentration, the pore size distribution exhibited a sharp peak centered at 15 nm with a full width at half maximum of 8 nm. The corresponding membrane featured a large mesopore volume V_{meso} of 0.93 cm^3/g, which equaled half of the total sample volume. The obtained result demonstrated the possibility of introducing large amounts of macrovoid-free mesopores by flash-freezing nanocrystallization. Since peak diameters continuously decreased with increasing polystyrene concentration, variation of the latter parameter allowed pore sizes to be controlled in the diameter range of 15–100 nm. Importantly, the developed method allowed the fabrication of pores nearly 100 times smaller than those obtained by conventional thermally induced phase separation methods.

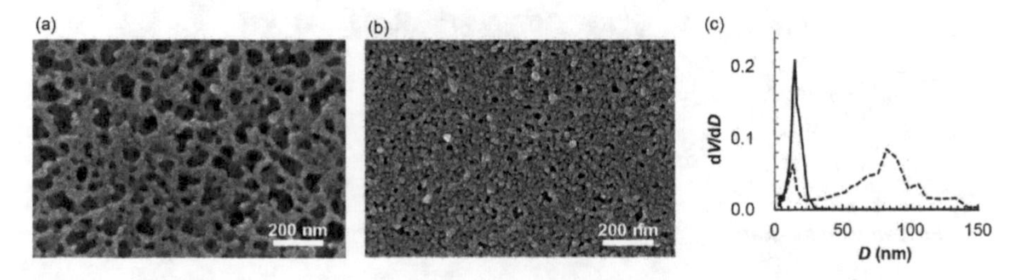

FIGURE 13.6 Cross-sectional SEM images of polystyrene NMs obtained at polystyrene concentrations of 20 wt.% (a) and 40 wt.% (b). (c) Pore size distribution of NMs (D = pore diameter). Dashed and solid lines represent membranes (a) and (b), respectively.

13.4.3 Membrane Materials

The described nanocrystallization method is versatile and directly applicable to a wide range of commercially available polymers such as polystyrene, poly(vinyl chloride), polycarbonate, polyacrylonitrile, polysulfone, PES, and poly(ether imide). Figure 13.7 shows SEM images of these mesoporous polymers, with the corresponding fabrication conditions listed in Table 13.2. Notably, nanofibers with average diameters of 12–21 nm were produced irrespective of polymer type, affording uniform mesoporous structures with pore diameters of 17–34 nm. The preservation of mesopores under large Laplace pressures during drying requires the T_g of polymers to exceed room temperature. In particular, engineering plastics with high T_g afforded thermally and mechanically stable mesoporous polymers. The above efficient and scalable method also allows the realization of different shapes such as bulks, membranes, fibers, and pellets by casting and spinning, thus being well suited for fabricating mesoporous polymers with industrial applications.

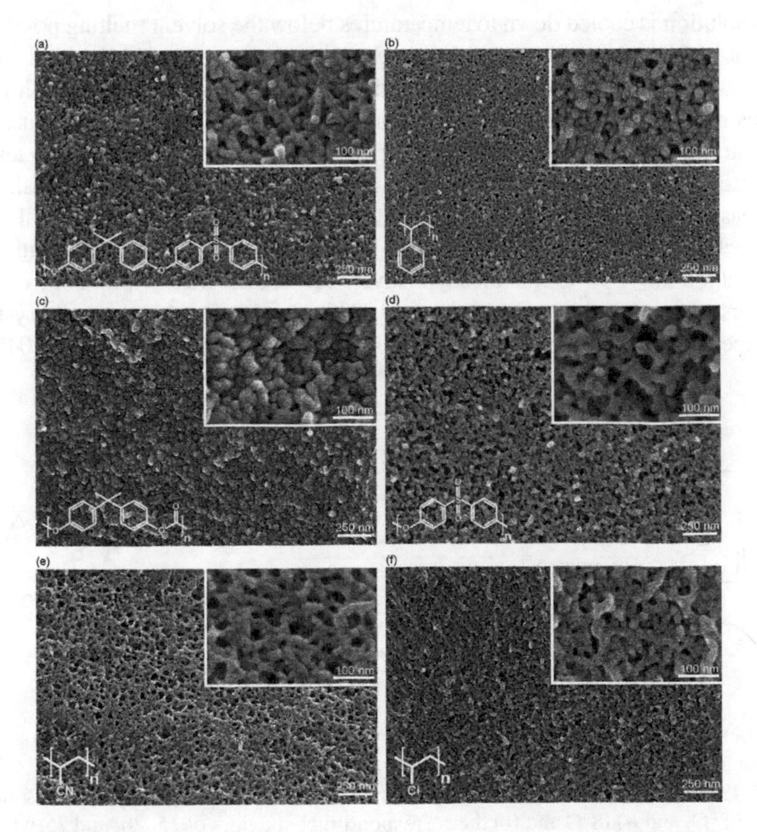

FIGURE 13.7 SEM images of mesoporous nanofiber networks. Insets show enlarged images. Detailed fabrication procedures are provided in Table 13.2. (Adapted from Samitsu, S. et al., *Nat. Commun.*, 4, 1–7, 2013. With permission from CC-BY license.)

TABLE 13.2
Fabrication Parameters for Mesoporous Polymer–Based Nanofiber Networks

Sample ID	Polymer	Concentration	Good Solvent	Freezing Temperature
NF1	Polysulfone	20 wt.%	DMF	−196°C
NF2	Polystyrene	40 wt.%	DMF	−196°C
NF3	Polycarbonate	20 wt.%	Cyclohexanone	−196°C
NF4	PES	20 wt.%	DMF	−196°C
NF5	Polyacrylonitrile	10 wt.%	DMF–DMAc	−196°C
NF6	Poly(vinyl chloride)	10 wt.%	DMF–DMAc	−196°C

Source: Samitsu, S. et al., *Nat. Commun.*, 4, 1–7, 2013.
Note: DMAc: *N, N*-dimethylacetamide. DMF–DMAc: a solvent mixture of DMF and DMAc at a weight ratio of 2:1.

13.4.4 Key Factors Influencing Membrane Fabrication

When a solution is cooled down to temperatures below the solvent melting point, nuclei of solvent molecules are stochastically generated in the supercooled liquid at a certain frequency. If nucleation occurs during cooling, the large crystal growth rate at high temperatures results in rapid nucleus growth on a macroscopic scale. Such unintentional nucleation should be completely avoided to perform crystallization at very low temperatures, which, based on the time–temperature–transformation diagrams of crystallization, can be easily achieved by flash freezing (Papon et al. 2006). The isothermal crystallization behavior of a 15-wt.% solution was probed by DSC, as shown in Figure 13.8a. Crystallization at ≥−96°C was mostly delayed because of the low probability of crystal nucleus generation, becoming faster at −98°C and slowing down as the crystallization temperatures was further decreased. For isothermal crystallization at −104°C, the

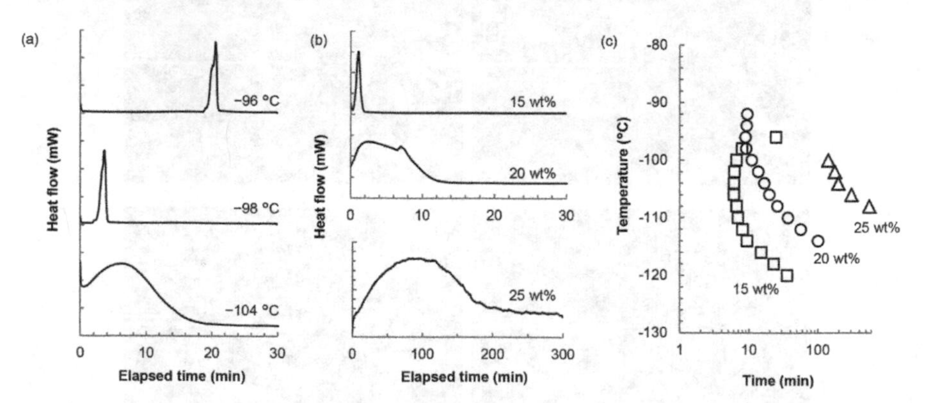

FIGURE 13.8 Isothermal crystallization behavior of (a) a 15-wt.% polystyrene solution at −96°C, −98°C, and −118°C and (b) the corresponding behaviors of 15, 20, and 25-wt.% polystyrene solutions at −104°C. (c) Time–temperature–transformation diagrams of polystyrene-DMF solutions obtained for isothermal crystallization: 15 wt.% (squares), 20 wt.% (circles), and 25 wt.% (triangles). Crystallization time indicates the end of crystallization.

crystal growth rate significantly decreased with increasing polystyrene concentration. The time–temperature–transformation diagrams obtained by isothermal crystallization are shown in Figure 13.8c. On slow cooling, the temperature profile intersects with the upper part of the crystallization region, and crystallization therefore occurs at high temperatures. On rapid cooling, on the other hand, the temperature profile reaches the target crystallization temperatures without intersection with the crystallization region, i.e., flash freezing allows polymer solutions to be cooled to very low temperatures without the formation of macroscopic crystals. However, the conventional ice-template method uses easily crystallizable water as a solvent, which makes nanocrystallization difficult to achieve even by freezing in liquid nitrogen (Gutiérrez et al. 2007).

Solvents with decreased crystallization capability are preferred for solvent nanocrystal formation. According to comprehensive studies on the glassy state of organic molecules, the boiling point/melting point ratio (T_b/T_m) empirically expresses the crystallization capability of organic molecules (Donth 2001), i.e., organic molecules having large T_b/T_m ratios are difficult to crystallize and are therefore recommended for the nanocrystallization method. Suitable examples include amide solvents (*N, N*-dimethylformamide, *N, N*-dimethylacetamide, *N*-methyl-2-pyrrolidone), aromatic solvents (*m*-xylene, *o*-xylene, nitrobenzene), halogenated solvents (carbon tetrachloride, chloroform, *o*-dichlorobenzene, 1,2-dichloroethane, 1,1,2,2-tetrachloroethane), and cyclohexanone. On the contrary, solvents with low T_b and high T_m (e.g., water, *p*-xylene, 1,4-dioxane, and dimethyl sulfoxide) are not suited for the above method due to readily forming macrocrystals even under flash-freezing conditions. In fact, the use of solvents with $T_b/T_m > 1.5$ resulted in the formation of mesoporous polymers with high S_{BET} (>200 m^2/g), small pores (diameters of 10–50 nm) and large mesopore volumes (>0.5 cm^3/g).

13.4.5 Applications of Mesoporous NMs as Separation Materials

An NM made of mesoporous polysulfone was obtained by spin-coating the corresponding precursor solution on a glass substrate followed by immersion in cold methanol (−80°C) to achieve simultaneous freezing and solvent exchange. The semi-transparency of the thus obtained membrane contrasted with the white appearance of conventional membranes (Figure 13.9a), and cross-sectional SEM imaging revealed a uniform mesoporous structure without macrovoids (Figure 13.9b). Such small structures exhibit a markedly reduced light scattering ability, as explained by the Mie scattering theory. The tensile modulus of the dried NM equaled 0.23 ± 0.02 GPa, and the average maximum stress and strain at break were determined as 7.4 ± 0.6 MPa and 26.5% ± 9.0%, respectively. Typically, non-porous bulk polysulfone features a Young's modulus of 2.5–2.6 GPa, a stress at break of ~70 MPa, and a strain at break of ~50%–100% (SpecialityPolymers Solvay 2017). Thus, the Young's modulus of the NM was ~10 times smaller than that of a typical non-porous one. Despite their large thicknesses (18–38 μm), NMs allowed water to pass through at high flux (11–92 L/m^2·h) at a reduced pressure of 80 kPa, which demonstrated that interconnected mesopores linked both sides of the membrane sheet. The fabricated membranes also showed rejection values of 92%–96% for 5-nm Au nanoparticles, thus being potentially useful for water purification.

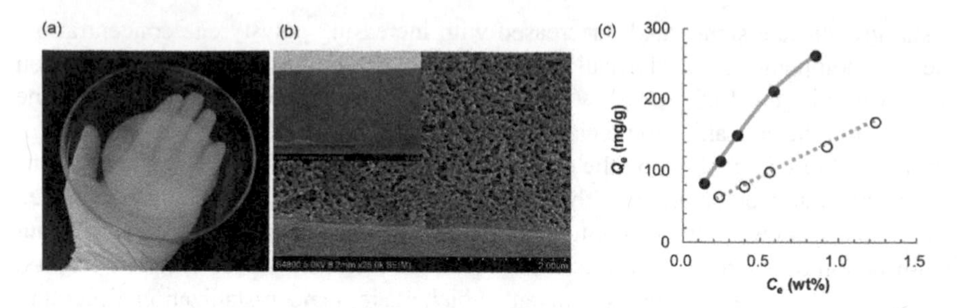

FIGURE 13.9 (a) Photograph of a large polysulfone nanofiber NM sheet. (b) Cross-sectional SEM image of a representative NM. (Inset) Low-magnification image of the whole thickness range. (c) Adsorption isotherms of *m*-cresol dissolved in water for a representative PES NM. Solid and open symbols represent data obtained at 20°C and 80°C, respectively, and lines are drawn as eye guides only. (Adapted from Samitsu, S. et al., *Nat. Commun.*, 4, 1–7, 2013. With permission from CC-BY license.)

The very high free surface area of NMs allows them to rapidly adsorb and desorb large amounts of CO_2 at near-room temperature. Within 1 min, the fabricated membranes adsorbed up to 90% of the saturation amount of CO_2, and the adsorbed gas was rapidly and completely desorbed under vacuum. The amount adsorbed at 10 atm and 25°C equaled 9.6 wt.%, being six times higher than the amount of N_2 adsorbed under these conditions. Interestingly, the Henry constant of the above membranes was about three times larger than that of the bulk polymer (Paterson et al. 1999) and very similar to that of liquid toluene (Wilhelm and Battino 1973). The free surface of polymers is known to significantly enhance the thermal fluctuation of polymer chains and thus increase their local mobility to soften the polymer surface layer (Forrest and Dalnoki-Veress 2001). The soft surface layer of the nanofibers probably facilitated the adsorption of gaseous molecules, and their continuous mesoporous structure enabled the capture of organic vapors by condensation. The condensation of organic molecules such as methanol, hexane, and acetonitrile in mesopores is regulated by pressure and temperature, which implies that the fabricated NMs can be utilized to remove volatile organic compounds from air.

The NMs effectively and rapidly adsorbed organic molecules dissolved in water and were thus viewed as promising materials for the purification of oilfield-produced effluent (Krishnan et al. 2014). Moreover, such membranes selectively captured toluene molecules from aqueous solutions at concentration of several tens of ppm and absorbed significant amounts of tetrahydrofuran from aqueous solutions. At 20°C, PES NMs captured *m*-cresol with a very high adsorption capacity of up to 262 mg/g (Figure 13.9c), which decreased to 128 mg/g at 80°C. The difference between these values was more than two times higher than the best values obtained for commercially available activated carbons and cross-linked polymer adsorbents. Since the above membranes could effectively release adsorbed molecules upon gentle heating, they were concluded to be promising adsorbents for the removal of oils by temperature swing adsorption/desorption.

13.4.6 Flash-Freezing Nanocrystallization: Conclusion

Herein, we have developed a flash freezing-based nanocrystallization procedure for fabricating mesoporous NMs made of commodity polymers. Typically, a polymer solution in a low-crystallization-capability solvent is rapidly frozen at a very low temperature to form solvent nanocrystals, and the subsequent microphase separation of solvent nanocrystals enclosed by the highly concentrated glassy polymer phase results in the formation of a reticular nanofiber network and is followed by solvent exchange at −80°C. The developed method allows the formation of nanofibers with diameters of 12–21 nm that provide high specific surface areas of 200–300 m^2/g. Moreover, the reticular network of thin nanofibers affords mesopores with diameters of 10–100 nm that can be controlled by variation of the polymer concentration, i.e., smaller mesopores are obtained at high polymer concentrations. Nanofiber networks can be obtained in the form of flat sheets, fibers, and monoliths by applying conventional polymer processing techniques. Thus, NMs featuring interconnected mesopores and high specific surface areas of >300 m^2/g are believed to be promising components of advanced separation materials such as filters and adsorbents.

13.5 FUTURE PERSPECTIVE

Despite the large amount of phase-separation phenomena-based procedures used for the fabrication of polymer membranes, the preparation of mesoporous structures comprising nanofiber networks remains challenging. The advanced phase separation techniques developed in this study allow one to obtain mesoporous NMs made of commodity polymers without the use of special polymer templates and elaborate processing. The above membranes are capable of selectively adsorbing certain molecules or letting them pass through and can facilitate vapor condensation, with the high specific surface area of nanofibers resulting in rapid absorption. The developed robust method is based on basic crystallization procedures and requires no specific templates for pore formation, thus being applicable to hybrid systems including those containing several polymers or nanofillers. The NMs that can be prepared by the described method exhibit unique porous morphology, and hence feature a number of medical, environmental, and energy-related applications.

REFERENCES

Abbott, S., and C. M. Hansen. 2018. "Hansen Solubility Parameters in Practice." https://www.hansen-solubility.com/HSPiP/.

Aubert, J. H. 1988. "Isotactic Polystyrene Phase Diagrams and Physical Gelation." *Macromolecules* 21 (12):3468–3473. doi:10.1021/ma00190a021.

Bell, L. 2011. "Top-Level Considerations for Planning Lunar/planetary Habitat Structures." *Journal of Aerospace Engineering* 24 (3):349–360. doi:10.1061/(ASCE)AS.

Biesmans, G., D. Randall, E. Francais, and M. Perrut. 1998. "Polyurethane-Based Organic Aerogels' Thermal Performance." *Journal of Non-Crystalline Solids* 225:36–40. doi:10.1016/S0022-3093(98)00103-3.

Blackadder, D.A., H. Ghavamikia, and A.H. Windle. 1979. "Evidence for Crystallinity in Poly(ether Sulphone)." *Polymer* 20 (6):781–782. doi:10.1016/0032-3861(79)90257-X.

Cho, T. H., M. Tanaka, H. Onishi, Y. Kondo, T. Nakamura, H. Yamazaki, S. Tanase, and T. Sakai. 2008. "Battery Performances and Thermal Stability of Polyacrylonitrile Nano-Fiber-Based Nonwoven Separators for Li-Ion Battery." *Journal of Power Sources* 181 (1):155–160. doi:10.1016/j.jpowsour.2008.03.010.

Daniel, C., S. Longo, R. Ricciardi, E. Reverchon, and G. Guerra. 2013. "Monolithic Nanoporous Crystalline Aerogels." *Macromolecular Rapid Communications* 34 (15):1194–1207. doi:10.1002/marc.201300260.

Debenedetti, P. 1997. *Metastable Liquids*. Princeton, NJ: Princeton University Press.

Deville, S. 2008. "Freeze-Casting of Porous Ceramics: A Review of Current Achievements and Issues." *Advanced Engineering Materials* 10 (3):155–169. doi:10.1002/adem.200700270.

Donth, E-. J. 2001. *The Glass Transition*. Vol. 48. Springer Series in Materials Science. Berlin, Germany: Springer. doi:10.1007/978-3-662-04365-3.

Ellison, C. J., A. Phatak, D. W. Giles, C. W. Macosko, and F. S. Bates. 2007. "Melt Blown Nanofibers: Fiber Diameter Distributions and Onset of Fiber Breakup." *Polymer* 48 (11):3306–3316. doi:10.1016/j.polymer.2007.04.005.

Forrest, J. A., and K. Dalnoki-Veress. 2001. "The Glass Transition in Thin Polymer Films." *Advances in Colloid and Interface Science* 94 (1–3):167–196. doi:10.1016/S0001-8686(01)00060-4.

Guenet, J.-M. 2008. *Polymer-Solvent Molecular Compounds*. Amsterdam, the Netherlands: Elsevier.

Gutiérrez, M. C., M. L. Ferrer, and F. Del Monte. 2008. "Ice-Templated Materials: Sophisticated Structures Exhibiting Enhanced Functionalities Obtained after Unidirectional Freezing and Ice-Segregation-Induced Self-Assembly." *Chemistry of Materials* 20 (3):634–648. doi:10.1021/cm702028z.

Gutiérrez, M. C., Z. Y. García-Carvajal, M. Jobbágy, F. Rubio, L. Yuste, F. Rojo, M. L. Ferrer, and F. Del Monte. 2007. "Poly(vinyl Alcohol) Scaffolds with Tailored Morphologies for Drug Delivery and Controlled Release." *Advanced Functional Materials* 17 (17):3505–3513. doi:10.1002/adfm.200700093.

Hansen, C. M. 2007. *Hansen Solubility Parameters: A User's Handbook*, 2nd ed. Boca Raton, FL: CRC Press.

Hassan, M. A., B. Y. Yeom, A. Wilkie, B. Pourdeyhimi, and S. A. Khan. 2013. "Fabrication of Nanofiber Meltblown Membranes and Their Filtration Properties." *Journal of Membrane Science* 427:336–344. doi:10.1016/j.memsci.2012.09.050.

Kalinova, K. 2017. "Resonance Effect of Nanofibrous Membrane for Sound Absorption Applications." In *Resonance*, edited by Jan Awrejcewicz, pp. 153–167. Rijeka: InTech. doi:10.5772/intechopen.70361.

Karan, S., Q. Wang, S. Samitsu, Y. Fujii, and I. Ichinose. 2013. "Ultrathin Free-Standing Membranes from Metal Hydroxide Nanostrands." *Journal of Membrane Science* 448:270–291. doi:10.1016/j.memsci.2013.07.068.

Krishnan, M. R., S. Samitsu, Y. Fujii, and I. Ichinose. 2014. "Hydrophilic Polymer Nanofibre Networks for Rapid Removal of Aromatic Compounds from Water." *Chemical Communications* 50 (66):9393–9396. doi:10.1039/C4CC01786B.

Lee, M. Y., J. W. Kim, N. S. Choi, A. J. Lee, W. H. Seol, and J. K. Park. 2005. "Novel Porous Separator Based on PVdF and PE Non-Woven Matrix for Rechargeable Lithium Batteries." *Journal of Power Sources* 139 (1–2):235–241. doi:10.1016/j.jpowsour.2004.06.055.

Li, W. L., K. Lu, and J. Y. Walz. 2012. "Freeze Casting of Porous Materials: Review of Critical Factors in Microstructure Evolution." *International Materials Reviews* 57 (1):37–60. doi:10.1179/1743280411Y.0000000011.

Liu, X., and P. X. Ma. 2009. "Phase Separation, Pore Structure, and Properties of Nanofibrous Gelatin Scaffolds." *Biomaterials* 30 (25):4094–4103. doi:10.1016/j.biomaterials.2009.04.024.

Lloyd, D. R., S. S. Kim, and K. E. Kinzer. 1991. "Microporous Membrane Formation via Thermally-Induced Phase Separation. II. Liquid-Liquid Phase Separation." *Journal of Membrane Science* 64 (1–2):1–11. doi:10.1016/0376-7388(91)80073-F.

Lloyd, D. R., K. E. Kinzer, and H. S. Tseng. 1990. "Microporous Membrane Formation via Thermally Induced Phase Separation. I. Solid-Liquid Phase Separation." *Journal of Membrane Science* 52 (3):239–261. doi:10.1016/S0376-7388(00)85130-3.

Lu, X., M. C. Arduini-Schuster, J. Kuhn, O. Nilsson, J. Fricke, and R. W. Pekala. 1992. "Thermal Conductivity of Monolithic Organic Aerogels." *Science* 255 (5047):971–972. doi:10.1126/science.255.5047.971.

Ma, P. X., and R. Zhang. 1999. "Synthetic Nano-Scale Fibrous Extracellular Matrix." *Journal of Biomedical Materials Research* 46 (1):60–72.

Mahler, W., and M. F. Bechtold. 1980. "Freeze-Formed Silica Fibres." *Nature* 285:27. doi:10.1038/285027a0.

Mukai, S. R., H. Nishihara, and H. Tamon. 2008. "Morphology Maps of Ice-Templated Silica Gels Derived from Silica Hydrogels and Hydrosols." *Microporous and Mesoporous Materials* 116 (1–3):166–170. doi:10.1016/j.micromeso.2008.03.031.

Mulder, M. 1996. *Basic Principles of Membrane Technology*. Dordrecht, the Netherlands: Springer. doi:10.1007/978-94-009-1766-8.

Notario, B., J. Pinto, and M. A. Rodriguez-Perez. 2016. "Nanoporous Polymeric Materials: A New Class of Materials with Enhanced Properties." *Progress in Materials Science* 78–79:93–139. doi:10.1016/j.pmatsci.2016.02.002.

Notario, B., J. Pinto, E. Solorzano, J. A. De Saja, M. Dumon, and M. A. Rodríguez-Pérez. 2015. "Experimental Validation of the Knudsen Effect in Nanocellular Polymeric Foams." *Polymer* 56:57–67. doi:10.1016/j.polymer.2014.10.006.

Papon, P., J. Leblond, and P. H. E. Meijer. 2006. *The Physics of Phase Transitions*. Berlin, Germany: Springer. doi:10.1007/3-540-33390-8.

Paterson, R., Y. Yampol'skii, P. G. T. Fogg, A. Bokarev, V. Bondar, O. Ilinich, and S. Shishatskii. 1999. "IUPAC-NIST Solubility Data Series 70. Solubility of Gases in Glassy Polymers." *Journal of Physical and Chemical Reference Data* 28 (5):1255–1264. doi:10.1063/1.556050.

Ramakrishna, S., K. Fujihara, W. E. Teo, T. Yong, Z. Ma, and R. Ramaseshan. 2006. "Electrospun Nanofibers: Solving Global Issues." *Materials Today* 9 (3):40–50. doi:10.1016/S1369-7021(06)71389-X.

Samitsu, S. 2018. "Thermally Stable Mesoporous Poly(ether Sulfone) Monoliths with Nanofiber Network Structures." *Macromolecules* 51 (1):151–160. doi:10.1021/acs.macromol.7b02217.

Samitsu, S., T. Shimomura, S. Heike, T. Hashizume, and K. Ito. 2008a. "Effective Production of Poly(3-Alkylthiophene) Nanofibers by Means of Whisker Method Using Anisole Solvent: Structural, Optical, and Electrical Properties." *Macromolecules* 41 (21):8000–8010. doi:10.1021/ma801128v.

Samitsu, S., T. Shimomura, and K. Ito. 2008b. "Nanofiber Preparation by Whisker Method Using Solvent-Soluble Conducting Polymers." *Thin Solid Films* 516 (9):2478–2486. doi:10.1016/j.tsf.2007.04.058.

Samitsu, S., R. Zhang, X. Peng, M. R. Krishnan, Y. Fujii, and I. Ichinose. 2013. "Flash Freezing Route to Mesoporous Polymer Nanofibre Networks." *Nature Communications* 4 (2653):1–7. doi:10.1038/ncomms3653.

SpecialityPolymers Solvay. 2017. "Technical Data Sheet Udel ® P-1700." http://catalog.ides.com/Datasheet.aspx?I=92041&U=0&FMT=PDF&E=25478.

Sundarrajan, S., K. L. Tan, S. H. Lim, and S. Ramakrishna. 2014. "Electrospun Nanofibers for Air Filtration Applications." *Procedia Engineering* 75:159–163. doi:10.1016/j.proeng.2013.11.034.

Wilhelm, E., and R. Battino. 1973. "Thermodynamic Functions of the Solubilities of Gases in Liquids at 25.deg." *Chemical Reviews* 73 (1):1–9. doi:10.1021/cr60281a001.

Witte, P. van de, P. J. Dijkstra, J. W. A. W a van den Berg, and J. Feijen. 1996. "Phase Separation Processes in Polymer Solutions in Relation to Membrane Formation." *Journal of Membrane Science* 117 (1–2):1–31. doi:10.1016/0376-7388(96)00088-9.

Xin, Y., T. Fujimoto, and H. Uyama. 2012. "Facile Fabrication of Polycarbonate Monolith by Non-Solvent Induced Phase Separation Method." *Polymer* 53 (14):2847–2853. doi:10.1016/j.polymer.2012.04.029.

Yoon, K., B. S. Hsiao, and B. Chu. 2008. "Functional Nanofibers for Environmental Applications." *Journal of Materials Chemistry* 18 (44):5326. doi:10.1039/b804128h.

Zhang, Q. G., C. Deng, F. Soyekwo, Q. L. Liu, and A. M. Zhu. 2016. "Sub-10 Nm Wide Cellulose Nanofibers for Ultrathin Nanoporous Membranes with High Organic Permeation." *Advanced Functional Materials* 26 (5):792–800. doi:10.1002/adfm.201503858.

Zhang, S. S. 2007. "A Review on the Separators of Liquid Electrolyte Li-Ion Batteries." *Journal of Power Sources* 164 (1):351–364. doi:10.1016/j.jpowsour.2006.10.065.

Index

Note: Page numbers in italic and bold refer to figures and tables, respectively.